THE UNIVERSE UNFOLDING

The spiral galaxy M81. (Hale Observatories.)

THE UNIVERSE UNFOLDING

IVAN R. KING

University of California, Berkeley

W. H. FREEMAN AND COMPANY
San Francisco

Cover photograph courtesy of Hale Observatories.

Library of Congress Cataloging in Publication Data

King, Ivan R.
    The universe unfolding.
    Bibliography: p.
    Includes index.
    1. Astronomy. I. Title.
QB45.K56   523   75-33369
ISBN 0-7167-0521-4

Printed in the United States of America

9 8 7 6 5 4 3 2 1

CONTENTS

PREFACE

The astronomical universe is a beauty to behold, a wonder to contemplate, and a challenge to understand. A study of astronomy should cultivate all of these. It should lead us to the beauties; and most of all, it should build an understanding that deepens and broadens our appreciation of the world in which we find ourselves.

The study of astronomy leads our minds in many different directions. It shows us the totality of what exists, and it marks our place in that immensity. Who are we, we ask, and where did we come from? Astronomy also leads us to examine the basic nature of the universe and the way of its working. What is a star? Why are there stars? Why must there be stars? Much of astronomy is devoted to finding those simple rules that make things as they are; indeed, one of our greatest satisfactions is to see, finally, that things could not have been otherwise.

The rules that lead us to an understanding of the universe are the rules of physics, and in a sense astronomy is simply an application of physics to problems on the largest scale of all. Thus a path into astronomy leads through much of the central territory of physics, and astronomy then serves as an introduction to that most basic of the physical sciences. The value of this experience lies, I believe, not in the material that is learned but in the experience itself. To follow physical processes requires a logical reasoning that complements in an invaluable way the processes of memory and intuition that one develops elsewhere. This is not to say that other intellectual areas lack logic, nor that physics can be carried on without intuition; but a study of physical science offers an incomparable opportunity to develop what for many people is a new strength of the mind.

Thus it is that the central theme of this book is understanding. Moreover, in a study of astronomy understanding has two facets that we can regard almost separately. One kind of understanding is to see why things behave as they do; this is common to all of science. The second kind of understanding concerns our own knowledge and how it develops. How do we know the things that we know, and how did we ever find them out? In this respect astronomy is extreme among the sciences: we cannot visit the stars, nor can we perform experiments on them; we can only look at them. All that we see is a scattering of points in a black sky; yet we know all sorts of detail about what they represent and why. Here is a unique opportunity that astronomy offers us, to follow one step after another in the unraveling of the clues of the universe. Surely, when we contemplate Man and the Universe, one of the greatest wonders is that we have come to know what we know about it. A humane study of astronomy should continually ask about our own ways of searching and knowing.

I have written this book to express this view of astronomy, which I do not see sufficiently emphasized elsewhere. The writing has been a stimulating experience, examining the basic questions of the universe and looking for words to communicate our attempts to answer them. It has not been easy, and I am grateful to many around me—particularly to my colleagues and students in the Berkeley Astronomy Department, who have tolerated all my impertinent questions, to Jane Rose, who carefully read the manuscript and helped me to remove many of its original obscurities, and to Robert Blanchard, who translated my crude scribbles into finished drawings. And I owe a special debt to my wife and children, who have endured so many ill-humored evenings and weekends. I hope that they think that it was worth it.

Berkeley, California
December 1975

Ivan R. King

THE UNIVERSE UNFOLDING

1 | THE SETTING

The realm of astronomy begins at the top of the Earth's atmosphere and does not end. It extends through the Solar System to the stars beyond, past the neighboring stars to the limits of the Milky Way, and outward, past ever remoter galaxies, to a limit imposed only by the limitations of today's telescopes. This is a diverse realm, whose successive ranges of distance pose different problems that in turn require very different approaches. It is thus natural that successive developments in astronomy have shifted the focus of interest and with it the whole tone of astronomical science.

It is equally natural that early modern astronomy should have begun in the Solar System, with the near and the familiar. The motions of the planets are easy to see, and the telescope makes much of their physical nature directly visible. The stars, by contrast, are incomparably remote; and never is their remoteness more vivid than when we face the problem of studying what they are. Even in the most powerful telescope, a star is only an infinitesimal point of light, to be studied only by analyzing the detailed make-up of that light and by measuring minute shifts of its position. Thus the centuries following Copernicus saw a one-sided development of astronomy. The workings of the Solar System were recorded and understood in increasing observational detail and mathematical complexity, whereas little was known about the stars, beyond the presumption that they were suns, faint only because of their tremendous distances.

Since 1900, however, astronomy has undergone a complete change. The stars have become a major concern and an extremely successful one. We know today what the stars are made of, how big they are, how and why they shine, how old they are, and how long they will live. We have traced the outline of our galaxy, probed other galaxies, and touched upon the question of why galaxies are what they are. In an elementary survey of astronomy today, most of the Solar-System part could have

VOLUME 198

THE ASTROPHYSICAL JOURNAL

1975 JUNE 1

NUMBER 2, PART 1

(Continued on inside page)

PUBLISHED BY THE UNIVERSITY OF CHICAGO PRESS FOR
THE AMERICAN ASTRONOMICAL SOCIETY

ASJOAB 198 (2, Pt. 1) 241-516 (1975)

FIGURE 1.1.
The tone of astronomy has changed with the centuries.
(Woodcut from Navin Sullivan, *Pioneer Astronomers*, 1964. Atheneum Publishers.)

been written in 1900; the stellar part could hardly have been dreamed of.

Of the factors that have contributed to the explosive advance of stellar astronomy in the twentieth century, three stand out. One factor is technological. Much has been written about the impact of science on technology, but here is the reverse process. Modern astronomy owes much of its success to the sharpening of its tools of observation. The construction of large telescopes is important, but even more important have been improvements in the light-receiving devices that can be attached to telescopes. The introduction of photography, nearly a century ago, was the first major step; it allowed the astronomer to record far more information than ever before. Later, the use of photoelectric devices introduced a new order of precision into measurements of brightness; we shall see how this increased accuracy has unfolded the basic nature of the stars. Furthermore, both photographic and photoelectric measurements have allowed the astronomer to observe fainter—and therefore more distant—stars than he could see directly. More recently, modern technology has opened new windows on the universe, by allowing observations of radiation other than visible light: radio, infrared, ultraviolet, X-rays, and gamma rays. Finally, spacecraft are beginning to add new and valuable modes of observation.

A second factor is the feedback of scientific knowledge; specifically, twentieth-century astronomy could not exist without twentieth-century physics. Today's astronomer is very much an applied physicist, and the rapid advance of physics has granted the astronomer powerful new insights.

The third factor is socioeconomic, and is in fact obvious enough: it is simply manpower. Today far more people than ever before are engaged in basic scientific research, and astronomy has its share of them. This year's crop of new astronomers could have staffed the whole astronomical world of a hundred years ago.

Along with its concerns and its techniques, modern astronomy has changed its tone. Last century's astronomer was a precise mathematician, absorbed in the elegant clockwork of the Solar System. His heir of today is an explorer of unknown territory, turning his knowledge of physical theory into a feeling for the ways of the universe. His explorations are often crude and his maps inaccurate, but they are maps of the new and hitherto unexplored. The age has passed in which astronomy was synonymous with precision; today's astronomer is often satisfied with a result that is uncertain by a factor of two—but this is where no number at all existed before. The nineteenth-century astronomer sought elegant approaches to clearly stated mathematical problems; the twentieth-century astronomer sketches a physical picture and is sometimes hard-pressed even to define the problems. The scene has changed, and with it the story: the courtly minuet of classical orbit theory has given way to the hurly-burly of the astrophysical frontier.

THE HIERARCHY OF DISTANCES

Today's astronomer turns his attention to a universe that has many
scales of distance and size, from the near to the very far. Contemplating
the universe is a continual series of readjustments, for each stage of
distance dwarfs the previous one. First, the Solar System still surrounds
us with unsolved problems. At the next level of distance come the nearby stars, close enough and few enough to be studied as individuals.
Farther away the stars are almost countless, but we can still study their
numbers in limited areas and investigate their statistics. On a larger
scale come problems of the size and structure of our galaxy, the Milky
Way, and then the structure of neighboring galaxies. Finally, with studies
of large numbers of distant galaxies, we approach the largest problem of
all, the structure of the whole universe.

To follow astronomical discussions one must first keep these ranges
of distance straight. The first problem is their very size; "astronomical
numbers" are proverbial. They are also confusing: one cannot always
remember whether the man said "billion" or "trillion," and it is easy to
lose one or two zeros in a string of twenty. Astronomers are no more
immune than anyone else to this confusion, and among themselves they
avoid the use of long numbers. One way of doing this is to measure
things in a unit of suitable size. We would not quote the distance from
New York to London in inches—nor would we quote the distance to a
star in miles. Astronomers quote stellar distances in terms of a unit called
a *parsec*, which is not far from the average distance between neighboring
stars. More precisely, a parsec is 206,000 times the distance from the
Earth to the sun. (Its origin will be explained in Chapter 12.) A distance
of 100,000,000,000,000 miles is unwieldy; 5.2 parsecs is much more
comfortable.

One other large unit of distance appears in many popular accounts of
astronomy, but is rarely used by astronomers. It is the light year, which is
the distance that light travels in a year. Since light travels 186,000 miles
(nearly to the Moon) in a single second, a light year is clearly a long,
long way. Aside from this obvious immensity, though, the light year has
no outstanding advantage as a unit, and in fact astronomers almost
always use the parsec. The distances in this book will therefore be given
in parsecs; anyone who wishes to convert them to light years can multiply
by 3.26.

(For most purposes of discussion it is convenient to use a unit, such as
the parsec, that is of an appropriate size for the subject at hand. Sometimes, however, one is obliged to compare quantities directly that are
of completely different orders of magnitude. For the very large or small
numbers that then result, it is convenient to use a notation based on
powers of ten. Thus a hundred is 10^2, a million is 10^6, 20 million is 2×10^7, etc. In this notation a parsec is 3.08×10^{13} kilometers, or 1.92×10^{13} miles.)

5

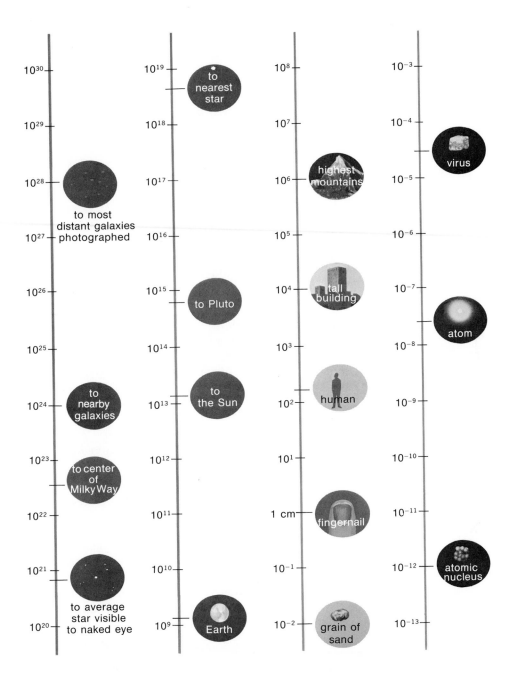

FIGURE 1.2.
Scales of distance, all expressed in the same basic unit, the centimeter. Each mark represents a factor of 10 in size or distance.

2

The first step in the hierarchy of distances is to make the jump from the Solar System to the stars. Here a scale model is useful. Imagine the Sun to be a marble and the Earth a grain of sand a yard away. On this scale the rest of the Solar System would fit into a good-sized garden, but the nearest other star would be a hundred miles away. On our real astronomical scale that star is just over a parsec from us. Within a few parsecs are several other stars, and with increasing distance the number goes up rapidly. Of the stars that we see easily at night, a few are near neighbors, while others are farther away; but a distance of about 200 parsecs will include nearly all the stars that are easily visible to the naked eye. Beyond this, the stars go on, but our eyes are too weak.

With such a thin sprinkling of stars—like marbles a hundred miles apart—the space around us is very empty. But it is not completely empty; spread through it is a very thin scattering of gas and dust. In any locality there is very little of this material—rather like a thimbleful of air spread over a cubic mile—but when it is added up over such large distances, there is almost as much material between stars as there is in the stars.

Our stellar neighborhood is just a tiny fraction of the whole galaxy called the Milky Way—the next step up in the distance scale. The Milky Way has the shape of a large, flat, spinning disc, made up of about a hundred billion stars plus a sprinkling of interstellar material. Our place in the Milky Way is not at all central; we are much closer to the edge. In actual distance we are some 10,000 parsecs from the center of the Milky Way. The outer edge is not clearly defined, but our galaxy certainly extends at least another 5,000 parsecs beyond us, giving a total diameter of 30,000 parsecs or more.

At the boundaries of the Milky Way stars become fewer and fewer, and beyond lies the emptiness of intergalactic space. In the distance lie other galaxies, and here is another step in the distance scale. A few parsecs separate stars, a few hundred parsecs span our neighborhood, and the Milky Way extends for tens of thousands of parsecs, but a typical distance between galaxies is a million parsecs.

Galaxies are scattered about us in all directions, as far as we can see into the distance. Taken all together, they make up the largest unit of all, the universe. Its size is unknown, for the limit of our vision is only *our* limitation—the limit of the vision that our imperfect instruments afford. The uniformity of our observed sample suggests that it is only a small part of the whole universe—if, indeed, the universe has any limited size at all. The one distance that can be safely quoted is that of the remotest galaxies that have been observed. It is several billion parsecs.

The twentieth-century explosion in astronomy has, step by step, sketched out this large, diverse realm. Along with the factual picture, moreover, modern astronomers are able for the first time to confront some of the really basic questions of the universe—not just what the

FIGURE 1.3.
The great Andromeda spiral galaxy, M31, with its two smaller companion elliptical galaxies. The Milky Way is a galaxy similar to M31. (Hale Observatories.)

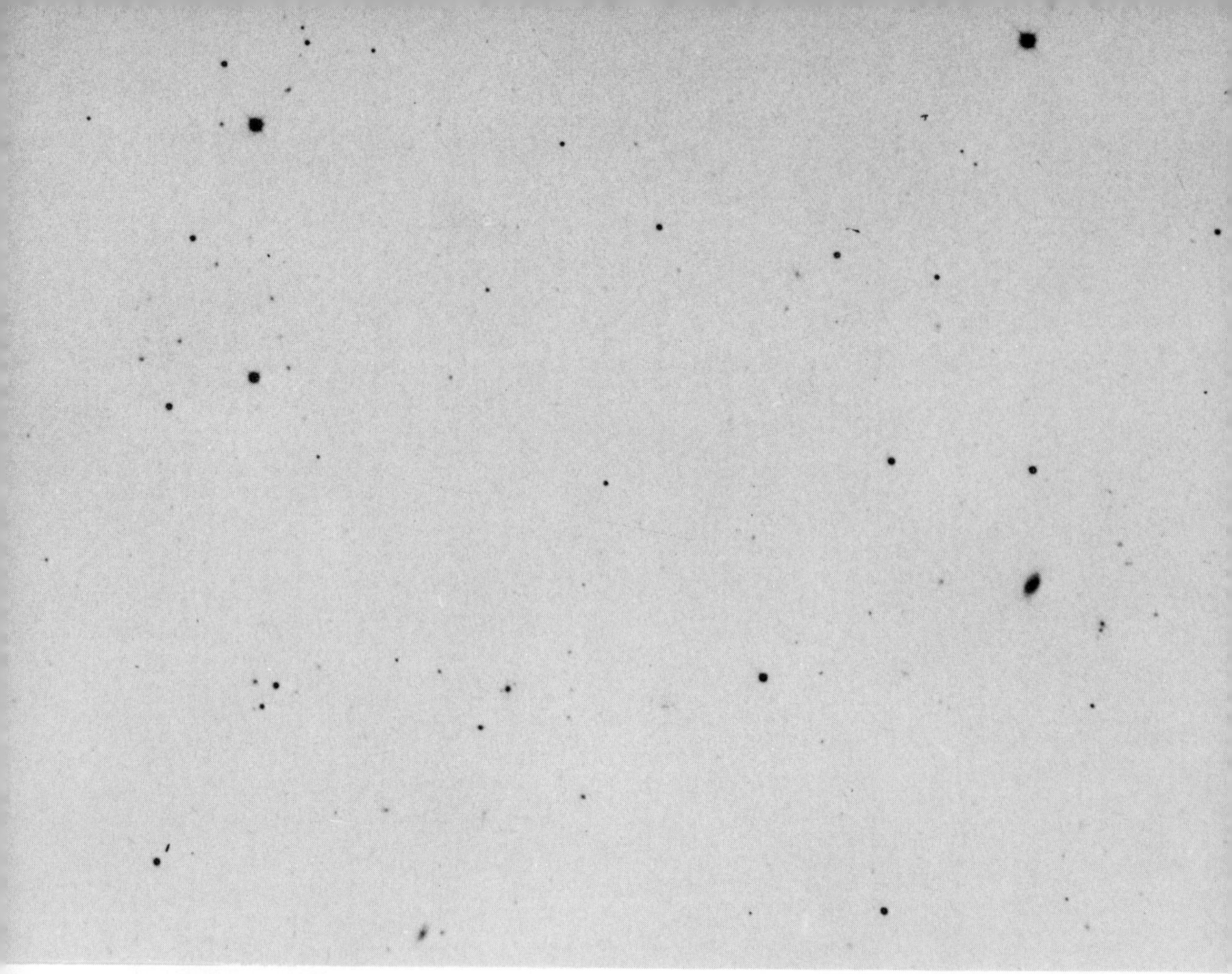

FIGURE 1.4.
In this long-exposure photograph of a small region of the sky, some of the faintest smudges are galaxies a billion parsecs away. The light that we see left them 3 billion years ago. (Lick Observatory photograph by the author.)

stars are, but how old they are, where they come from, and what will become of them. And we can even approach the question of why the universe contains such things as stars.

GETTING AT THE ANSWERS

The modern picture of the universe is fascinating and often exciting, but not all the interest lies in the answers themselves. The process of studying the universe is like the solution of an involved mystery and carries its own kind of interest. Like any good detective story, stellar astronomy starts with the slimmest of clues. Consider the plight of the astronomer.

FIGURE 1.5.
NGC 2264, a region in our Milky Way in which stars are being born. (Hale Observatories.)

With his naked eye he sees a star as a point of light, remote and uninformative. Even when seen through a telescope, that star is still only a point of light—much brighter, but still an enigmatic point. Yet from that point of light and others like it, the astronomer must deduce the nature of the universe. This he does by careful and reasoned measurements of minute shifts in the position of a star and above all by detailed analysis of the quality of the light that it sends us. All is measurement and deduction, and little is seen directly. Nothing could be farther from the truth than the traditional picture of the astronomer peering through his telescope for long hours to see what he can see.

A modern astronomer is indeed very like the detective in an intricate mystery story. The detective is given a few scattered clues, and the plot

9

begins to develop only when he quietly forms some notion of the skein of occurrences that might have left such clues behind. At this stage he keeps his own counsel, using his hypothesis only to suggest new lines of inquiry, until finally the picture begins to fit together. The spreading lines of deduction converge and cross and mesh, and in the last chapter he announces a solution that no one can then doubt. Only the inspired detective could have followed that first path, but his final picture is so clear that we wonder why we did not see it sooner.

So it is in astronomy. The opening scene shows a pitiful set of clues—points of light in a black sky. The astronomer begins asking questions and making measurements, not broadside or at random, but logically and incisively. Each successful step suggests new directions, and the chains of fact and deduction reach outward. Finally some large circle closes, and what was a shaky hypothesis becomes a firm structure as another episode of detection closes.

In building one deduction on another, the astronomer has two basic elements: observation and theory. Neither can succeed alone; every astronomical structure is made of the bricks of observation held together by the mortar of theory. Observations are generally of no interest taken alone; for instance, the brightness and color of a star, and the changes in its position, are of no great significance in themselves, but connected properly by theory they reveal the size of the star. Similarly, theory by itself is an empty game; it tells us what might be, but only observation can transform "might be" into "is." Even together, though, observation and theory can build a meaningful structure only when vitalized by one more factor: a design. This design is the story of modern astronomy. It is a story that can be told in two ways, just as astronomy itself can be looked at from two points of view. So, for that matter, can any science. On the one hand there are the results, the coherent picture of the universe or of one of its parts. These are the objectives gained, the fruits of victory, for which the scientist has spent his labor. Yet they are not all. The scientist, like the athlete, does not play the game purely for the sake of winning; the game itself is often satisfaction enough. And so it should be with the layman examining science. He should contemplate not just the universe but, more important, the way in which it can be explored, not just what we know, but how we could ever have come to know it.

In short, science is not just a set of answers; it is the process of finding out. To present astronomy in this light is the theme of this book.

THE UNITY OF THE UNIVERSE

Our starting point has already been stated: in studying the nature of the stars, all the astronomer has is the light that they send us. He must then use his knowledge of the properties of light and proceed to the characteristics of bodies that emit it. Soon he is concerned with all the properties of material bodies and their interactions with each other—in a word, he is concerned with physics. The laws of physics supply the connection between the bare observed facts and the rich reality that underlies them— between the color of a star and the temperature of its surface, between the rotation of a galaxy and the gravitational forces that hold it together. Even more, they connect this first level of deduction with the remote and inaccessible. The nature of a star's surface can be understood fairly directly from the light it emits, but we have no observational contact with the star's interior. Nevertheless, astronomers know what the inside of a star is like. To keep the surface as it is, the deeper layers can only be just as they are. This *must* be so, because this is how matter always behaves, everywhere.

Buried in this phrase is an idea that is central to modern astronomy: underlying all the phenomena of the universe is a single set of rules. The laws of physics that we discover in the laboratory here on Earth apply equally well to the deepest heart of the farthest star. This assumption is rarely stated explicitly, for it seems quite obvious to citizens of the modern scientific world; yet an assumption it is, and without it modern astronomy could not exist. Indeed, it has not always been thought obvious. For many centuries, quite to the contrary, it was considered to be unthinkable; and during that long era astronomy languished. In the firmly entrenched religious outlook of the Middle Ages, the heavens were as different from the Earth as God is from man. What is obvious in the twentieth century was a daring step in the sixteenth.

I have called this basic premise of uniformity an assumption, because in fact it cannot be proved in any logical and deductive way. To put it as bluntly as necessary, we believe in it because it works, because reliance on this principle has led to such splendid successes in understanding the universe.

With his faith in the principle of the unity of the universe, the astronomer thus weaves together observation and theory into a picture of the unseen that is just as sharp and clear as what we see directly. But here we come to the second crucial point: how does he know that this picture is correct? Surely not by faith again, for then we would find that faith in the method and faith in the results were each holding the other up. True support can come, on the contrary, only from a new contact with the firm reality of observed fact. In short, the picture is accepted only when it leads to observational confirmations. Only rarely are these the dramatic single tests; probably never again will a Leverrier write, "Point your telescope to longitude 326° and you will discover a new planet." Instead the confirmations are generally a host of small points that did not have to come out right but in fact did. Building up an astronomical picture is like filling in a crossword puzzle. When only "down" words are filled in, we are not sure they are right; but if the "across" words also read correctly, then the puzzle has been correctly solved.

This, then, is the role of the astronomer, building a large picture of the universe out of many small circles that close on themselves and lock together as strong checkpoints. And underlying it all is a faith that the workings of the universe can be understood logically, in terms of the same neat set of rules that cover the workings of familiar objects.

REVIEW QUESTIONS

1. Why has astronomy advanced so rapidly during the twentieth century?

2. Outline the various scales of distance that characterize the realms of astronomy. Give examples and relative sizes.

3. What is a parsec? What is a light year? What is the relation between them?

4. What is the relation between astronomy and physics?

2 | LOOKING AT THE STARS

Consider the stars. Go out on a clear night and look at them. Some are bright, some are faint, but all are remote and inscrutable. Here is the first problem of astronomy: how are we to discover the large, bright reality behind those little glimmers in the black sky?

Clearly, our approach must be through the evidence at hand: the light that the stars send us. Their whole story—insofar as we have yet been able to discover it—is written in those thin beams. The first question to ask is, What is starlight? The answer holds a wealth of information, but to understand it, we must expend some effort; to read the message of starlight, we must first learn the language of light—or, more properly, of electromagnetic radiation.

Electromagnetic radiation is indeed a language; and our eyes, excellent as they are for our animal purposes, can decipher only a very few of its words. The narrow sample that we call *visible light* is just a small section of the broad spectrum of what physicists call electromagnetic radiation. This spectrum includes such apparently diverse emanations as gamma rays, visible light, heat rays, and radio waves. The ability to discover unity in apparent diversity is, of course, characteristic of physics—or of any other science, for that matter. In this case, the physicist understands that all these types of radiation are oscillations of electric and magnetic fields traveling through space. The differences between these types of radiation are all differences in one characteristic: the wavelength. Physically, the wavelength is directly related to the electric and magnetic fields that make up the radiation; but, for many purposes,

14

FIGURE 2.1.
Stars are everywhere, even in this photograph showing a relatively smooth
section of the Milky Way. (Kitt Peale National Observatory.)

one can dispense with such complexities and regard the wavelength
merely as an identifying number that labels the properties of the radiation. Be it essence or label, however, the wavelength determines how the
radiation interacts with matter. In terms of everyday experience, this
means that the wavelength governs the circumstances under which we
may encounter any given type of radiation, the ways in which we may
generate it, and the instruments with which we may detect it. The eye
is one of these instruments, and the radiation within the narrow range
of wavelength that it can detect has thus been given the name "visible
light."

The shortest waves in the electromagnetic spectrum are produced in
nuclear reactions and are called gamma rays. Somewhat longer waves,
called X-rays, can be generated in specially designed electronic tubes.
Still longer in wavelength is ultraviolet radiation, which is produced
in much the same way as visible light and is received by many of the
same detectors—but not by the human eye. Next, marked off only by
the particular characteristics of the eye, comes visible light, where wavelength corresponds to the sensation of color. Visible light is delimited by

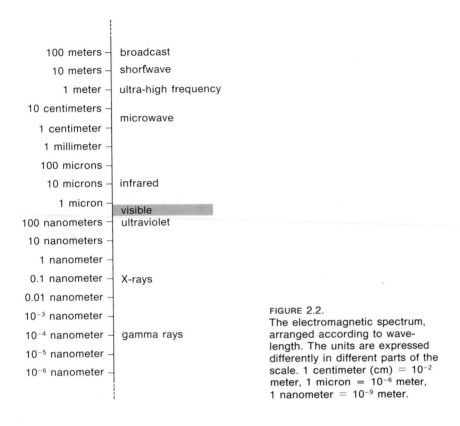

FIGURE 2.2.
The electromagnetic spectrum, arranged according to wavelength. The units are expressed differently in different parts of the scale. 1 centimeter (cm) = 10^{-2} meter, 1 micron = 10^{-6} meter, 1 nanometer = 10^{-9} meter.

the ends of the well-known rainbow spectrum, which stretches from violet at the short-wavelength end to red at the long-wavelength end. Waves too long to be detected by our eyes are called infrared; we are most often aware of them as the heat waves from such things as a bed of coals or an electric iron. Longer than infrared are the waves called radio: the shortest of these, microwaves, are used in radar; next in length are the high-frequency waves used for television and FM radio; somewhat longer are the so-called "short waves," improvidently named at a time when they were the shortest used; longer still are the waves in the narrow band devoted to ordinary AM broadcasting; and the longest waves are used in certain long-distance communications systems.

These are the bands of the electromagnetic spectrum. The divisions between them are often vague, for they lack any sharp physical distinction, reflecting only our own habit and convenience. The important fact is that they are all examples of the same kind of radiation, differing only in wavelength.

The stars emit radiation throughout the electromagnetic spectrum, from the longest waves to the shortest. Parts of their story are to be read

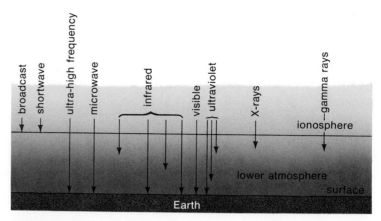

FIGURE 2.3.
Radiation of different wavelengths penetrates to different levels of the
Earth's atmosphere. Only in a few ranges of wavelength does the
radiation reach the surface.

in all parts of this array, and it is the task of the astronomer to use as
much of it as he can. Unfortunately, between the astronomer and the
stars lies a thick wall: the Earth's atmosphere. The atmosphere is fairly
transparent to visible light (a statement that is really backward in its
logic, because it was the transparency of the atmosphere that determined
the wavelengths for which sight organs would be of any use), but most
parts of the electromagnetic spectrum fare badly in their attempt to
penetrate to the Earth's surface. In only two major parts of the spectrum
can our instruments "see" the stars. One wavelength region to which
the atmosphere is transparent includes visible light with a little infrared
and ultraviolet on either side of it; this is referred to as the "optical window." All shorter waves, including all the gamma rays and X-rays and
most of the ultraviolet, are blocked by the atmosphere—an essential
thing for the astronomer's health, but frustrating to his science. On the
other side of the optical window, most of the longer infrared waves fail
to penetrate the atmosphere, but part of the radio spectrum does come
through: microwaves and the wavelengths used for FM and TV. All
longer waves are blocked.

It is perhaps confusing to learn that the Earth's atmosphere is opaque
to radio wavelengths that are commonly used for communication over
long distances. The resolution of the paradox lies in realizing that the
atmosphere consists of separate layers. The lower layers are transparent
to all radio wavelengths; but higher up (between 50 and 300 miles) lie
electrically charged layers called the ionosphere, which not only block
the longer waves but reflect them, too. The existence of this "radio roof"
in the atmosphere resolves another potential paradox. If radio waves,

like light, travel in straight lines, how can we receive them far beyond the horizon? The answer is that these waves bounce off the ionospheric ceiling; for communication over very long distances, they must ricochet several times between the ionosphere and the surface. Shorter waves, on the other hand, pass right through the ionosphere; hence, television waves fail to bounce back and are limited to the direct line of sight. For radio astronomy, however, the values are reversed. The shorter waves reach us unhindered, while longer waves from the stars bounce off the *outside* of the ionosphere and never reach the earthbound astronomer.

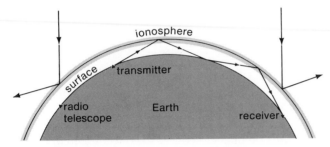

FIGURE 2.4.
Just as the ionosphere reflects communications signals around the Earth, it prevents cosmic radiation of the same wavelengths from reaching us.

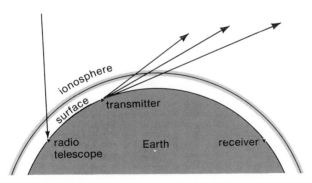

FIGURE 2.5.
The wavelengths used in radio astronomy pass readily through the ionosphere and thus cannot be used for communication beyond the horizon.

Thus, observations made from the Earth's surface are restricted in wavelength. The greater wealth of information lies in the traditionally exploited optical window, but recent developments in radio techniques have revealed the rich treasures available through the radio window. In addition, equipment has been greatly improved for those portions of the infrared spectrum to which the Earth's atmosphere is at least partly transparent. But large parts of the electromagnetic spectrum still remain inaccessible to the ground-based astronomer.

Thus the real value of the space program to astronomy is that it opens new wavelength windows on the universe. Of greatest interest are the short wavelengths: ultraviolet, X-rays, and gamma rays. Only a tiny fraction of the space program has been devoted to scientific observations of such radiation, but the results have already done much for our understanding of the universe, particularly of the high-energy phenomena with which these short wavelengths are so often associated.

THE USE OF TELESCOPES

Submerged in a sea of murky air, the astronomer observes the stars. His basic tool is the telescope, which one ordinarily thinks of as a device for magnifying distant objects. Magnify it does—and in a few astronomical situations the magnification is indeed crucial—but in most of astronomy the main function of a telescope is to collect the light of faint objects, in order to make them look brighter. Rarely does the astronomer have enough light; his ability to study a star is almost always limited by its pitiful faintness. The telescope serves him as a sort of light-funnel; the beams of starlight that enter its entire aperture are gathered together at the focus. Thus the main value of a telescope lies in its collecting area; so it is natural that astronomers should rate telescopes by their diameters. (The light-gathering power increases, of course, as the area, which goes as the square of the diameter.) Thus, compared with the human eye, whose iris can open to about a third of an inch, the 200-inch Palomar telescope collects $600 \times 600 = 360,000$ times as much light.

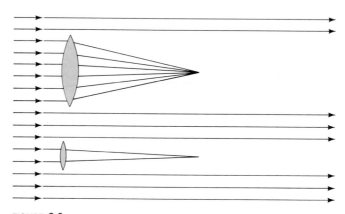

FIGURE 2.6.
A telescope lens (or mirror) collects light at its focus. A telescope of larger diameter gathers in a larger bundle of light.

The 200-inch (5-meter) telescope, at Palomar Observatory, is the world's largest. (Hale Observatories.)

This is not to say that magnification has no value; it does indeed. The trouble, however, is that magnification is limited not by the telescope but by the Earth's atmosphere. Air currents blur our vision—as anyone can see by looking down the road on a hot day—and a telescope magnifies the blurring. Even at the best observing sites, the air is rarely steady enough to allow us to see more than a few hundred times as sharply as we see with the unaided eye. In fact, it is for this characteristic, steadiness of the air, that observatory sites are chosen. Astronomers call it "good seeing."

The telescope makes starlight available, but it is up to the astronomer to use it. He could do this by simply looking through the telescope, but he rarely does such a crude thing. The human eye is too poor an instrument, poor in sensitivity, in versatility, in accuracy, and in the ability to record large amounts of data. For some of his recording, the astronomer uses photographs. At the focus of the telescope, he sets his photographic plate (most astronomical pictures are taken on rigid glass plates

19

FIGURE 2.8.
The 4-meter (158-inch) reflector of Kitt Peak
National Observatory is one of the world's
largest and most modern telescopes. (Kitt Peak
National Observatory.)

FIGURE 2.9.
The 120-inch (3-meter) telescope of Lick
Observatory. (Lick Observatory.)

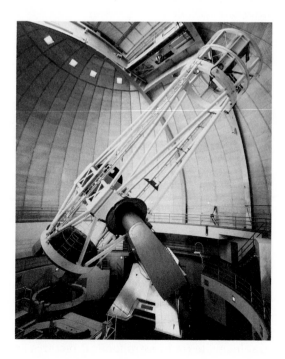

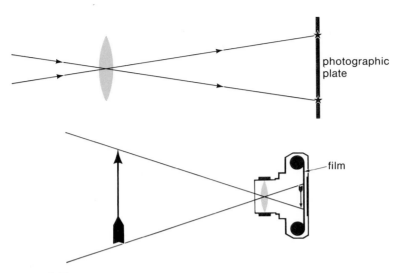

FIGURE 2.10.
A photographic telescope takes a picture by forming an image on a
photographic plate placed at its focus, just as a camera forms an image
on its film.

rather than on flexible films), just as he might place a film at the focus
of an ordinary camera lens. Thus, the telescope becomes a big camera,
even though the proportions are different. The exposure times are very
different, too, for the astronomer relies on the marvelous ability of photography to build up an image over a long exposure time. Many of his
exposures last for an hour or more, during which time he "guides" the
telescope to keep it sharply pointed at the stars whose photograph he
must not blur. The only limitation is the brightness of the sky itself;
ultimately, the photograph picks up a smooth haze of night-sky background, a faint atmospheric glow, along with sunlight scattered from the
widespread dust of the Solar System.

Not only does the photograph register, by untiring accumulation of
light, objects much too faint to see with the unaided eye, but it also
covers a sizeable area of the sky in a single picture. Thousands of stars
appear, the position and brightness of each shown as a tiny dot. For
mapping the stars and for many types of measurement, direct photographs are ideal. More detailed analysis, however, requires separating
the starlight into the different wavelengths that make up its spectrum.
This separation can be effected in a crude way by filters that select light
of certain wavelengths, or it can be done more precisely by optical devices that send the different colors in slightly different directions, usually
by either a glass prism or a grating of fine grooves ruled several thousand
to the inch. Mounted with suitable lenses or mirrors and a photographic
plate to record the results, such a device is called a spectrograph. It

records a streak of light in which every point corresponds to a different wavelength—in other words, a picture of the spectrum. Again the telescope acts as a light funnel, but now at its focus is placed the spectrograph, its narrow opening adjusted to select the light of a single star for spectral study.

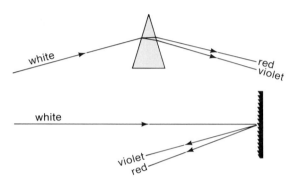

FIGURE 2.11.
Dispersion of colors of light by a prism and by a reflection grating. For the grating, the separation of the colors depends on the spacing of the grooves; their slope concentrates the light in a particular direction.

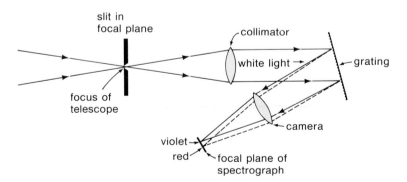

FIGURE 2.12.
An astronomical spectrograph. The focal-plane slit isolates the star that is to be studied; the collimator makes the light parallel; the grating disperses it; and the camera focuses it into a spectrum.

Just as direct photography yields to spectroscopy for detailed analysis of starlight, it must again step aside when brightness needs to be measured with the highest accuracy. For this, the astronomer uses a photoelectric cell, which converts starlight into an electric current that can be amplified into a precise measure of the brightness of the star. With a photocell, however, only one star is measured at a time, so that observations are slow and laborious. What is clearly needed is a device that combines the accuracy and sensitivity of the photocell with the wide-field capability of the photograph. Such devices exist. One type, called an image tube, produces pictures directly; it is really just a photographic device of improved speed and sensitivity. There is a quite different class of "panoramic detectors," however, that operate very much like television cameras. Here the picture is converted into an electronic signal that can be recorded and then played back, perhaps as a picture, perhaps just as a string of numbers for analysis.

If electronic detectors are so efficient, one might ask why astronomers ever take photographs at all. The answer is that an electronic detector is able to cover only a relatively small area at a time. If a large region of the sky needs to be surveyed, photographs are still the only efficient way.

Even so, some panoramic detectors are able to record the brightness at a hundred thousand points simultaneously, and it is thus natural that the astronomer should use a computer to control his detector and to record the results. Every modern telescope has at least one computer tied directly into its operations.

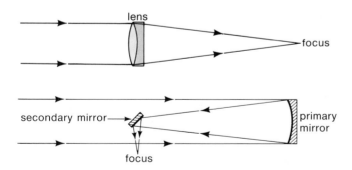

FIGURE 2.13.
Refracting and reflecting telescopes. The refractor has a compound lens, in order to make all colors come to the same focus. The reflector is color-free, but needs a secondary mirror to remove the focus from the middle of the incoming beam.

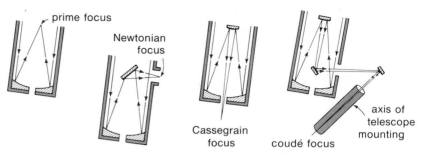

prime focus

Newtonian focus

Cassegrain focus

coudé focus

axis of telescope mounting

FIGURE 2.14.
With the use of different auxiliary mirrors, the same large primary mirror can be made to focus light at many different points.

The telescope itself has changed, too, since the day of the long tube with a lens at its upper end. All large telescopes today are reflectors, collecting light by reflection from a large concave mirror, which is much easier to make than a lens of the same size. There is a difficulty inherent in the use of a curved mirror, however: it sends the light back in the direction from which it came. (Oblique reflection from a curved mirror produces too blurred a picture.) The focus is then in the middle of the incoming beam, where the astronomer can use it only by blocking out part of the light. There are several solutions to this dilemma: In the largest telescopes, the astronomer himself is too small an obstruction to matter; he sits in a small observing cage at the "prime focus," high in the middle of the telescope tube, carried about with the telescope as it is pointed. In smaller telescopes, a diagonal flat mirror is placed just in front of the prime focus; it sends the converging beam of light to a "Newtonian focus" just outside the top of the telescope tube. The astronomer then uses the telescope while standing on an elevator platform hung from the dome that encloses the telescope. Large instruments, however, are too heavy for the lightweight top end of a telescope tube. They are hung behind a hole in the main mirror, at the "Cassegrain focus," which light reaches after being reflected back by a small secondary mirror. Finally, the largest instruments of all are mounted in a fixed position at what is called the "coudé focus," to which an ingenious arrangement of mirrors sends the starlight down the hollow axis on which the telescope pivots. An astronomer working underground at the coudé focus of a large telescope may spend the entire night "observing" without ever seeing the sky!

INVISIBLE LIGHT

In his investigation of the mystery of the universe, the astronomer uses every clue that he can find. As we have seen, the Earth's atmosphere restricts the range of available wavelengths, but even further limitations are set by our ability to discriminate and to detect radiation of different types. Originally the astronomer used his eye, and later, photographic materials and photoelectric cells. All of these have their sensitivity confined to the same general spectral region, around the wavelengths that the eye itself defines as "visible." For other spectral regions, detectors have not been nearly so easy to find, and progress in astronomy has had to wait for developments in technology.

The name "infrared" covers a large part of the electromagnetic spectrum, from the long-wavelength edge of the visible out to the beginning of what we call the radio spectrum. It is the spectral range in which we find the major part of the radiation of objects cooler than stars. Although the Earth's atmosphere blocks most parts of the infrared wavelength region quite solidly, there are a few wavelength "windows" through which we can observe. The major problem, however, has always been the availability of detectors of adequate sensitivity. These have become available in the past decade or two, largely as a result of military developments. It is poetic and peaceful to make swords into plowshares, but the cold fact is that infrared astronomy has grown out of a technology whose original aim was to seek out enemies in the jungle night.

Radio astronomy also came from a military ancestry: the development of radar during the Second World War. Earlier, there had been little incentive to develop electronic circuitry for the very short radio wavelengths that do not bounce off the ionosphere and are therefore of no use for long-distance communication. But with the desire to "see" planes and ships in the darkness or through clouds, large engineering efforts went into broadening the radio spectrum. Then, with the return to peacetime pursuits, television became an important commercial activity, and radio technology continued to be improved.

Astronomers have taken advantage of all this technological development to "watch" the radio-wavelength radiation that comes from the objects of the Universe. To do so, however, they have to make use of telescopes of a special kind. A "radio telescope" is in a sense a radio antenna, but it has to have a special characteristic: the ability to point at a particular part of the sky. An ordinary antenna does not have this property; it receives radiation from a wide range of directions. Your radio always receives the signals from all the broadcast stations around you, but each of the stations transmits only in its own narrow range of wavelengths, and by tuning your receiver to that wavelength, you select the station to which you wish to listen. Interference between stations is avoided by government regulation, which carefully insures that no two stations near each other broadcast on the same wavelength (or frequency, which is a different but equivalent measure of the species of radiation).

Radio astronomy cannot use this kind of discrimination, since all the cosmic sources radiate over similar wide ranges of wavelength. Instead,

FIGURE 2.15.
The 140-foot (43-meter) reflector at Green Bank,
West Virginia. Note the hole in the main dish; this
radio telescope uses the Cassegrain focus.
(National Radio Astronomy Observatory, Green
Bank, West Virginia.)

sources must be discriminated according to their differing directions. In this sense radio astronomy is more similar to optical astronomy than to radio reception; we "look" by pointing the telescope, rather than "listen" by tuning it.

A radio telescope must be tuned nevertheless, because there are man-made signals at nearly all wavelengths and the radio astronomer must avoid interference from them. He therefore tunes carefully to one of the few "clear channels," reserved for radio astronomy through the political efforts of radio astronomers, who have had to plead their needs against those of other special-interest groups.

But the basic problem of radio telescopes is, of course, directivity. The ability to achieve this depends on the physical nature of the radio waves themselves, and specifically on the wavelength. What matters is the size of the telescope compared with the wavelength. A telescope that is about 70 wavelengths across has a directional sharpness of 1 degree; to resolve a tenth of a degree, a telescope would have to be ten times as large. Thus radio telescopes need to be big, but fortunately they do not need to be constructed as accurately as optical telescopes. Again, the quantity that matters is the wavelength; the reflecting surface needs to be accurate only to a modest fraction of a wavelength, typically an inch or so, com-

FIGURE 2.16.
Lining a valley near Arecibo, Puerto Rico, is a stationary reflector 1,000 feet (305 meters) across, the world's largest radio telescope. The radio receivers are in the cage suspended at the prime focus, above the "dish." (National Atmospheric and Ionospheric Observatory.)

pared with a few millionths of an inch needed for precision in an optical mirror.

Along with its high directivity, a large radio telescope has, like a large optical telescope, a high sensitivity. It collects more radiation than a small telescope, and this enables it to observe fainter sources. As with optical telescopes, the limitation on size is set by economics and engineering problems. The largest radio telescopes are imposing structures. Imagine a football field, raised in the air and capable of being tilted in any direction!

The wavelengths shorter than visible light pose a quite different set of observing problems. The foremost problem is the Earth's atmosphere,

27

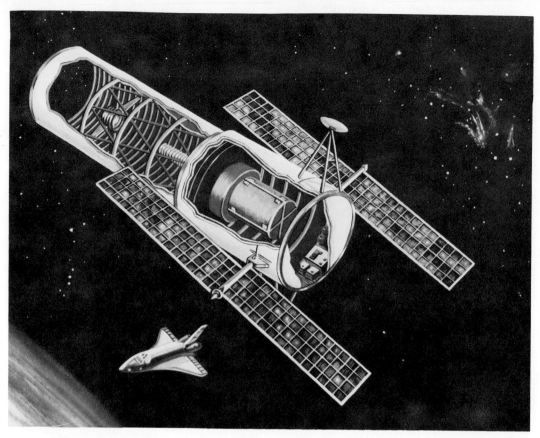

FIGURE 2.17.
In the 1980's, observations will be made by remote control with the Large Space Telescope. (National Aeronautics and Space Administration.)

which completely blocks all these short wavelengths except for a small fringe of ultraviolet. Observations must be made from above the atmosphere, in a limited way from rockets and balloons, or at greater length from spacecraft. Thus observations are expensive and scarce. A further complication is the lack of good telescopes for short wavelengths. In the extreme ultraviolet, X-ray, and gamma-ray regions, lenses are opaque and mirrors do not reflect. A typical "X-ray telescope" achieves its directivity by simply looking through long, narrow holes in an opaque block of metal.

PUTTING IT ALL TOGETHER

The tools of the astronomer are various, and the scientific problems that he faces are even more varied. It is important that we should think of the observations in terms of the underlying problems, however, because that is just what the astronomer does. Observation of the heavens is anything but indiscriminate mapping. Most often, the astronomer makes a particular set of observations because he believes that the results will illuminate some general question. Sometimes it is more of a shot in the dark, where he feels intuitively that measurements of a particular kind are bound to produce an answer that is interesting in some way. But always, as he gazes into the distance, his eye is set on the basic problems of the universe.

REVIEW QUESTIONS

1. List the names given to the different parts of the electromagnetic spectrum. For each region, give an example (if possible) from everyday experience. In which regions can we observe the heavens from the Earth's surface?

2. Explain the role of the ionosphere in radio communication. Contrast this with its role in radio astronomy.

3, Explain why a telescope is usually more valuable as a light-collector than as a magnifier. How does light-gathering power depend on the size of a telescope?

4. What are the important kinds of auxiliary instruments that are used at the focus of a large telescope? What is the purpose of each?

5. What are the different foci of a reflecting telescope? Why should several of them be available on the same telescope?

6. In Chapter 1, it was stated that the progress of astronomy has depended on developments in technology. Give specific illustrations.

3 | THE EARTH

It is natural that the study of astronomy should begin at home, with the Solar System, even though in a broader sense this is a narrow and parochial view. Our Solar System consists, after all, of a collection of minor bodies going around an ordinary star among billions of galaxies. (For this reason, this book is arranged so that a reader who so chooses can skip from here to Chapter 12 and continue immediately with the study of the stars.) Yet the Solar System is our home, and in our view of things it bridges the gap from the familiar, solid surface of the Earth to the black depths of the universe. Even more important, *we* are in it. Surely, when we consider the universe and its mysteries, one of the greatest wonders of all is that you and I are here to contemplate it. We will indeed move our attention toward the question of life in the universe, but our starting point is an astronomer's view of the Earth.

The Earth is our planet. It is the solid ground beneath our feet, and its gravity holds us to that ground. It is the air around us, and indeed, it is the very material that we are made of. But it is a planet, and our study of astronomy begins by regarding it as one.

SIZE AND SHAPE

Educated people have known for thousands of years that the Earth is round, and about 200 B.C. Eratosthenes of Alexandria succeeded in measuring its size. He reasoned simply and clearly that if he could measure

FIGURE 3.1.
The Earth, viewed as a planet, from just above the surface of the Moon.
(National Aeronautics and Space Administration.)

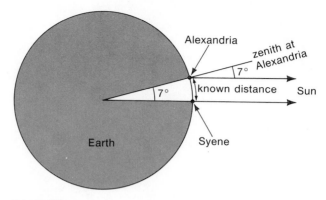

FIGURE 3.2.
Eratosthenes' measurement of the size of the Earth. By knowing the angle of the Sun from the zenith at Alexandria, and the distance to Syene, he was able to calculate the circumference of the Earth. (The angles are exaggerated in this diagram.)

a known fraction of the Earth he could then deduce the size of the whole. He had been told that at a point 500 miles to the south of him sunlight went straight down a well at noon on June 21, whereas he knew that at Alexandria on that day it missed the overhead point by seven degrees. Since the whole circumference is 360°, he reasoned correctly that the circumference of the Earth must be 360/7 of 500 miles, or about 25,000 miles. Eratosthenes did not press for the highest accuracy, but his method was correct, and in principle it is the method that we use today. Two measurements must be combined: an angular distance determined by observing the same astronomical bodies from two places, and the linear distance between those places as directly surveyed. If these measurements are made accurately enough, we can accurately calculate the size of the Earth. Its radius is 6,371 kilometers (3,959 miles).

But accurate measurements of the size of the Earth raise new questions—as accurate measurements so often do. The Earth turns out not to be perfectly round; its diameter through the equator is about 1/298 greater than the diameter from pole to pole. This is a very small bulge at the equator—far too small to notice in a scale drawing, for instance—but it makes a difference. It matters to the engineer who wants to predict the flight of a satellite or a missile; and it also matters to the student of science, because he can understand it as a natural consequence of the Earth's rotation. Rotation tends to throw things outward, and this "centrifugal force" causes the equatorial bulge. Dynamical theory connects the amount of the bulge with both the speed of the Earth's rotation and the extent to which its material is concentrated toward a dense center.

When theory establishes such a connection, it can be used in any direction that is convenient to the scientist. Thus from a knowledge of the Earth's rotational speed and its equatorial bulge, both of which are easily observed, geophysicists deduce useful information about the distribution of material inside the Earth, which they cannot observe directly. Similarly, instead of using the Earth's shape to predict the detailed orbit of a satellite, the geophysicist can use refined observations of the orbit to deduce the amount of the Earth's equatorial bulge. In this way the bulge has been measured with higher accuracy than ever before, and more detailed departures of the Earth from symmetry have been detected, down to differences on the scale of a few yards.

MASS AND DENSITY

Another physical quantity that expresses a different kind of "size" is the mass of a body. Mass is a quantity that lies at the base of all dynamics, but it is hard to define in a clear way. Some elementary books describe it as "the amount of material in a body." This does indeed express intuitively one of the basic properties of mass, although it is not the kind of rigorous statement on which science builds its structure. We shall see more of the role of mass in Chapter 6, in discussing forces and motion, and in Chapter 8 we shall try to understand how the origins of modern science depended on recognizing and suitably defining such concepts.

In everyday circumstances, the mass of a body is closely connected with its weight. When we weigh bodies, in fact, our real intent is to find out their masses—again, in an intuitive sense, the amount of material in them. Weight is a direct consequence of one of the basic properties of mass, gravitation; and it is a step into astronomy to recognize how the weight of an object in your hand is a manifestation of a more general law that controls a large part of the workings of the universe.

The law is gravitation. It says that all bodies attract each other, and that the force of attraction is proportional to the mass of each body. The force decreases, however, with the square of the distance between two bodies. Thus the large bodies of the universe exert very little gravitational force on us, just because they are so far away. The one gravitational force that we see strongly in everyday life is the force between the Earth and objects near its surface. This is what we call the force of gravity; it is responsible for the weights of objects. From the law of gravitation, it is clear that the weights of two bodies that are attracted by the Earth must be proportional to their masses. Thus it is that we can use weight as a measure of mass. Removed from the Earth, however, and put in a different gravitational field, the bodies would have different weights—but their masses still remain the same.

In applying the law of gravitation to the attraction between the Earth and objects near its surface, the question immediately arises, what is the "distance" of the object from the Earth? Should we use its distance from the surface of the Earth, its distance from the center, or what? The answer can be found only by adding up the attractions due to all the infinitesimal bits that make up the whole Earth. The calculation is annoyingly difficult—it held up Newton's researches for a long time—but the answer is fortunately simple: any spherical body attracts as if all of its mass were concentrated at its center. Since the Earth is nearly spherical, the distance that we should use is therefore the distance to the center of the Earth; as we have seen, the equatorial bulge gives a small additional attraction.

Because the Earth is so much more massive than everyday objects, its attraction on them is tremendously greater than their attraction for each other. Thus the only gravitational force of which we are directly aware is that of the Earth. If we use sensitive enough measuring apparatus, however, the forces between objects of everyday size can be measured in the laboratory. By means of such an experiment, in fact, it is possible to measure the mass of the Earth itself, by comparing its force on an object with the gravitational force exerted on that same object by another object of known mass. The mass of the Earth is 6×10^{27} grams, or 6×10^{21} tons. (A gram is about 1/28 ounce; 454 grams make a pound.)

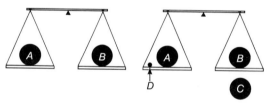

FIGURE 3.3.
Measuring the mass of the Earth. Masses A and B are adjusted to balance exactly. The attraction of C on B unbalances them, and the attraction of the Earth on D just restores the balance.

Merely knowing this astronomical number does not give us any insight into the properties of the Earth, but from it we can calculate another number that does. The *density* of a body is the ratio of its mass to its volume; this depends on the material of which a body is made but not on its size, so that the density is a clue to the nature of the material. (In everyday language, density is often loosely confused with weight; we say that lead is heavier than aluminum, when what we really mean is that it is denser.) Densities can be expressed in units of grams per cubic centimeter (g/cm^3), which automatically compares each substance with water, whose density is exactly 1 g/cm^3. (This last is not an accident; the gram was defined to make this so.)

The mean density of the Earth is 5.5 g/cm^3. Since the surface rocks have an average density of only about 3 g/cm^3, there must be much denser material inside, to bring up the average. As we shall see later, this is one of the facts upon which our picture of the Earth's interior is built.

THE EARTH'S ATMOSPHERE

Between the solid Earth and the emptiness of space is a layer of air. Although vestiges of the atmosphere extend hundreds of miles up, nearly all its material is within a few miles of the surface—a layer relatively as thin as the skin of an apple. Like any other material, air has weight in the Earth's field of gravity; and in spite of the low density of air, miles of depth add up to a substantial weight. This weight is the atmospheric pressure. When we quote a pressure of 15 pounds per square inch, we are saying that the column of air standing on each square inch of the Earth's surface actually weighs 15 pounds. Air is compressible—as everyone knows from stuffing it into a tire or a balloon—and the weight of the air on top compresses the bottom of the atmosphere. Thus the air is densest near the surface and becomes rapidly less dense at higher altitudes, where less atmosphere is standing on top of it. Half the atmosphere is within 3½ miles of the surface, and most people begin to feel the thinning out of the air after the first mile or so of altitude.

FIGURE 3.4.
The atmosphere thins upward, because its upper layers have less weight to bear.

TABLE 3.1   THE COMPOSITION
OF THE EARTH'S ATMOSPHERE

Gas	Per cent[a]
Nitrogen	78
Oxygen	21
Argon	1
Carbon dioxide	0.03

[a]Of dry air. Amount of water vapor varies from 0 to 5 per cent, depending on temperature and humidity.

The Earth's atmosphere is a mixture of gases, in nearly the same proportions everywhere. This composition is given in Table 3.1. It does not include a firm number for water vapor, because the proportion of water vapor is highly variable. (This variation, with its accompanying clouds and rain, is one of the principal problems of meteorology.)

The existence and composition of the Earth's atmosphere are a consequence of the material from which the Earth was made and the way it has developed since. It is very likely that the Earth was made out of the same material as the Sun, itself a typical sample of the material of the universe. All the most common elements in that sample take a gaseous form at the temperature of the Earth's surface, or else they form compounds that are gaseous. They make up the atmosphere. The composition of our atmosphere, however, is strikingly different from that of the universe; in our atmosphere the two elements most abundant in the universe, hydrogen and helium, are almost completely missing. But this too is a natural consequence of the physical conditions: hydrogen and helium escape from the Earth's gravitational force.

The escape of gases from a planetary atmosphere is a result of the motions of the atoms and molecules. (When two or more atoms are joined together, as in water, H_2O, or carbon dioxide, CO_2, we call them a molecule.) The particles of a gas are in constant motion, according to the temperature of the gas; the motion is, in fact, a measure of the temperature. Typical speeds are half a mile per second, and at this speed a molecule collides with other molecules millions of times per second. As one might expect, the lighter molecules bounce off faster than the heavy ones. Very quickly an equilibrium is set up, in which lighter molecules move faster on the average, and heavier molecules have a lower average speed. The other factor that governs speeds is the temperature; at higher temperature all the molecules move faster than they do at low temperature.

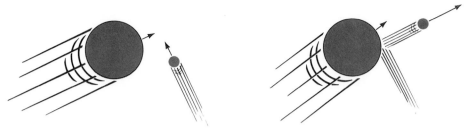

FIGURE 3.5.
When a small and a large molecule collide, the small molecule rebounds hard, but the large molecule is hardly affected. Thus collisions tend to give small molecules high speeds.

Near the Earth's surface these mean speeds merely determine how fast the molecules bounce around, but high in the atmosphere a molecule goes much farther between collisions, and very high up it has almost no collisions at all. Thus near the top of the Earth's atmosphere a molecule that is moving fast enough can escape unimpeded from the Earth's gravitation. The question is, how many molecules have more than the minimum velocity needed for escape? The answer clearly depends on the relation between the escape velocity from the Earth and the mean velocity of the molecules—and in fact it depends critically on the ratio of these two velocities. Many molecules have a little more than the mean velocity, but very few move with speeds several times the mean. In practice, molecules of a particular kind will escape in an astronomically short time if their mean speed is one fourth of the escape velocity, while they will last for the age of the Earth if their mean speed is a sixth of the escape velocity.

At the temperature of the Earth's surface, hydrogen, the lightest of molecules and therefore the fastest moving, escapes easily. Helium is also able to escape, but all the heavier atoms and molecules are slower moving and are unable to escape. Thus the Earth differs from the typical material of the universe mainly in having lost nearly all its hydrogen and helium.

Other planets have more or less atmosphere than the Earth, according to their ability to hold it gravitationally. We will return to this question, and to differences of atmospheric composition, when comparing the planets with each other.

Just as the density of the atmosphere depends on height, so also does the temperature. The differences of temperature are more complicated, however. They depend on two phenomena. One is vertical mixing; the

38

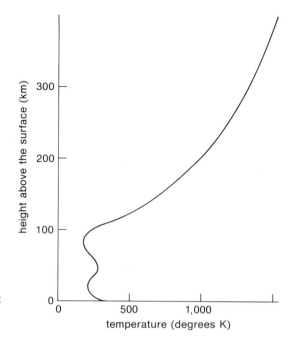

FIGURE 3.6.
Variation of temperature with height in the Earth's
atmosphere. Each of the warm regions is a level at
which some significant part of the Sun's radiation
is absorbed.

warm air near the ground rises, and as it rises it cools; similarly, when
cool air sinks toward the ground, it warms up. This vertical interchange
thus establishes a temperature gradient, which typically sees a drop of
about 1° C for each 110 meters increase in altitude (about 4° F per
thousand feet).

The other phenomenon that determines the temperature distribution
in the Earth's atmosphere is the input of solar energy. Most of it is absorbed at the Earth's surface or in clouds near the surface, but a few per
cent of the Sun's radiation is at short wavelengths that are absorbed in
the upper atmosphere. The shortest ultraviolet is absorbed at altitudes
above 100 kilometers (60 miles), and some longer wavelengths are absorbed in a layer of ozone (oxygen molecules consisting of three atoms
each) at 30 kilometers altitude (20 miles). The highest temperatures are
at the levels where the energy is absorbed; and in between, the temperatures are lower. Figure 3.6 shows how temperature varies with height.

It is this same absorption of solar radiation that determines the temperature of the Earth as a whole. There is a balance between the energy
that the Earth absorbs from the Sun and what it radiates into space.
The balance is a stable one, so that the Earth keeps the same temperature
from year to year. If the Sun gave us more energy, our gain of energy
would exceed the loss and the Earth would warm up; but it would then

radiate more—as all warmer bodies do—and a balance would soon be restored, at a higher temperature.

A solid body without atmosphere, bathed in the same sunlight that the Earth receives, would reach equilibrium at a temperature of 4° C (about 40° F). The Earth is actually a number of degrees warmer because of its atmosphere, which lets in the warming sunlight but partially blocks the outgoing infrared radiation by which the Earth cools itself. This warming is called the "greenhouse effect," by analogy with the similar effect that the glass of a greenhouse produces.

It is natural to give different names to the different layers of the atmosphere. Thus the first five or ten miles, in which extensive mixing occurs, are called the troposphere (Greek *tropos*, turn). The stable layers above are the stratosphere (*stratos*, layer). Then comes the ionosphere, which contains many electrically charged atoms (ions), produced when the solar ultraviolet radiation removes electrons from atoms. The uppermost layer, in which the atoms fly up and fall back down almost like free projectiles, is called the exosphere (*exo*, outside).

The ionosphere, because of its charged particles, has a strong effect on the longest electromagnetic waves, radio. The longer radio waves are reflected off the ionosphere, and it is their successive bounce between ionosphere and surface that make it possible to communicate around the Earth with waves that themselves travel in straight lines.

THE SURFACE OF THE EARTH

We live our lives in only a tiny fraction of the Earth's volume, on the surface and immediately above and below it. Air travel takes us only a few miles above the surface, and mines and wells go only a few miles deep. The heights and depths of the surface itself are also small, about six miles for the highest mountains and seven miles for the greatest ocean depths. Relative to the radius of the Earth, these are very small irregularities. In comparative terms, the Earth is much smoother than an orange, although it is not quite as smooth as a billiard ball.

It is the small irregularities in height that make the land. If the surface of the Earth were level, the water that is in the oceans would cover it more than two miles deep everywhere. Instead, there are high areas, the continents. These are made of a lower-density rock than the rock that is beneath them and in the ocean-bottom areas, so that the continents are in a sense floating in a sea of denser rock. The supporting rock does indeed move and flow, and pieces of the surface slide about. The surface consists, in fact, of a number of large, thin "plates," a few miles thick and thousands of miles across, which carry continents or parts of continents and ocean bottom. It had long been conjectured that the continents were in motion ("continental drift"), but only recently has study of ocean-

40

FIGURE 3.7.
This picture, taken from an altitude of 400 miles, shows a large part of India, and the island of Ceylon.
The white areas are cloud-covered; the picture gives some idea of the thinness of the surface layer in
which our weather occurs. (National Aeronautics and Space Administration.)

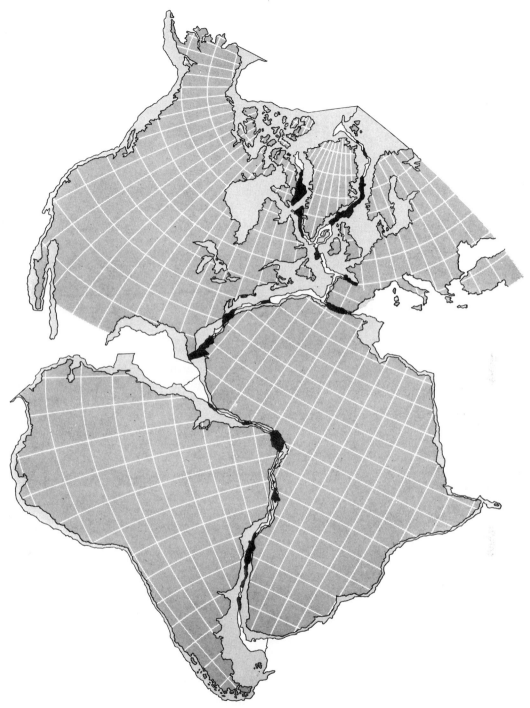

FIGURE 3.8.
The Atlantic Ocean is widening as the continental plates move apart. This diagram shows how well we can fit them back into their original positions, nearly 200 million years ago. (From P. M. Hurley, "The Confirmation of Continental Drift." Copyright © 1968 by Scientific American, Inc. All rights reserved.)

bottom rocks shown clearly that the plates are spreading away from several mid-ocean underwater ridges, where rock gradually wells up and pushes the plates apart. The motion is slow, but over the millions of years it leads to large drifts. Thus South America was once in contact with Africa but has now drifted 1,500 miles away. In other places, plates are driven against each other; these are regions of mountain-building and frequent earthquakes. In California, for instance, the East Pacific ocean-bottom plate is driving against the North American plate, sliding northward as it goes. Its forces have raised the Sierra Nevada over the past few million years, and frequent slippages occur along the San Andreas fault, which is the surface boundary between the plates. Similarly, the impaction of the Indian plate against the Asian plate is responsible for the greatest uplift in the world, the Himalaya mountains.

Tectonic activity, as this pressure-driven mountain building is called, is one of the two processes that continually rebuild and mold the Earth's surface. The other process is erosion. Rain, wind, and ice grind material off the Earth's surface, and streams and rivers then carry it away. Much of it ends as mud under the sea, whose pressure compacts the mud into new layers of sedimentary rock. Lifting and lowering, grinding and layering, these geologic processes have turned the layers of the Earth's surface into a chaotic jumble. If we could live 30 million years, instead of our 30 million minutes, we could watch the churning and renewal of the Earth's surface. As it is, the geologist examines the structure and the layering today, and he can often deduce the history that built it.

INSIDE THE EARTH

The study of the Earth's interior belongs to geophysics, but the results have astronomical implications. As in astronomy, the principal methods are indirect. Most important is seismology, the study of how earthquake waves travel through the Earth. The speed of these waves depends on the density of the material through which they are traveling, and the relative speeds at different depths determine in turn the path that the waves take. As in any material in which waves travel, the waves always bend toward regions where their speed is lower, and any abrupt boundary also reflects some fraction of the waves. Thus after an earthquake shock, seismologists consider where the waves are felt and how long they take to arrive, and they are able to deduce the density of the material at various depths and to locate the boundaries between layers of different properties. In putting the picture together, they are also aided by two quantities that are measured astronomically: the mean density of the Earth, which is the ratio of its total mass to its total volume, and its degree of central concentration, which can be deduced from the way in which the equatorial bulge perturbs artificial satellites.

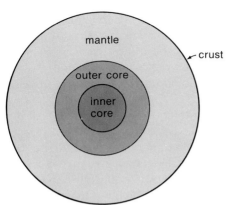

FIGURE 3.9.
The interior of the Earth. On this scale, the thickness of
the ink line represents the thickness of the crust.

In the resulting picture of the Earth's interior, there are four distinct regions. The crust, which is only about 30 kilometers (20 miles) thick, is made of rocks similar to those on the surface, with a density of about 3 grams per cubic centimeter (g/cm³). The next 2,900 km (1,800 miles) of depth make up a region called the mantle. Its upper boundary is a little denser than the crust, whereas at greater depths the rock is compressed to a density as high as 6 g/cm³ by the great weight of the overlying layers. Below the mantle is a sudden change to much denser material, called the core. The density in the outer core is about 10 g/cm³; the deeper region called the inner core is even denser.

The passage of seismic waves reveals a surprising property of the outer core. Of the two kinds of waves that can pass through a material body,

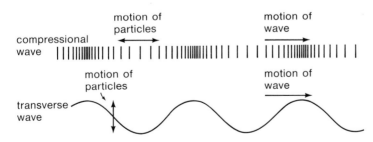

FIGURE 3.10.
Compressional and transverse waves. In both kinds the wave travels along through the material, but the motion of the particles is different for the two kinds of wave. Compressional waves can be carried by solids, liquids, or gases (for instance, sound waves in air). Transverse waves can be carried on the surface of a liquid (ocean waves), or inside a solid (the shaking of jelly.)

the compressional (in-and-out) waves go through the core, but the transverse (side-to-side) waves do not. This behavior is just that of liquids, which lack the rigidity that is required for the transmission of the side-to-side waves; hence we deduce that the outer core is liquid. At the boundary of the inner core is another discontinuity, which may indicate that the inner core is again solid.

Below the Earth's surface the temperature goes up, and this heat determines the shallow depth to which men can work in mines. Deeper in the Earth the temperature does not continue such a rapid rise, but it does increase slowly, reaching many thousands of degrees at the center. Although such a temperature would melt the rock under ordinary circumstances, the high pressures inside the Earth keep the material solid. The liquid layer in the core is somehow a result of the temperature and pressure, but we do not know enough about the properties of matter under such high pressures to be sure of just how it happens.

THE COMPOSITION OF THE EARTH

The basic material of the Earth is rock. Just as with the atmosphere, the nature of the Earth's solid parts depends on the composition of the material from which it was made. To see this, let us look in a little more detail at the composition of the universe, as it has been determined from spectral analysis of stars (Chapter 16) and interstellar gas clouds (Chapter 19). Most of the material of the universe consists of the elements hydrogen and helium. All the other elements add up to only about 2 per cent of the total mass. Of this residue, most of the material is carbon, nitrogen, oxygen, and neon; all the remaining elements add up to less than a tenth of the total of these four.

In the Earth, however, the most abundant elements are oxygen, silicon, aluminum, and iron—far from the typical composition of the material of the universe. Why this strange composition? The answer is to be found in the behavior of each of the elements as the material condensed to form the Earth. In the hot gas of a star, or in the very low density of interstellar space, the atoms go their separate ways, and very few of them join into the bound pairs or sets of atoms that are called molecules. When material comes together at a cool temperature like that of the Earth, however, the atoms begin to join into molecules. The atoms of each chemical element—that is, the atoms of each different type—behave in their own characteristic way, but the result depends on the density, which determines how often atoms encounter each other, on the temperature, which determines how hard they hit, and especially on what other atoms are present.

Hydrogen, the most abundant element, forms molecules readily, whereas helium, the second most abundant, does not form molecules at all. Thus the helium remained gaseous as the Earth formed. Because hydrogen was present in such excess, the other atoms combined with hydrogen, and the remaining hydrogen atoms combined with each other. The carbon produced methane (whose chemical formula is CH_4), the nitrogen produced ammonia (NH_3), and the oxygen produced water (H_2O). At Earth-like temperatures and moderate densities, these substances are all gases, as is molecular hydrogen. Since neon behaves like helium in not forming molecules, it also remained gaseous. Thus the six most abundant elements, making up to 99.9 per cent of the material, remained in a gaseous state as the Solar System began to form.

The seventh most abundant element, however, is silicon. It combines readily with oxygen, and the resulting molecular groups then combine with metallic atoms to form the solid substances called silicates. These are the rocks of the Earth's interior and of much of its surface. It thus appears that the existing Earth is a tiny solid residue, from which the volatile majority of the original material has been lost. Since the Earth's gravitation today is strong enough to hold methane, ammonia, and water vapor, either it was once heated enough to lose its envelope of these gases, or else it never had them. It is quite possible that the first stage was the formation of small rocky bodies, each of which had far too little gravitation to hold any gases, and that these then collected and coalesced to form the Earth.

In short, it was the affinity of silicon for oxygen, along with the inability to hold gases, that made the Earth the rocky lump that it is. But just as the silicon bound a part of the original oxygen, so a small part of the other gaseous elements was bound up in solid molecules. It is possible that the water of the oceans, which is one two-thousandth of the Earth's mass, and also the gases that now make up the atmosphere, which has only a millionth of the Earth's total mass, came from those compounds.

One further differentiation apparently took place in the interior of the Earth. The sharp difference between the mantle and the core indicates that at some time the whole interior of the Earth was fluid enough for the iron to separate from the lighter rock and sink to the center. This is another of the clues to the early history of the Earth, but large parts of the story still elude understanding.

THE EARTH'S MAGNETIC FIELD

It has been recognized since ancient times that the Earth has a magnetic field. The field is not strong enough to affect daily life in any serious way, but it is easy enough to detect, with as crude an instrument as a dime-store compass. The magnetic lines of force extend through the Earth, from pole to pole; and outside the Earth they loop around in wide arches from pole to pole, through and beyond the atmosphere. The

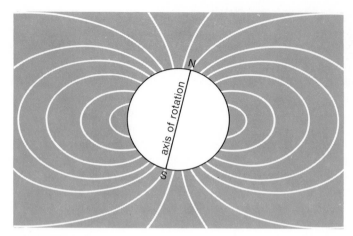

FIGURE 3.11.
The Earth's magnetic field arches through the space surrounding
the Earth.

field is probably due in some way to currents—fluid, electric, or both—
in the Earth's core; but the reason for its existence is basically not
understood.

The magnetic field is most important in the Earth's exosphere, particularly in its outermost parts. So far above the Earth's surface, there is
so little air that the magnetic field is completely dominant and controls
the motions of the particles (which at that level are electrically charged
and extremely sensitive to a magnetic field). At two different levels, high
up, are zones that are filled with high-speed charged particles, trapped
along the magnetic field lines and oscillating back and forth from pole
to pole. These zones are called radiation belts (according to a sloppy
terminology that extends the word "radiation" to any high-speed particles); another name for them is the Van Allen belts, after the geophysicist whose satellite-borne experiments first discovered them in the late
1950's.

At the Earth's surface, the most spectacular effect of the magnetic
field is the *aurora borealis,* or northern lights (in the southern hemisphere, *aurora australis*). These are a glow that sometimes flickers over
the sky in northern latitudes; it is caused by charged particles that gyrate
down the lines of magnetic force and make the upper levels of the atmosphere glow just like the gas in an electric advertising sign. Since the
magnetic field lines are concentrated near the poles, so is the aurora.
The Earth's magnetic poles do not coincide with the geographic poles,
however. The north magnetic pole is in northeastern Canada; hence
auroras are commonly seen much farther south in Canada and the eastern United States than they are in Europe or Asia.

FIGURE 3.12.
A bright aurora, as seen in Alaska. (V. P. Hessler.)

Bright auroral displays are usually caused by a "dumping" of charged particles from the radiation belts into the atmosphere, as a result of a burst of charged particles from the Sun. Hence their frequency is connected with the 11-year cycle of solar activity, which is discussed further in Chapter 17.

THE EARTH'S MOTIONS

Among the many and complicated motions of the Earth, the two simplest and most obvious are its daily rotation on its own axis and its annual revolution around the Sun. (It is customary to use different names for these two kinds of turning: a rotation is the turning of a body around an axis within itself, whereas a revolution is a motion around a distant point.)

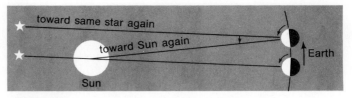

FIGURE 3.13.
The solar day is longer than the sidereal day. A day later, the
Earth has moved along in its orbit by 1/365 of a circle, and a
point on the Earth must turn an extra 1/365 before it again faces
the Sun. (The angles are exaggerated here.)

The Earth spins on its axis at the rate of one turn per day. It is this
rotation, of course, that makes the day and night; as the Earth turns from
west to east, the Sun and all the other celestial scenery appear to move
the other way. Thus a day is, for everyday purposes, the period from
noon to noon; and it is this *solar day* (averaged over some variations
that will be discussed in the following chapter) that we divide into hours,
minutes, and seconds for timekeeping purposes. From a more physical
point of view, however, this is a poor way to measure the rotation of the
Earth, because the reference direction to the Sun keeps changing as the
Earth moves around the Sun. It is better to use a standard fixed in the
distant stars. As is clear from Figure 3.13, a point on the Earth will come
around to face in such a fixed direction again a little before it again
faces toward the Sun. Thus a *sidereal day,* measured with respect to the
stars, is a little shorter than a solar day. The difference adds up to one
full day per year, so the difference in the lengths of the two kinds of day
is 1/365 of a day, or 3 minutes 56 seconds.

The Earth's rotation has a number of measurable effects on objects
on its surface. In all cases it is not the speed of the motion that we feel,
but rather the change of direction. (This is an example of the distinction
between velocity and acceleration that will be discussed in more detail
in Chapter 6.) This change of motion, or acceleration, manifests itself
in two ways: a centrifugal force and a deviation of moving objects.

The centrifugal force leads to a small variation of gravity with latitude.
Although simple in principle, this effect is complicated to work out in
detail, since gravity is also affected by the Earth's equatorial bulge,
which is itself related to the centrifugal force. The net result is that gravity is 1/200 less at the equator than at the poles.

Much more striking are the effects of our continuously turning reference system. For those of us who live in the northern hemisphere, the
ground is always turning counterclockwise under our feet. A moving
object tends to continue in a straight line as the Earth turns; hence relative to the moving surface it seems to be deviated clockwise, or to its own

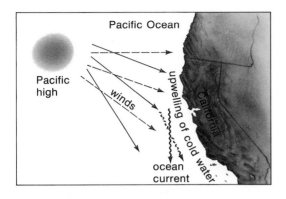

FIGURE 3.14.
California summer climate, as a result of Coriolis forces on winds and ocean currents. Because of Coriolis forces the motions are in the direction of the solid rather than the dashed arrows. The onshore winds are cooled as they blow across the cold coastal waters.

right. This effect, called a Coriolis force, affects projectiles, winds, and ocean currents. Thus the winds blowing away from high-pressure areas and toward the lows are all deviated into a pattern of clockwise circulation around the highs and counterclockwise circulation around the lows. In the southern hemisphere the directions are reversed. Similarly, the ocean currents follow a clockwise pattern in the oceans of the northern hemisphere and go counterclockwise in those of the south.

On a much smaller scale, a freely swinging pendulum feels a Coriolis force and slowly changes its direction of swing. Many science museums have such a demonstration, which is called a Foucault pendulum.

Although the Earth's yearly motion around the Sun has measurable effects on the apparent positions and motions of the stars, by far the dominant practical effect is the changing of the seasons. This phenomenon is due to the fact that the Earth's axis of rotation is not perpendicular to the plane of our orbit around the Sun. Thus our axis of rotation looks tilted—by 23½ degrees—as seen from the Sun. Because the axis maintains the same direction in space—that is, with respect to the stars— the Earth's northern hemisphere seems to lean toward the Sun during one half of the year and the southern hemisphere toward the Sun for the other half of the year.

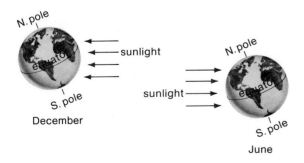

FIGURE 3.15.
The constant tilt of the Earth's axis of rotation is responsible for the seasons.

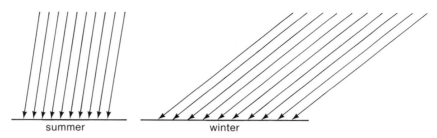

FIGURE 3.16.
The low-angled winter sunshine is spread over more ground and therefore warms it less.

Since time immemorial, men have noticed the annual change in the height of the noonday Sun, and they have kept their calendars by it. The time when the Sun is highest (for observers in the northern hemisphere) is called the summer solstice; it falls on June 21. The winter solstice, when the noonday Sun is lowest, is December 21. In between, the year is further divided by the dates when the noonday Sun is halfway between its extreme altitudes; these are the vernal equinox, March 21, and the autumnal equinox, September 23. (These dates can come a day early, on account of leap years.)

The hemisphere that leans toward the Sun receives more sunlight in two ways. First, the Sun is up for more than half of the day. Second, the Sun's rays strike the surface in a more nearly vertical direction in the summer, whereas in the winter the same rays, at a low angle, are spread over a larger surface. At latitude 40 degrees a unit area of surface actually receives three times as much sunlight at the summer solstice as it does at the winter solstice.

In addition to this result of the Earth's axial tilt, the amount of radiation received by the whole Earth varies with our distance from the Sun. In January we are slightly closer than in July. This effect somewhat moderates the northern-hemisphere seasons and makes those of the southern hemisphere more extreme, but these differences are small compared to those produced by local configurations of land and ocean.

Arguments like these tell us how much heat each hemisphere of the Earth receives, but they do not tell how warm it gets. The temperature of each part of the Earth is determined, like that of the whole Earth, by the balance between gains and losses of heat. Thus at the time of the summer solstice, when we receive the most sunlight, we are still warming up, and the warmest time of year is a month or so later, just as the warmest time of day is not at noon but rather an hour or two later.

REVIEW QUESTIONS

1. "How big is the Earth?" Answer this question in as many different ways as you can.

2. Distinguish between mass, weight, and density. What does the density of a body tell us about its composition? What does it not tell us?

3. What makes atmospheric pressure? Why is the pressure much less on top of a high mountain?

4. Why does an atmosphere escape from a body whose gravity is too low? Why do hydrogen and helium escape most readily?

5. State three important facts about
   (a) the Earth's atmosphere,
   (b) the surface of the Earth,
   (c) the interior of the Earth.
   In each case, explain why you consider these facts to be important.

6. Contrast the composition of the Earth with that of most of the remaining material of the universe. Why is the Earth made mostly of rock?

7. Explain, in as much detail as you can, why summer is warmer than winter. Why do the extremes of temperature not occur at the solstices?

4 | TIME, PLACE, AND THE SKY

We inhabitants of the Earth look out at the celestial scene. We see the unchanging background of the stars, and the Sun, Moon, and planets move in front of it. But our whole view is modified and modulated by the moving Earth and our position on it. The sky changes throughout the night and throughout the year, and its appearance is different from different parts of the Earth. Our first task in this chapter is to understand these complications; then we will see how they are used to solve the practical problems of navigation and timekeeping.

THE CELESTIAL SPHERE

Because we do not directly perceive the distances of the celestial bodies, they look as if they were all on the inner surface of a large sphere at whose center we stand. Some ancient cosmologies really believed in this sphere—which we do not—but even as a fiction it is a great convenience. Positions on the celestial sphere correspond to directions from us, and the relations between spatial directions are more easily seen when they are represented by points on a sphere.

   In specifying directions on the celestial sphere, two different systems of coordinates naturally arise: on the one hand we need to map the skies in a system fixed among the stars, while on the other hand it is natural to choose a system that is fixed with respect to the observer's horizon.

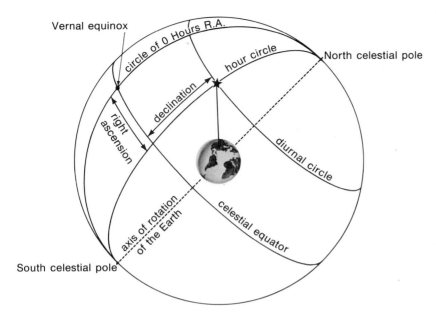

FIGURE 4.1.
The directions of stars in the sky are marked as "positions" on the "celestial sphere."

For a system fixed in the stars, the most convenient is one called the *equatorial system,* based on the Earth's poles and equator. The Earth's axis of rotation, extended upward from the North Pole, cuts the celestial sphere at a point called the celestial North Pole. Similarly, the point diametrically opposite is called the celestial South Pole. Halfway between the poles, the celestial equator runs completely around the sphere. In specifying the position of an object on the celestial sphere, one step is to locate it with respect to the equator and the poles. This coordinate is called *declination;* it runs from +90° at the North Pole of the celestial sphere down to −90° at the South Pole, 0° being at the celestial equator.

As the Earth rotates on its axis, the heavens appear to rotate in the opposite direction around the same axis. Since each star always keeps the same distance from the celestial equator and the poles, it appears to follow a circle of constant declination in its daily rotation, and for this reason such a circle is called a *diurnal circle.* All stars on a given diurnal circle have the same declination.

The other coordinate in the equatorial system locates a star on its diurnal circle. Since a circle has no beginning or end, this requires an arbitrary starting point on all the diurnal circles. For this marker we use the plane of the Earth's orbit, which is inclined 23½° to the equator,

FIGURE 4.2.
To a stationary camera, the stars have appeared to follow circles around the North celestial pole in this picture, which was exposed for a third of a day. (Lick Observatory.)

and so intersects the equator at two points. One of the two intersection points of these planes on the celestial sphere is called the vernal equinox, because it marks the direction of the Sun on the day of the vernal equinox each year. We draw a circle that connects the vernal equinox with the poles; then we specify the location of a star on its diurnal circle relative to the vernal-equinox circle by a coordinate called *right ascension*. It is counted continuously from zero to 24 hours, toward the east. The pole-to-pole arc on which a star lies is called its hour circle.

Right ascension and declination together give the position of a star on the celestial sphere. Since the coordinate lines are taken to be fixed on

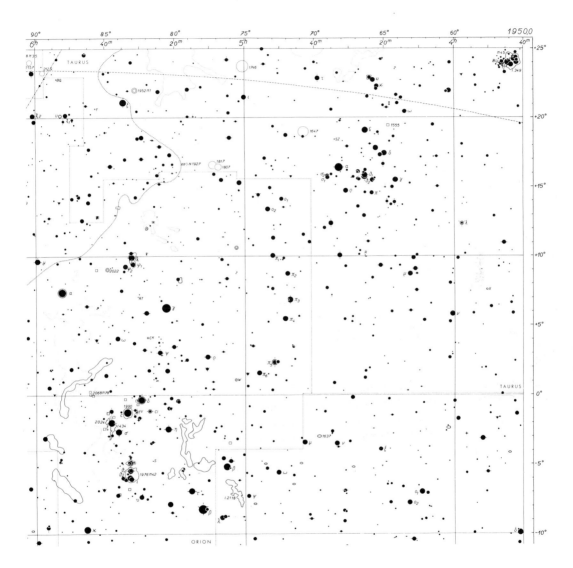

FIGURE 4.3.
Part of a star map, showing lines of right ascension and declination and constellation boundaries.
(From A. Bečvař, *Skalnate Pleso Atlas of the Heavens.* Copyright 1949 by Sky Publishing Corp.)

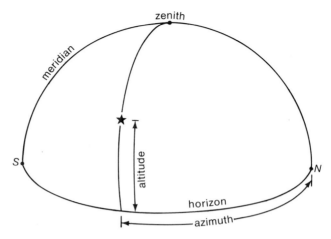

FIGURE 4.4.
In the horizon system, the position of a star in the sky is
given by its altitude and its azimuth.

the celestial sphere, the right ascension and declination of a star do not
change as the Earth rotates.

The rotation of the Earth does, however, make the stars appear to
rise in the east, move diagonally across the sky, and set in the west. Thus
the unchanging right ascension and declination do not tell an observer
where to look for a star at any given moment. Directions for pointing in
the sky are most easily given in terms of the observer's horizon and the
direction directly above him, which is called the zenith. In the horizon
system the angle of an object above the horizon is called its *altitude*; at
the zenith the altitude reaches 90°. The other coordinate is measured
around the horizon, and again it requires a starting point. For this, the
north point is chosen; the coordinate is called *azimuth* and is measured
to the right, from zero to 360°.

Right ascension and declination tell where a star is on a fixed map of
the heavens, whereas altitude and azimuth place it in our sky at the
moment. Calculating the relation between the two systems is one of the
basic tasks of spherical astronomy. The relation depends on the observer's latitude and on the time, and conversions from one system to the
other require tedious calculations.

Because the celestial poles represent the extension of the Earth's axis
of rotation, their place in the observer's sky depends in a simple way on
his latitude: the altitude of the pole is equal to his latitude. The hour
circles of the equatorial system all radiate from the poles, and the diurnal
circles have their centers at the poles.

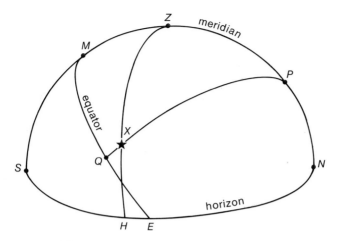

FIGURE 4.5.
Relation between the equatorial and horizon systems. *N, E,*
and *S* are north, east, and south points, respectively; *Z* is
zenith, *P* celestial north pole; $NP = MZ =$ latitude;
$XH =$ altitude, $NH =$ azimuth; $MQ =$ hour angle,
$QX =$ declination.

Uniting the two coordinate systems is one circle, the *celestial meridian.*
It runs from the north point of the horizon up through the zenith and
down to the south point; it divides every diurnal circle in half, and every
object reaches its highest point in the sky as it crosses the meridian. Since
the meridian passes through the celestial pole, it coincides with one of
the hour circles of the equatorial system. But the rotation of the Earth
brings each hour circle past the meridian in turn. Thus, on the one hand,
any stage in the Earth's daily rotation can be identified by specifying
which hour circle is on the meridian, and, on the other hand, the location
of any hour circle can be specified by stating where it is in relation to
the meridian. The latter is expressed as the *hour angle* of a star: if it is
2 hours west, it crossed the meridian two hours ago, whereas stars that
are 1 hour east will cross the meridian an hour from now. The final step
in the relationship between the equatorial and horizon systems is to
specify which hour circle is on the meridian. At any given moment, the
right ascension of the hour circle that lies along the meridian is called
the *sidereal time.* Hour circles of higher right ascension lie east of the
meridian, whereas hour circles of lower right ascension are in the west.
As the sidereal time ticks on, the celestial sphere turns from east to west,
and the hour angle of each star keeps changing. At a given moment,
the hour angle of a star is given by the difference between its right ascension and the sidereal time. The hour angle is east if the right ascension
is greater than the sidereal time and west if the right ascension is less.
As explained earlier, the Earth's motion around the Sun makes the

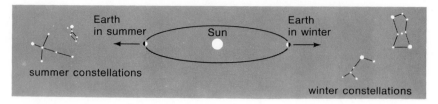

FIGURE 4.6.
As the Earth goes around the Sun, we see different groups of constellations in our nighttime sky at different times of year.

sidereal day shorter than the solar day, and it is of course the solar day that is used for ordinary "civil" time. The result is that each star crosses the meridian four minutes earlier each day; or, from the other point of view, the sidereal time is four minutes later each successive midnight, or two hours later each month. As the year passes, successive parts of the heavens dominate our nighttime sky. The reason for this annual change of the nighttime stars is even clearer if we look at a diagram of the Earth, Sun, and stars.

CONSTELLATIONS AND STAR MAPS

Anyone who looks at the stars sees patterns among them, and as time passes the patterns become traditional. Our own constellations probably date from the third millennium B.C.; they have been handed down through the cultures of Mesopotamia, Greece, the Arabic world, and western Europe. In recent centuries astronomers have filled in constellations in the extreme southern sky that were not visible from the classical lands, and the whole sky is now divided into 88 constellations. Early in this century, astronomers agreed on precise constellation boundaries, so that every point on the celestial sphere now falls within the area of some constellation. Many types of star are named or numbered by constellation. (For charts of the constellations, see pp. 481–484.)

The stars too have names. Many of these also date from ancient times. A few are Greek, but most star names come from the centuries when our classical heritage was preserved by Arabic-speaking peoples. They keep their Arabic flavor, if not their authentic Arabic forms. Modern astronomers use star names only for convenience, however, and most of the old names have been dropped in favor of a system that assigns to successive stars in a constellation the letters of the Greek alphabet. The ordering is usually by brightness, but in a few constellations it goes according to position. Fainter stars are designated by their serial numbers in some catalog; since many sorts of catalog exist, a learned paper on a particular star often begins with a list of its aliases. If all else fails, an accurate position in the sky is given—preferably along with a marked chart.

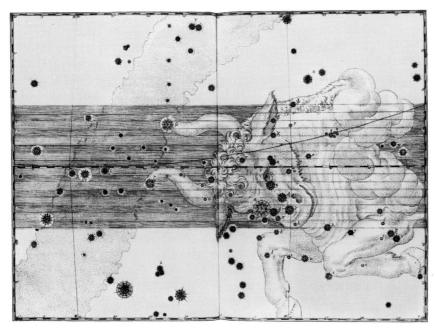

FIGURE 4.7.
Johann Bayer's seventeenth-century star atlas was illustrated with fanciful
drawings of the constellation figures. (Courtesy of Bancroft Library, University of
California, Berkeley.)

Star positions are quoted in right ascension and declination, since
the equatorial system, based only on the Earth's equator and ecliptic
planes, does not change with time of day or time of year. Even these
coordinates, however, are not completely invariant. The Moon's gravitational attraction shifts the direction of the Earth's pole of rotation,
and the other planets change the plane of our orbit. The amounts of
these shifts are known, and their effects can be calculated, but the result
is that the right ascension and declination of every star slowly change
with time; this effect is called precession. A star's position changes about
a degree in a century. Thus with a star position a date, or epoch, must
be quoted; and to set a telescope on it accurately, the position must be
precessed up to the current year.

EFFECTS OF LATITUDE AND LONGITUDE

Since the zenith, the straight-up direction, is the direction away from the
center of the Earth, then at any moment of time each point on the Earth

has its zenith in a different direction. Thus the sky looks different according to an observer's longitude and according to his latitude.

The effect of longitude is directly related to time, since the Earth's daily rotation carries successive longitudes around successively through the same orientations. Thus observers at the same latitude but at different longitudes will have the same views of the heavens—but at different times. These sights include the passage of the Sun across the meridian, so different longitudes have different times of day at the same moment. At 40 degrees latitude, the local apparent time changes by one minute for each 13 miles of travel east or west. It would be utter chaos for everyone to keep strictly local time, of course, so by agreement every region has adopted a standard time that differs by a whole number of hours from the time at the reference location, the Greenwich Observatory in London. Thus the United States is divided into four standard time zones; Eastern, Central, Mountain, and Pacific Standard Times are respectively 5, 6, 7, and 8 hours behind Greenwich Mean Time. The standard meridians of longitude pass approximately through Philadelphia, Chicago, Denver, and Los Angeles.

A traveler who went westward completely around the Earth would have to set his watch back successively by twenty-four hours; to make up for this there must be a point where the time jumps forward a whole day. This is the International Date Line, which zigzags down the Pacific Ocean, avoiding all land so as not to assign different dates to closely neighboring places.

FIGURE 4.8.
Time zones of the world.

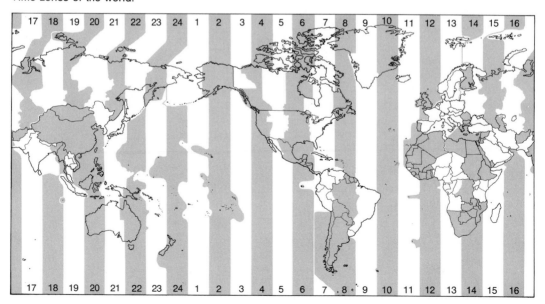

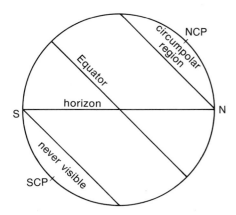

FIGURE 4.9.
Circumpolar regions for an observer
in midnorthern latitudes.

Latitude, by contrast, has a quite fundamental effect on one's view
of the sky. Figures 4.9 and 4.10 show how an observer's latitude determines the height of the celestial pole above his horizon; its altitude is,
in fact, equal to his latitude. Each celestial object appears to move, of
course, in a diurnal circle around the pole. Diurnal circles near the celestial pole are completely above the horizon. Such stars are said to be
circumpolar; they are always up and never rise or set. There is a corresponding area around the opposite celestial pole whose stars never
appear above the horizon. In between these polar caps all stars rise and
set, and the slope of their paths across the sky depends on the observer's
latitude.

At the equator the horizon cuts all diurnal circles in half, and every
object is above the horizon for just half a day. In particular, the Sun is up
for 12 hours regardless of the time of year. At other latitudes the length
of the day varies. At 40 degrees north the June Sun is up for 15 hours,
whereas in December, at its extreme southern declination, the Sun is up
for only 9 hours. North of latitude 66½ degrees; the Sun at its extreme
northern declination is circumpolar and does not set; this is the "midnight Sun." By way of compensation—if it can be called that—the mid-
winter Sun does not rise at all.

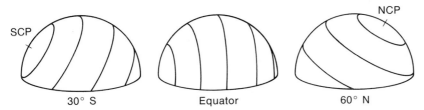

FIGURE 4.10.
Diurnal circles on the celestial sphere at three different latitudes.

TIMEKEEPING

The stars appear to move regularly throughout the night and throughout the year. Not only is it natural to keep time by them, it is also a necessity, since their apparent motions define what we mean by a day and a year.

A day should clearly correspond to the time between successive passages of the Sun across the sky. This definition, however, cannot be applied exactly, because the exact interval between the Sun's meridian crossing on one day and the next is not a constant. As was explained in Chapter 3, the solar day is shorter than the more fundamental sidereal day by an amount that depends on the Earth's motion around the Sun. Since our rate of motion around the Sun varies with the time of year, so also does the length of the solar day. It would be totally impractical to regulate clocks by this variable day, so instead the standard time is the *mean solar day,* in which the annual variations are averaged out.

Along with the variation caused by the Earth's varying orbital speed, the tilt of our axis causes a secondary variation. The resultant of the two is the "equation of time." Apparent solar time runs alternately fast and slow, with the result that the Sun crosses the meridian as much as 16 minutes before or 14 minutes after mean noon, the amount varying regularly with the time of year.

Even mean solar time is not constant, because the fundamental standard, the sidereal day, varies in length. In addition to small seasonal effects, there are random unexplained changes. Over short times such changes can be monitored by very accurate laboratory clocks. Over years and decades the motions of the other bodies in the Solar System have

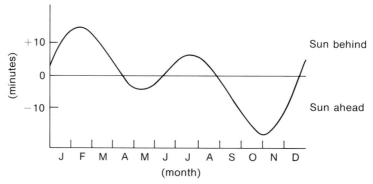

FIGURE 4.11.
The equation of time. This graph shows the difference between apparent solar time and the mean solar time to which our clocks are adjusted.

served as a check. These observations indicate that the Earth's rotation is slowing down—but so gradually that it will take 100,000 years for the day to lengthen by one second. This is the change in the length of a single day; because each day is then a little longer, the error in the time of day accumulates surprisingly fast. The oldest records of eclipses indicate, by the locations in which the Moon's shadow fell on the turning Earth, that 2,500 years ago the Earth was hours ahead of its current schedule of rotation.

The gradual lengthening of the day is a direct consequence of the friction of the tides, which is steadily slowing the Earth's rotation. The details of this process will be discussed in Chapter 5.

NAVIGATION

A scientific relationship can usually be used in either direction. Navigation is an example. Just as we saw how one's position on the Earth determines the appearance of the sky, so we can turn the question around and ask how, from the appearance of the sky, to discover one's position on the Earth. The answers are implicit in our previous discussion.

Latitude is easily calculated from stars of known declination. Figure 4.12 illustrates two simple rules: the observer's latitude is equal to (1) the altitude of the celestial pole, or (2) the declination of stars that pass through the zenith. Or, at the cost of a simple subtraction, one's latitude can be found from the altitude at which any star of known declination crosses the meridian.

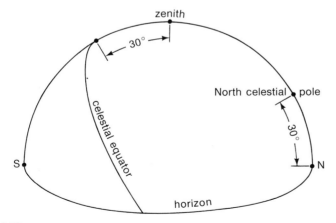

FIGURE 4.12.
At latitude 30 degrees north (corresponding to extreme southern parts of the United States), the north celestial pole is 30 degrees above the horizon, and stars of declination +30° pass through the zenith.

Longitude is calculated by means of a different principle. Observers at different longitudes have the same view of the sky, but at different times; so a longitude calculation requires a measurement of local time. The most straightforward way of doing this is to mark the meridian passage of the Sun or of some other celestial body of known right ascension. A longitude calculation requires one piece of auxiliary information, however; the observer must know the Greenwich time, either from a radio or from an accurate clock that is kept on Greenwich time. His longitude is simply the difference between his local time and Greenwich time.

The navigational methods just described are very simple, but they are not necessarily the most convenient. Practical celestial navigation involves more complicated trigonometry, but the observations are quick and need not wait for any particular object to cross the meridian.

THE CALENDAR

Practical necessity has saddled us with two incompatible units of time: the day and the year. There is, of course, no reason why they should be simply related, depending as they do on the two quite independent rates at which the Earth turns on its axis and travels around the Sun. The exact number of days in a year is not a whole number, and it can be made a whole number only by the device of leap years, where an extra whole day is added in just enough years to make the average come out right.

The ancient Egyptians knew that a year should have 365 days, but the first systematic rule for leap years was established by Julius Caesar, who decreed that every fourth year would be a leap year. The Julian calendar remained the calendar of the western world for sixteen centuries. Its year, of average length 365.25 days, is not quite correct, however; astronomers now know that the year should average 365.2422 days. Thus the Julian calendar is wrong by an amount that accumulates at the rate of one day in 128 years. By the sixteenth century the error had accumulated to a point where the vernal equinox was coming nearly two weeks earlier than it had when Caesar established the calendar. In 1582 Pope Gregory XIII corrected most of the accumulated error by dropping ten days from that year; in doing so, he set the vernal equinox at March 21, which had been its date in the fourth century A.D., when the Church set the rules for the date of Easter. He also provided for the future by adjusting the rule for leap years: thenceforth a year that was divisible by 100 would not be a leap year unless it was also divisible by 400. Thus 1700, 1800, and 1900 were not leap years, but 2000 will be. The Gregorian calendar was not accepted until much later in non-Catholic countries (in 1752 in Great Britain and its colonies, including America), but now it is used in all countries. Its year has an average length of 365.2425 days, so that the error is only one day in 3,000 years.

The confusion of the calendar has been compounded by the custom of reckoning in months. Aside from the mythological or religious significance of the Moon, it is a convenience to keep track of which nights will have moonlight. Ancient calendars reckoned in months that began regularly at new moon, but such a calendar has complications. The month averages 29.53 days; so extra days have to be added frequently. More important, the number of months in a year averages 12.37, so that extra months have to be added too. The Romans used a lunar calendar, but never kept it very accurate. Eventually, after politicians had gerrymandered the calendar out of shape to extend their own terms of office, Julius Caesar established our present system, in which the months are arbitrary divisions that bear no relation to the actual phases of the Moon.

Lunar calendars still exist. The Jewish calendar uses lunar months, and adds a whole month often enough to keep in step with the year—actually, seven extra months in every 19 years. The Christian church calendar uses this same 19-year cycle for setting the date of Easter, which is the next Sunday following the "Paschal" full moon. The vernal equinox is considered to be March 21, and the first full moon on or after that date is the Paschal full moon. Easter can come as early as March 22, or as late as April 25.

The Mohammedan calendar is also lunar, but it avoids the extra-month problem in a bizarre way—it ignores it! Its year has an average length of 354 days, and its months march around the seasons in a 33-year cycle.

REVIEW QUESTIONS

1. Identify the quantities that correspond to each other between these pairs: latitude, longitude; right ascension, declination; altitude, azimuth.

2. Why is it convenient for some purposes to identify a point in the sky by its altitude and azimuth and for other purposes to identify it by its hour angle and declination?

3. Distinguish clearly between right ascension, hour angle, and sidereal time. What is the quantitative relationship between them?

4. Describe (or sketch) the path of the Sun across the sky on December 21 and on March 21 as seen from (a) 40° north latitude, (b) the Equator, and (c) 20° south latitude.

5. If you were marooned in an unknown location on the Earth, how could you determine your latitude? Why would it be much more difficult to determine your longitude? What would you need in order to do the latter?

6. What does it mean to say that a calendar is "right" or "wrong"? What are the differences between the Julian and Gregorian calendars? Why is the Gregorian calendar not perfect?

5 | OUR SATELLITE, THE MOON

The Moon is much smaller than any of the planets, but it is the most conspicuous object in our nighttime sky, just because it is so close. Although the Moon has often been of more interest to poets than to professional astronomers, the space program has brought it into the forefront of research again. In our own outlook, as in the space program, it can serve as a starting point for a study of the universe. It is near but at the same time far. On the one hand, we see its apparent motion and its changes in the sky; on the other hand, we can view it objectively as an astronomical body and learn to understand why it looks to us as it does.

THE MOTIONS OF THE MOON

The Moon revolves around the Earth in a period of about a month; it is this repetition, of course, that is responsible for the use (and the naming) of the month as an interval of time. The average distance of the Moon from the Earth is 384,000 kilometers (239,000 miles). This is 30 times the diameter of the Earth, but is only 1/400 of the distance from the Earth to the Sun. From radar and laser measurements (both of which measure how long light or radio waves take to travel to the Moon and back), the distance to the Moon is known with high accuracy.

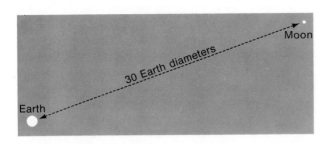

The relative sizes of the Earth and Moon,
and their separation.

The Moon's orbit around the Earth is an oval curve called an ellipse.
(Orbits will be discussed in more detail in the following chapter.) At
its closest point to us, *perigee*, the Moon is almost 5 per cent closer than
average; at its farthest, or *apogee*, point, it is 5 per cent farther than
average.

The length of time that the Moon takes to go around the Earth depends on how we look at its motion. When marked against the background of the stars, which serve as a fixed reference standard for all
motion, its period is 27⅓ days; this is called the *sidereal month.* More
obvious to us, however, is the Moon's position relative to the Sun, since
this is what determines its phases, from new moon to full moon to new
again. Because the Earth moves around the Sun, the directions of successive new moons shift relative to the stars, as is clear from the accompanying figure; and the interval between new moons is thus longer than
a sidereal month. This period, 29½ days, is called the *synodic month.*
This is the month on which lunar calendars are based. (Note the similarity here to the distinction between the sidereal and solar days.)

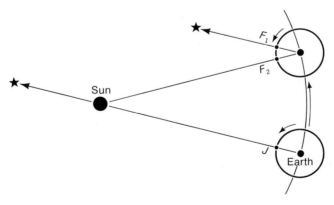

FIGURE 5.2.
New moon in January occurs when the Moon is at *J.* One
sidereal month later, at F_1, the Moon is in the same direction
with respect to the stars; but new moon occurs later, when
the Moon is at F_2. (The orbits are not drawn to scale.)

68

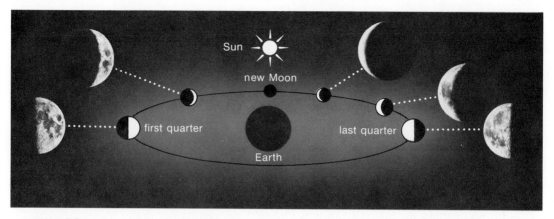

FIGURE 5.3a.
The phases of the Moon during the "darker" half of the month. The pictures show how the Moon looks from the Earth at each of the marked phases.

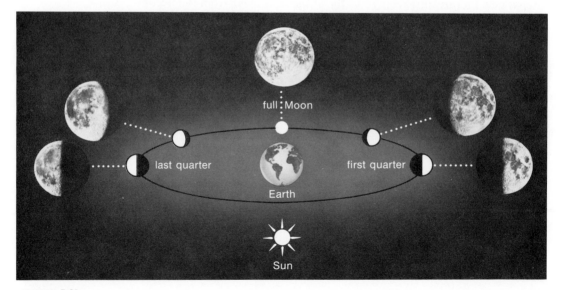

FIGURE 5.3b.
The phases of the Moon during the "bright" half of the month. In this picture, the sunlight comes from behind the observer. The pictures show how the Moon looks from the Earth at each of the marked phases.

As the Moon moves through the synodic month, its appearance changes completely, as we see it from the Earth. These changes are called its phases. Only the side of the Moon that faces the Sun is illuminated by sunlight; we see more or less of that side, depending on the Moon's position relative to the Sun and the Earth. Near *new moon* (as we call the phase when the Moon is between the Sun and the Earth) we are looking at the dark side of the Moon, and only a small edge of the bright side shows, as a narrow *crescent*. The crescent grows, and seven days after new moon, at the phase called *first quarter,* we see half of the bright side and half of the dark side; so the Moon looks then like a half-filled circle. The bright part continues to grow after first quarter, through the phases called *gibbous*. Halfway round the month, when the Moon itself has gone halfway round the Earth, we look directly at its illuminated side and see a *full moon*. After this two-week period of *waxing*, the Moon then *wanes*, as it moves through positions in which we see progressively less of its bright side. During the week following full moon, the Moon is gibbous again, until at *last quarter* we again see a half moon. During the last week of the synodic month, the Moon passes through its waning crescent phases, until it is again at new moon.

The phases of the Moon are thus just the changing appearance of a dark body that is illuminated by the Sun, as this body moves around us. It is clear from this picture that the times of day when we can see the Moon must be closely connected with its phases. Thus at new moon the

TABLE 5.1   THE PHASES OF THE MOON

Time	Shape	Position	Rises	Sets	Visible
New moon	Little or no crescent	Very near Sun	Sunrise	Sunset	Not at all
3½ days after new moon	Crescent	45° east of Sun	Midmorning	A few hours after sunset	Early evening
First quarter	Half circle	90° east of Sun	Noon	Midnight	Afternoon and evening
3½ days after first quarter	Gibbous	135° east of Sun	Midafternoon	After midnight	Late afternoon and most of night
Full moon	Full circle	Opposite Sun	Sunset	Sunrise	All night long
3½ days after full moon	Gibbous	135° west of Sun	A few hours after sunset	Midmorning	Most of night and early morning
Last quarter	Half circle	90° west of Sun	Midnight	Noon	Second half of night and all morning
3½ days before new moon	Crescent	45° west of Sun	A few hours before sunrise	Afternoon	Before dawn

Moon is so close to the Sun in the sky that we do not see it at all, while our night is completely dark. At full moon, on the other hand, the Moon is opposite the Sun; it rises at sunset and floods us with light all night long. Table 5.1 shows in more detail the relationship between phase and visibility.

Not only does the amount of bright area vary with the phase of the Moon; the apparent brightness of the Moon's surface changes too. Because the Moon reflects much more light back in the direction toward the Sun than it does in other directions, it looks much brighter at full moon than the mere size of its disc would suggest. Correspondingly, the surface brightness of the crescent moon is very low; this explains why we fail to see the crescent moon in the daytime, even when it is not very close to the Sun. At quarter and gibbous phases, on the other hand, the Moon is bright enough to be conspicuous in the daytime sky.

A few days away from new moon, when the Moon is a narrow crescent, the remainder of its disc can often be seen dimly. What we are seeing then is earthlight on the Moon. At this phase, while we look at a small crescent moon, the Moon sees a nearly "full earth." Because the Earth is much larger than the Moon and also reflects sunlight better, this bright earthlight makes the Moon quite easy to see.

Just as the Moon revolves around the Earth, it also rotates on its own axis. This rotation has one very strange characteristic, however; it takes exactly the same length of time as the Moon's orbital revolution. The effect of this synchronism is that, as the Moon moves around us, it turns on its axis by just the same amount, and as a result it always keeps the same face turned toward the Earth. The synchronism remains perfect throughout the centuries; we see the same face of the Moon that our ancestors saw. During each month there is a slight oscillation, however, because the Moon moves alternately faster and slower in its orbit, while its rate of rotation remains absolutely constant. Thus we see alternately a small amount around each edge. Also, the Moon's axis of rotation is not exactly perpendicular to our line of sight, so that when it is on one side of its orbit we see a little bit over its north pole, while half a month later we see a little bit beyond its south pole. All these small variations are called *librations;* in all, they allow 59 per cent of the Moon's surface to be seen from the Earth at one time or another.

The exact agreement of these two periods, the Moon's revolution and rotation, cannot be a matter of mere accident; something must have made them the same. The cause is the gravitational forces between the Earth and the Moon; they lead to a process called tidal friction, which we shall examine in some detail later in this chapter. Long in the past, tidal friction locked the Moon's rotation into synchronism with its orbital revolution, and it has remained this way ever since.

FIGURE 5.4.
When the Moon is a narrow crescent, we can see Earthshine illuminating the remainder of the surface that faces us. (Yerkes Observatory.)

FIGURE 5.5.
Libration slightly changes the face that the Moon turns toward us, as can be seen in these two photographs taken at the same phase but at different times. (Lick Observatory.)

THE MOON AS A PHYSICAL BODY

Measuring the size of the Moon is simple and straightforward. We measure its angular size, and knowing its distance from us we can calculate its linear size. The radius of the Moon is 1,738 kilometers (1,080 miles), about a quarter that of the Earth. The mass of the Moon is calculated from its gravitational effect on spacecraft that have passed near it; it is about 1/82 of the mass of the Earth. From the radius we can calculate the volume; mass divided by volume gives the mean density of the Moon. It is 3.4 times the density of water—considerably lower than the Earth's value of 5.5. The force of gravity on the Moon's surface, proportional to mass divided by radius squared, is about a sixth of that on the Earth's surface; thus a 180-pound astronaut weighs only 31 pounds on the Moon. The escape velocity from the surface of a body is proportional to the square root of its mass-divided-by-radius; for the Moon this is 2.4 km/sec. At the temperature of the Moon's surface, average-speed gas molecules would move about at a considerable fraction of this speed; hence the faster-moving molecules would escape easily, and an atmosphere on the Moon would soon be gone. If the Moon ever had any atmosphere, it has long since disappeared.

Thus the surface of the Moon is a barren, airless expanse of rock, covered with the finely broken rock that we call soil. It is not the same everywhere, however. Even a glance at a map of the Moon shows that there are areas of two quite different sorts. One kind of area is relatively smooth and dark; these regions, which are usually round, are called *maria* (singular, *mare*). Their name comes from the Latin word for sea, but they are actually low-lying plains of dark gray rock that is very similar to the basalt found in many volcanic regions on the Earth. The remainder of the Moon's surface is quite rugged, consisting of a succession of mountains and craters. These latter regions have come to be called the lunar highlands.

Craters are the most characteristic feature of the lunar landscape. Numbers of them dot the maria, and in the highlands they are everywhere. They range in diameter from 150 miles down to a few feet. The larger craters, which are easy to recognize on a photograph of the Moon, are great circular plains surrounded by walls several miles high. Often there is a mountain in the center, also some miles in height.

The lunar craters are scars of past impacts, when other small bodies struck the Moon at high speed. The craters are round because of the speed of the impact. When a small body strikes a solid surface at high speed, instead of plowing into the surface it releases the energy of its motion in a single, sudden explosion, which blows a round hole in the surface. At the speed at which a vagrant body of the Solar System would strike the Moon, a one-foot chunk of rock would have the energy of several tons of high explosive.

FIGURE 5.6.
The Earthward side of the Moon. This is a composite of photographs taken at first quarter and last quarter, when the angle of sunlight makes features stand out more clearly than they do at full moon. (Lick Observatory.)

FIGURE 5.7.
Mare Serenitatis is typical of lunar maria: round, with well-defined edges, flat, and relatively free of craters. (Lick Observatory.)

It seems likely that the maria are also the result of impacts, but much larger ones. Here a large portion of the lunar surface is smooth and level; the area is so large that there must have been a general flow of molten rock that filled up a large basin. Whatever the details may be, however, the maria must have been formed at a later time than most of the craters, because relatively few craters have marked their surfaces since they solidified.

The chronology of lunar features has been greatly improved by the analysis of rock samples brought back by the U.S. Apollo astronauts and the unmanned Soviet Lunik vehicles. Both highland and mare areas have been sampled; the analyses have shown what type of rocks compose the surface and how old the rocks are. (The dating of rocks from study of radioactive materials in them will be discussed in Chapter 11.) The rocks from the maria are indeed basalts, and their ages are between 3 and 4 billion years. Highland rocks are quite different, however. They are a type of rock called breccia, which is a compacted mass of broken fragments, in good agreement with the picture of the highlands as areas that were heavily bombarded and never remelted again. These rocks are older than those from the maria—a little over 4 billion years, on the average—but they are not as old as astronomers had expected them to be. Similar study of asteroids that have collided with the Earth shows clearly that the Solar System is 4.6 billion years old. It is therefore surprising to find that the rocks of the lunar surface did not solidify until nearly half a billion years later. Did the Moon form at that late date, in a hot state, or did it form earlier and cool slowly? Or did some great event, much more energetic than a mere crater-forming bombardment, heat the lunar surface to the melting point at that time? Unfortunately we still know far too little about the early history of the Moon.

At the same time as the lunar craters were being formed, the Earth must have received a similar bombardment. On the Earth, however, the marks have been almost totally erased. First, all but the largest bodies are slowed down and burned up by friction, as they plow their way down through the atmosphere. Second, scars are quickly wiped out by erosion or covered by deposits of soil or sediments. Such a conspicuous feature as the Arizona meteor crater was formed only a few tens of thousands of years ago; a few million years from now, little of it will be left to recognize. Finally, the drifting motions of the continental plates of the Earth's surface have repeatedly sunk, lifted, and folded the surface features until they are buried or changed beyond recognition.

Thus on the Earth we should expect to find only the marks of the few most recent impacts. Furthermore, these should be very few indeed, because most of the bombardment of the Earth and Moon took place during a relatively brief period in the early history of the Solar System, and the rate has since been much lower, as is clear from the relative

FIGURE 5.8.
The Arizona meteor crater, the best-known impact scar on the Earth, is nearly a mile across.
(From *Geology Illustrated* by John S. Shelton. W. H. Freeman and Company. Copyright © 1966.)

absence of craters in the lunar maria. They have lain exposed for three quarters as long as the highlands, but that whole period has left only a few marks on their smooth surfaces.

The Moon seems to be free of the tectonic activity that has so rumpled the surface of the Earth. The maria and craters lie as the last impact left them. A very sensitive test of this constancy comes from the systems of *rays* that surround some craters. These are trains of debris thrown out by the impact that formed the crater. Some of them run for hundreds of miles in straight lines over the lunar surface, undisturbed throughout the ages.

The total map of the Moon has one characteristic that is very puzzling. The far side almost completely lacks the bare, flat maria that cover so much of the side of the Moon that faces the Earth. Large circular structures do exist, but they are heavily cratered over. Apparently the history of bombardment was different on the two sides. Although the reason for this difference is not yet understood, it may indicate that the Earth had a strong influence on the paths of the bodies that hit the Moon. As we shall see, calculations of tidal friction indicate that the Moon was much closer to the Earth at that time; the difference between the two present-day faces of the Moon would then indicate that the Moon was already locked in its present orientation at the time the bombardment took place.

On the other hand, it is possible that the asymmetry in the Moon came first. Perhaps what is now the far side was simply a region of thicker

FIGURE 5.9.
Most of this photograph, taken from the Apollo 16 spacecraft, covers the far side of the Moon, but three of the near-side maria are at the upper left. The contrast between the terrain of the near side and the far side is striking. (National Aeronautics and Space Administration.)

FIGURE 5.10.
The surface of the Moon. The rock in the foreground is about 2 feet across.
(National Aeronautics and Space Administration.)

crust, less vulnerable to punctures by *mare*-forming impacts. On this hypothesis, the asymmetry determined the orientation into which the Moon's rotation would eventually be locked by tidal friction. The possibility of such differing explanations shows how hazy is our view of the early history of the Solar System.

On the small scale, the lunar surface consists largely of a soil that is made of broken particles of the surface rock. Although occasional bare rocks stick up, most of the surface is covered with soil to a depth of several feet. This porous soil is a poor conductor of heat; thus on the slowly rotating Moon the immediate surface swings through a temperature range of several hundred degrees each month, as it absorbs the heat of the Sun during its two-week-long "day" and radiates into space during its two weeks of "night." A few feet below the surface, the rock varies very little in temperature throughout the month, but the surface has so little thermal contact with the interior that the surface heats and cools almost as if the two were not connected at all.

Deeper below the surface, the interior of the Moon can now be studied in the same way as we study the interior of the Earth. Several of the Apollo expeditions set up seismographs on the lunar surface, and these have continued to radio their records back to Earth. The shocks whose reverberations are measured originate from moonquakes, from impacts of large meteors, and occasionally from intentional crash-landings of rocket vehicles. As can be done for the Earth, the timing and strength of the various waves can be analyzed to show how the properties of the lunar material vary with depth. The Moon's crust is about 60 km thick, with a gradual transition from the broken rock of the surface to a more solid rock below. The solid rock of the mantle extends about halfway to the center, but the material of the core, below the mantle, is much less rigid. Transverse waves pass through the Moon's core only with difficulty; so it must be at least partially molten.

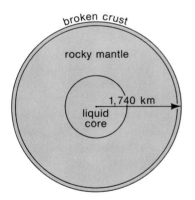

FIGURE 5.11.
The interior structure of the Moon, as shown by the transmission of moonquake waves.

Moonquakes are very much rarer than earthquakes, because the Moon lacks the tectonic activity that keeps the Earth's surface moving and reforming. The Moon is a passive body, and its quakes are probably induced by changes in the strain of the Earth's tidal force, just as an old frame house groans in the shifting wind. Moonquakes reverberate longer, however. The lunar rocks, devoid of water or gases, do not damp out vibrations nearly as heavily as do the rocks of the Earth's interior; they respond with a ring rather than a thud.

The Moon does not have any general magnetic field over its surface, although magnetizations of rocks indicate that there must have been some appreciable field at the ancient epoch when those rocks were formed. The disappearance of this field is probably connected with the way that the temperatures deep inside the Moon have changed with time.

The history of the Moon's interior is not well understood, but we can make some educated guesses. The Moon must have begun its life hot, and the surface then solidified during the next few hundred million years; this would account for the ages of the surface rocks in the highlands. Below the thin solid surface, however, the rock must have remained molten for a long time, so that the surface could be punctured by impacts and flooded by molten rock from below. Many of the large craters in the highlands are clearly flow-filled; and even a billion years later, maria were formed by much larger flows. But the seismic data indicate that at the present time solidification extends far in toward the center.

It is fortunate that the Moon is such a simple body, lacking the infinite complexities of the Earth's surface. It was possible to understand a great deal of its nature from earthbound observations, and now the first fruits of exploration have filled in the picture even further.

LUNAR EXPLORATION

Throughout human history, the Moon symbolized the unreachable. In a few decades of space technology, however, it has come within our grasp. The space age had its seeds in the German effort, near the end of the Second World War, to launch destructive rockets against Britain. After the end of the war, the remaining V-2 rockets were shot upward for scientific purposes, to explore the Earth's upper atmosphere. More importantly, the German rocket engineers became the nucleus of newly developing space programs in the United States and in the Soviet Union. The real space age began on October 4, 1957, with the launching of the first *Sputnik* (the Russian word for "satellite"). This Soviet achievement was a shocking blow to American prestige; it gave such political impetus to the space program that it was only a dozen years until Americans stood on the Moon.

In both nations, developments have been rapid. The first Soviet astronaut circled the Earth in 1961, and an unmanned Soviet vehicle had already photographed the far side of the Moon in 1959. Series of vehicles examined the Moon during the 1960's: Rangers took close-up photographs before crash-landing, a Soviet Luna and several American Surveyors made soft landings and photographed and tested the soil, and

FIGURE 5.12.
Locations from which lunar samples have been returned to the Earth.

Lunar Orbiters mapped the surface in detail. But the greatest day in space exploration was July 20, 1969, when an Apollo vehicle first landed men on the Moon: "One small step for a man, one giant leap for Mankind." Since then, four more Apollos have landed American astronauts to explore the lunar surface, and two unmanned Soviet Luniks have returned samples of lunar material to the Earth.

The issue of manned versus unmanned vehicles has been controversial throughout the history of the space program. Advocates of the manned program argue that the human astronaut has judgment and flexibility; he can react to what he sees, discuss it with his ground-based colleagues, and change his plans accordingly. Advocates of the unmanned program point out that most of the cost and complexity of a manned spacecraft goes into merely keeping the astronaut alive; for a fraction of this effort, incomparably better instruments could be flown in an unmanned vehicle. Without doubt, much of the reason that the balance has swung to the manned program is emotional. Scientific understanding is a step in human progress, but no advances of science appeal to the imagination as much as the sight of a human being standing on the Moon.

FIGURE 5.13.
On July 20, 1969, Man first left his footprints on the Moon.
(National Aeronautics and Space Administration.)

THE TIDES

Along with giving us light at night, the Moon has one other effect on everyday life: the tides. These are caused by its gravitational attraction on the Earth—or, rather, by the way in which that attraction varies with distance from the Moon. This last distinction, between the amount of attraction and its variation with distance, is an important one. The chief effect of the Moon's gravitational force on the Earth is just to move the Earth. The resulting motion is small and slow, however (a displacement back and forth every month by less than one Earth-radius), and only careful measurements can detect it. If the Moon attracted all parts of the Earth with the same force, this displacement would be its only effect. The actual force decreases with distance, however, so that the Moon pulls harder on the near side of the Earth than it does on the far side. The over-all result, then, is not just the displacement already referred to; there are also additional small forces that act to pull the near side of the Earth toward the Moon and the far side of the Earth away from the Moon. This kind of differential force, which acts as a stretch, is called a *tidal force*.

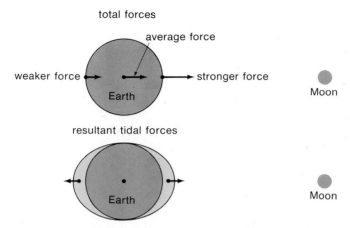

FIGURE 5.14.
Because the Moon attracts the near side of the Earth more strongly than the far side, its net effect is a stretching force on the Earth. The average force, as felt at the Earth's center, moves the whole Earth in a small monthly orbit around the common center of mass of the Earth and Moon; whereas the difference in the forces on the near and far sides of the Earth displaces the water of the oceans. The displacement is nearly equal on both sides.

The solid body of the Earth stretches somewhat under the Moon's tidal force, but the oceans yield much more, since their water has no rigidity. A flow of water takes place, raising the level of the ocean surface by about 3 feet at the points facing and opposite to the Moon. At the same time, the displacement lowers the water level at places around a circle that goes around the Earth halfway between these points. This is the way, at least, that the water would behave on an Earth whose surface was all ocean. On the real Earth, the tidal flow laps against the shores of the land, and the tides arrive at different times at successive points of each shoreline. Their heights differ from point to point, too, depending on the configuration of the shoreline; there are places where tidewater funnels up the narrowing shoreline of a bay and produces a rise and fall of 20 feet or more.

Because there are high-tide zones both on the side of the Earth that faces the Moon and on the side opposite, the Earth's daily rotation carries each point of the shoreline through two high tides per day, with low tides in between. During the half-day interval between tides, however, the Moon has moved somewhat in its orbit, so that successive high tides actually occur 12½ hours apart.

The Moon is not the only body that produces tides on the Earth; the Sun also contributes to the tides. Although the Sun is much farther away than the Moon, it also has a much greater mass, and its tidal force (which is proportional to mass divided by distance cubed) is almost half as great as that of the Moon. The net tidal force that we feel is the sum of the tidal force of the Moon and that of the Sun, and the resultant depends on how the two individual forces are directed. Near new moon and full moon, the two are pulling along the same line, and both the high tides and the low tides are more extreme than usual. At quarter moon, on the other hand, the solar high-tide region occurs where the Moon produces low tides, and vice versa; and the net effect is to reduce the size of both the high and the low tides.

The actual tides are even more complicated. The distance of the Moon from the Earth is continually varying, and this changes the strength of the tidal force. Perigee tides are almost 40 per cent stronger than apogee tides. The height of the tide at a particular location also depends on the latitude of the location and on the position of the Moon in relation to the Earth's equator. In practice, tide tables must be computed for various locations, taking all these factors into account and giving the times and heights of the high and low tides for each day.

In addition to the daily ebb and flow, the tides have an important long-term effect on the evolution of the Earth-Moon system. This is the phenomenon of tidal friction, which gradually slows down the Earth's rotation and increases the Moon's distance from us. The friction is between the ocean water and the bottom over which it must flow, particularly in the shallow seas over the "continental shelf" that extends beyond many parts of the shoreline of the continents. As a result of the friction, the Earth's rotation carries the tidal bulge around by a small amount, displacing it from the line connecting the center of the Earth with that of the Moon. The Moon's attraction on the displaced bulge then acts as a drag that slows down the rotation of the Earth. At the same time, the attraction of the tidal bulge on the Moon pulls it forward in its orbit, and this tends to enlarge the size of the Moon's orbit.

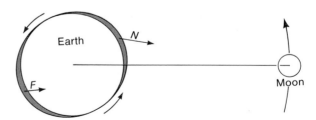

FIGURE 5.15.
The action of tidal friction on the rotation of the Earth. Friction carries the tidal bulge in the direction of the Earth's rotation. The force on the near-side bulge, N, which acts to slow the Earth down, is stronger than the force on the far-side bulge, F, which tends to speed the Earth up.

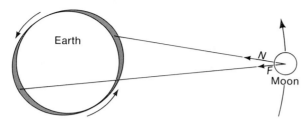

FIGURE 5.16.
The action of tidal friction on the motion of the Moon.

The effect of tidal friction is very small and slow, but over the ages it accumulates. At the present rate, the day becomes one second longer with the passage of about 100,000 years. It is very difficult to trace the past history of tidal friction, however, because its amount depends on the configuration of the shallow seas, which changes over relatively short geologic times. With any reasonable rate at all, however, tidal friction must have had a profound effect on the Earth-Moon system during the 4 billion years or more that the two bodies have lived together. In the remote past, the Earth must have turned much faster, and the Moon must have been much closer to the Earth.

At present the Earth produces no tidal friction on the Moon. The Moon always keeps the same face toward us, and the tidal bulge in the body of the Moon is thus "frozen" into one place. Since the bulge does not move around, there is no friction. If the Moon was rotating faster at some time in the past, however, the tidal friction that the massive Earth produced in it must have been quite strong, especially if it was much closer to the Earth at that time. (Even though the Moon has no oceans, the deformations in its solid material would produce considerable friction nevertheless.) This is undoubtedly what actually happened. Early in the Moon's history, tidal friction by the Earth dragged the Moon's rotation to a stop, relative to the Earth; and ever since then the Moon has rotated synchronously with its revolution, keeping the same face toward us.

ECLIPSES

Every month, at new moon, the Moon passes approximately between the Sun and the Earth, and there is a chance that its shadow will fall on the Earth, causing an eclipse. Similarly, at full moon the Moon passes on the opposite side of the Earth from the Sun, and there is a chance of its passing through the Earth's shadow. Because the Moon's orbit is inclined to that of the Earth, however, its shadow usually passes above or below the Earth at new moon, and the Moon usually passes above or below the Earth's shadow at full moon. But when the Moon happens to pass through its new or full phase near one of the points where it is crossing the plane of the Earth's orbit, an eclipse will take place. The line connecting these points is called the *line of nodes* of the Moon's orbit. This line stays relatively fixed in direction (relative to the stars) as the Earth and Moon go around the Sun; so eclipses take place at the two times of year when the Earth and Sun are oriented along the Moon's line of nodes. However, the direction of the Moon's line of nodes shifts slowly backward (as a result of the way that the Sun perturbs the Moon's orbit); so the eclipse seasons occur about 19 days earlier each year.

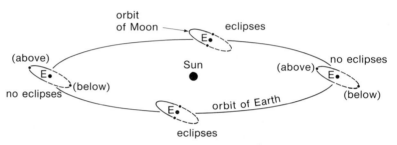

FIGURE 5.17.
Because the orbit of the Moon is inclined to that of the Earth, it is sometimes above the plane of the Earth's orbit (solid arc) and sometimes below the plane (dotted arc). Eclipses occur only when the Moon is both new or full and near its crossings of the Earth's orbital plane.

The Earth and Moon cannot get through an eclipse season without at least one eclipse occurring, and with suitable timing there can be an eclipse of the Sun at new moon, an eclipse of the Moon two weeks later at full moon, and another eclipse of the Sun again at the following new moon.

The circumstances of eclipses are easy to understand if we look at the profile of the shadows that the Earth and the Moon cast. Behind each body is a region called the *umbra,* in which the Sun is completely hidden, and a surrounding region called the *penumbra,* within which an observer would see the Sun's face partly hidden. Because the Sun is larger than the Earth or Moon, the umbra of each body tapers to a point, beyond which there is only penumbra.

When the Moon's shadow sweeps across the Earth, only observers in the path of the shadow see an eclipse of the Sun. When the Moon passes through the Earth's shadow, however, anyone on the nighttime side of the Earth can see the Moon eclipsed. The two events are fundamentally different; at an eclipse of the Sun it is really the observer who is being eclipsed. There is also a great difference between the size of the shadows of the Earth and the Moon. Where the Earth's umbra passes the Moon,

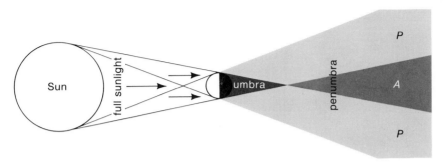

FIGURE 5.18.
The umbral and penumbral parts of a shadow. An observer in region *A* would
see an annular eclipse; in either of the regions *P* he would see a partial eclipse.
Only in the umbra would he see the Sun totally eclipsed.

it is several times as large as the Moon; during a total eclipse of the
Moon, the entire Moon is immersed in the Earth's umbra. The Moon's
umbra, by contrast, can reach the Earth's surface only when the Moon is
closer than average, and it never covers a spot more than about 150
miles in diameter. As this spot sweeps across the Earth's surface, observers within its narrow path see a brief total eclipse of the Sun, never last-
ing longer than 8 minutes. If the umbra fails to reach the Earth, observers
in the center of the Moon's shadow see an *annular eclipse,* in which the
Moon, at mid-eclipse, blots out the center of the Sun's disc but leaves
a bright ring all around. Annular eclipses of the Sun are more common
than total eclipses. In each case, however, the Moon's penumbra sweeps
over a path about half as wide as the Earth, so that a large part of the
Earth then sees a partial eclipse of the Sun.

FIGURE 5.19.
During a total eclipse of the Moon,
the Moon passes totally into the
Earth's umbra.

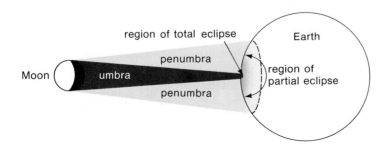

FIGURE 5.20.
The shadow of the Moon on the
Earth during a total eclipse of the
Sun.

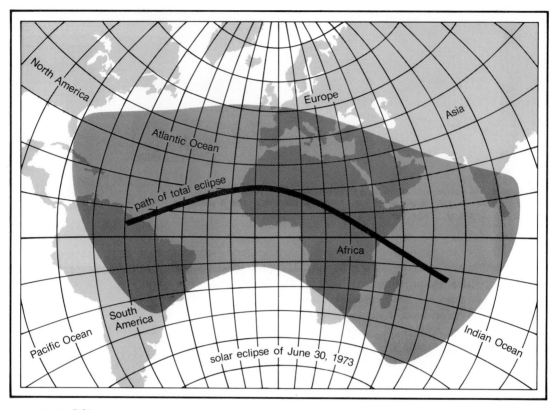

FIGURE 5.21.
Large parts of the Earth can see a partial eclipse of the Sun, but the total eclipse is seen only within a narrow track. This map shows the area in which the solar eclipse of June 30, 1973, could be seen. (Adapted from *American Ephemeris and Nautical Almanac.*)

A total eclipse of the Sun is a never-to-be-forgotten event. The hours of the partial phase are like any other partial eclipse, but the few minutes of total eclipse are a completely unique experience. As totality approaches, the Sun dwindles to a narrow crescent. The sunlight is much less bright than usual, but even the last edge of the Sun keeps the illumination at the semblance of a daylight level. But then, when totality begins, daylight turns into night in an instant. Stars are easily visible, and around the edge of the darkened Sun shines its corona, the pale outer fringes of the Sun's atmosphere, which are too faint to be seen in the ordinary bright skies of the daytime. The whole spectacle persists for the few brief minutes of totality; then daylight rushes back as the Moon uncovers the other edge of the Sun. All that is left is the few remaining hours of partial eclipse, plus the memory of seeing the sky's most spectacular event.

FIGURE 5.22.
The totally eclipsed Sun. The Sun's corona, surrounding the black disc of the Moon, is too faint to
see at other times. (Hale Observatories.)

Because the path of totality is so narrow, one who wishes to view a total eclipse of the Sun must almost always travel to do so. At any given place on the Earth, one can expect to see a total eclipse of the Sun only about once every 300 years.

A number of factors determine whether an eclipse of the Sun is total or annular, and if total, how long it lasts. Most important is the distance of the Moon from the Earth at the time of the eclipse; the longest total eclipses happen when the Moon is at perigee. The duration of totality also tends to be longer when the eclipse track is near the Earth's equator, where the Earth's rotation is fastest and therefore best carries the observer in pursuit of the rapidly moving shadow of the Moon. The point on the Earth that is closest to the Moon is also near the equator. Finally, total eclipses are longer when the Earth is farthest from the Sun, because the Sun then looks smaller.

Opportunities are much more frequent to see a total eclipse of the Moon, but the event itself is much less interesting. Observers on the whole nightime side of the Earth can watch the Moon slowly move into the Earth's umbra. (The slight dimming by the penumbra is hardly noticeable.) Totality can last more than an hour and a half, and the partial phases before and after take about an hour each. The eclipsed moon does not disappear completely; it is illuminated by a pale reddish light. This is sunlight that has passed through the Earth's atmosphere and has been bent around into the umbra. Just as the glass of a lens bends, or *refracts,* light, so does the Earth's atmosphere refract light by a small amount. This light is red because the atmosphere filters the mixture of colors that make up sunlight, and the red is the part of the spectrum that gets through best. This filtering is greatest when the light follows a long path through the atmosphere; it is what makes the setting Sun look red to us.

Eclipses do not occur with any obvious regularity, because the length of the month does not bear any simple relation to the interval between eclipse seasons. There is one long-period repetition, however, that is very striking. After 18 years and 11⅓ days (or 10⅓ days, depending on the number of leap years in between), the circumstances of an eclipse repeat almost exactly. By a chance coincidence, during this length of time 223 synodic months coincide almost exactly with the length of 38 eclipse seasons. The Earth is within a few days of the same place in its orbit, so that the angular size of the Sun is nearly the same as it was. More important, the Moon's perigee, which shifts around its orbit in a nine-year cycle, has gone twice around and is at the same place again, so that the Moon has the same angular size. This 18-year periodicity, called the *saros,* was known to the ancient Babylonians, who used it to predict the occurrence of eclipses. All the solar eclipses of this century that have the longest durations of totality belong to the same saros cycle (May 18, 1901; May 29, 1919; June 8, 1937; June 20, 1955; June 30, 1973; July 11, 1991).

REVIEW QUESTIONS

1. Explain the relationship between the Moon's phase and its position in its orbit. Why is it that eclipses of the Sun can occur only at new moon and eclipses of the Moon can take place only at full moon?

2. How do the lunar maria differ from the highlands? How were the maria formed? How do we know that they are younger than the highlands?

3. Contrast the number and the nature of craters on the Moon and on the Earth. What accounts for the differences?

4. Describe the interior structure of the Moon. How do we know about it?

5. How has the space program contributed to knowledge about the Moon?

6. Why do tides occur simultaneously on both sides of the Earth, rather than just on the side facing the Moon?

7. Why do lakes not have tides?

8. What causes tidal friction? Explain how it operates on the rotation of the Earth and on the orbit of the Moon.

9. Draw a cross section of the shadow of the Moon, showing both umbra and penumbra. Explain the difference between them. What causes an annular eclipse?

10. Why is it so unusual to see an eclipse of the Sun, and not at all unusual to see an eclipse of the Moon?

6 | ORBITS AND MOTION

The "whirling heavens" is a vivid metaphor. It is also a bewildering spectacle. The Earth turns on its axis daily and circles the Sun yearly; the Moon goes around the Earth each month. Other planets go round the Sun at different rates, and we watch them from our own careening platform. Small wonder that the workings of such a system are confusing to a casual observer—indeed, they are not at all obvious even to an observer who charts all the positions carefully. Yet underlying it all, there is really a beautiful simplicity, a few rules from which everything else flows as a logical consequence. Everyone knows, of course, that it is the force of gravitation that is responsible, but in fact this law is able to provide a solution to the problem only because of the existence of an even more basic set of laws: those that govern motion in general. We shall see in this chapter how planetary orbits illustrate the general laws of motion—or how the laws of motion explain the behavior of planetary orbits, which is just another way of saying the same thing.

The *apparent* motions of planets, as seen from the moving Earth, are a poor place to start, because our own motion so much complicates what we see. Instead, let us use some foreknowledge of the answers and look at the planets as they would appear to an observer who could magically hold himself fixed at a point above the plane of the Solar System. He would see the planets circling the Sun, all in the same direction and all in nearly the same plane. If he looked carefully he would notice that the orbits are not exactly circular, but that nevertheless each orbit repeats itself rather closely at each revolution. He might also notice that a planet moves a little faster in the part of its orbit that is closest to the Sun. He would certainly be struck that the planets closest to the Sun go around much faster than those farther out.

KEPLER'S LAWS

Three and a half centuries ago, Johann Kepler managed to put this picture together clearly, in spite of the complication of observing it from the moving Earth. He summarized it all in three laws:

(1) *The orbit of each planet is an ellipse, with one focus placed at the Sun.*

(2) *As a planet moves around its orbit, the straight line drawn from the planet to the Sun sweeps over equal areas in equal times.*

(3) *The squares of the periods of revolution of the planets are proportional to the cubes of their average distances from the Sun.*

Thus the first of these laws tells what the path is, the second law tells how the planet moves faster and slower in different parts of its path, and the third law tells how fast the average speed of each planet is.

Kepler's laws exemplify a number of things about scientific laws in general. First, they are not statements of what *ought* to be, like moral laws; they are statements of what *is*. Second, they simplify the whole picture, because they are generalizations—statements about the properties of all planetary orbits, rather than enumerations of the properties of each. Finally, they are able to simplify the picture just because they deal in new concepts and new language.

In Kepler's laws the new language is mathematical. (As we shall see later, it is quite different with the laws of motion, where the new language is physics.) Kepler's first law will tell us very little—except that orbits are not true circles—unless we first know what an ellipse is. There are various ways of defining an ellipse. Even mathematics books use various definitions, and none of the more common ones are particularly easy to visualize; here we will simply note a few properties of an ellipse. First, an ellipse can be thought of as a circle that has been stretched uniformly along one direction—or, what is equivalent, squashed the other way. We see an ellipse every time that we look at a circle obliquely. Alternatively, as an operational definition, a string whose two ends are pinned at different points will constrain a pencil that pulls against it; and as the pencil slides along, it will trace an ellipse. The two pinned points are called the foci. They have this name because among their properties is the fact that a light source at one focus will have its rays reflected by the ellipse to the other focus. (The same is true for sound; this is the basis of the

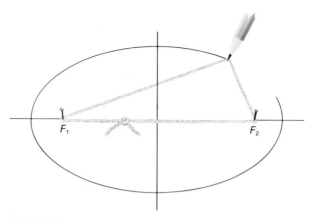

FIGURE 6.1.
How to draw an ellipse. The pins at F_1 and F_2 mark
the two foci of the ellipse. (From A. Baez, *The New
College Physics*. W. H. Freeman and Company. Copyright
© 1967.)

"whispering galleries" that focus a voice from one point to another in some vaulted corridors.)

Aside from its placement, any particular ellipse is described by two properties: a size and a shape. As a measure of size we use a quantity analogous to the radius of a circle, but measured along the longest diameter; it is gracelessly called the semi-major axis. To specify the shape of an ellipse, we measure how far out along the semi-major axis the focus lies. The quantity that measures this is called the eccentricity, a name that makes good etymological sense: "out-of-center-ness." The size of the eccentricity ranges from zero, for a circle, to 1, for an infinitely narrow ellipse.

One might think it more natural to describe the shape of an ellipse by the ratio of its minor to its major axis, but mathematicians find that the more sophisticated properties of ellipses and other related curves are better studied in terms of the eccentricity than of the axial ratio. (For an ellipse of eccentricity e, the axial ratio is $\sqrt{1 - e^2}$.)

Astronomers add another pair of labels to the elliptical orbit of each planet. One end of the major axis, where the planet passes closest to the Sun, is called the *perihelion,* and the other end of the major axis, farthest from the Sun, is called the *aphelion.* (These come from the Greek *peri,* "close to," and *apo,* "away from," joined to *helios,* "Sun." For orbits around the Earth, the corresponding terms are *perigee* and *apogee,* whereas double-star astronomers refer to *periastron* and *apastron.*)

Kepler's second law is stated in terms that are even less familiar; none of us have spent much time watching rotating lines as they sweep over areas. In Kepler's time geometry was the language of mathematics, but

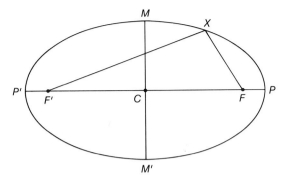

FIGURE 6.2.
An ellipse, with its foci, *F* and *F'*, its major axis
PP', and its minor axis *MM'*. The eccentricity is the
ratio *CF/CP*. For any point *X* on the ellipse,
FX + F'X = PP'.

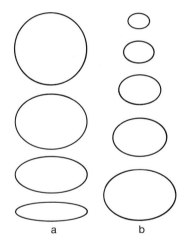

FIGURE 6.3.
Ellipses of different sizes and shapes. Those in column a have
the same major axis but different eccentricities, whereas those
in column b have the same eccentricity but major axes of
different length.

today we would find it more comfortable to state his second law in more
modern terms:

(2') *As a planet moves around its orbit, its direction as seen from the
Sun changes at a rate that is inversely proportional to the square
of its distance from the Sun.* (In more physical terms, the angular
speed goes inversely as the square of the distance.)

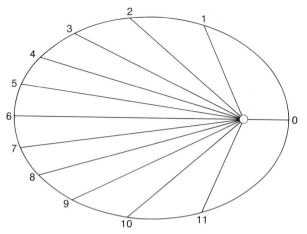

FIGURE 6.4.
Kepler's second law. Each of the 12 sectors has the same area, and a planet going around the Sun will traverse each of the 12 arcs in the same length of time. (From P. van de Kamp, *Elements of Astromechanics.* W. H. Freeman and Company. Copyright © 1964.)

Notice that this law does not tell us directly the speed of the planet *along* its orbit; that has to be figured from the angular speed and the direction in which the orbit is taking the planet at that moment.

The third law is fortunately straightforward in its mathematical statement. Only one small improvement is needed in the simple statement that was given above: for the imprecise term "average distances," we should say, more specifically, "semi-major axes."

THE LAWS OF MOTION

Kepler's laws tell *how* the planets move, but they say nothing about *why*. We think of motions as being governed by forces, but nowhere yet has a force appeared in our discussion. Indeed, even if we eruditely cite the force of gravitation, already mentioned in a previous chapter, it does nothing to make the particular behavior of orbits understandable in any direct way. What is lacking so far is an understanding of the *way* in which forces govern motion. This understanding comes from studying simpler situations and arriving at generalizations—laws—about motion. It is surprisingly hard to find these simple situations, however; nearly every motion that we observe is subject to a mixture of forces, some of which are hard to identify.

The greatest hindrance to a clear understanding is friction, which slows moving bodies and makes stationary bodies hard to set in motion.

If bodies moved on perfect, frictionless bearings, then we would see, clearly and often, the operation of Newton's first law of motion:

(1) *A body remains at rest, or else it continues to move with the same speed in the same direction, unless it is acted on by a net force.*

The law says "a net force," because a force has no effect when it is balanced by an equal opposing force—like gravity pulling down on a body while the ground underneath holds it up.

Contained in this law is an interesting combination of quantities: speed and direction. Many quantities in physics have both a size and a direction; they are called vectors. This particular vector, speed-and-direction, is called velocity. In terms of velocity we can state Newton's first law more economically. In fact, if we recognize that "at rest" is just a particular value of velocity—namely zero—we can say simply:

(1′) *A body moves with constant velocity unless it is acted on by a net force.*

The next step will be to discuss changes in velocity. Changes in velocity vectors are easy to handle, if we remember that the change itself is a vector and has a direction. To add two vectors we simply lay them end to end (preserving their lengths and directions) and replace the pair by a single arrow that goes straight from the beginning to the end.

FIGURE 6.5.
The vectors $\vec{V_1}$ and $\vec{V_2}$ can be added by placing the beginning of one at the end of the other. The resultant $\vec{V}$ is the vector sum of the two.

Force is a vector too, and in terms of it we can begin to state Newton's second law of motion:

(2a) *The velocity of a body changes at a rate that is proportional to the net force acting on it.*

An important idea lies hidden in the vector nature of the quantities: the change of velocity is also in the same *direction* as the force. Thus a force in the direction of a body's motion increases its speed, and an

98

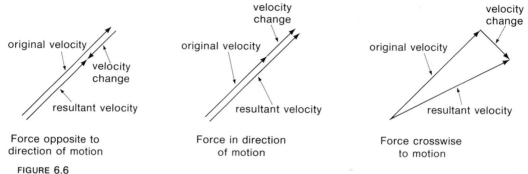

Force opposite to
direction of motion

Force in direction
of motion

Force crosswise
to motion

FIGURE 6.6
A force always produces an acceleration, and hence a change of velocity, that is in the direction of the force.

opposing force slows it down, but a crosswise force changes the direction of its velocity.

Still another vital fact is included in this simple wording: the law talks about the *rate* at which the velocity changes. As long as the force keeps acting, the velocity keeps changing; and the stronger the force, the faster the change. The rate of change of velocity is called the acceleration, and it is of course also a vector. Hence the simpler statement:

(2a′) *Acceleration is proportional to force.*

For the sake of brevity, we have dropped the word "net."

It is necessary to say more about the effect of forces. Clearly, a given force—say, the force exerted by a push of your hand—does not accelerate all bodies the same. Hence the remaining statement contained in Newton's second law:

(2b) *A given force accelerates different bodies in inverse proportion to their masses.*

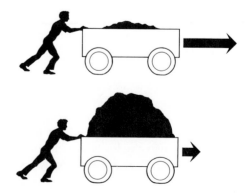

FIGURE 6.7.
The same force will accelerate a larger mass less.

Such a statement assumes, of course, that we know what mass is; this is a subtle point, and we shall postpone it till a later discussion. For the time being, let us content ourselves with an intuitive understanding of what we mean by more massive bodies and less massive bodies.

We can now combine the two parts of Newton's second law:

(2) *The acceleration of a body is proportional to the force acting on it and inversely proportional to its own mass.*

Physicists prefer to write this as an equation, in the neat form $f = ma$, instead of in the above order, which corresponds to the completely equivalent equation $a = f/m$. Note that in writing the law as an equation they have set the constant of proportionality equal to 1; this really comes down to agreeing that they will always express forces in units that give the constant that simple value.

Newton's second law, in this form, says many things indeed. In particular, it includes all the information in the first law, which is no longer needed as a separate statement. Constant velocity, referred to in the first law, is the same as zero acceleration. Thus the first law simply states that zero force gives zero acceleration—which is just a special case of the second law.

Thus far, we have been imagining the motion of a "test body," under the action of forces that come out of an unspecified nowhere. But in the real world, forces are exerted by one body on another—and this works both ways. Newton's third law of motion states:

(3) *When two bodies exert forces on each other, the two forces have the same strength but are opposite in direction.*

(The most common statement of this law is, "To every action there is an equal and opposite reaction"—a pernicious piece of obscurity that has done more to retard the teaching of science than to advance it. Here, I have tried instead to express what the law really means.)

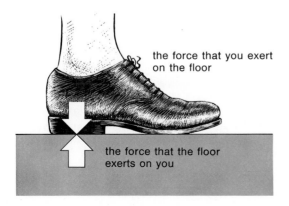

the force that you exert on the floor

the force that the floor exerts on you

FIGURE 6.8.
Why your feet hurt from standing. Your weight presses down on the floor, but the floor presses up against you with a force of equal strength.

Newton's third law takes on its practical significance when combined with the second law: when two bodies interact, since the forces are equal and opposite, the accelerations must be in opposite directions, but they must also be in inverse proportion to the masses. Everyone who has bounced off a much more massive person is aware of this phenomenon. The recoil of a gun is another example. The bullet, with its low mass, has a large acceleration and therefore reaches a high velocity before it loses contact with the gun. The gun experiences a force of the same strength, but because of its large mass it is accelerated very much less and therefore does not reach a high velocity.

The details of this process are worth emphasizing. What we see is the end result: the high velocity of the bullet. What made this velocity, however, was the acceleration that took place while the bullet was still in the gun. The acceleration in turn resulted from the force that the exploding gas exerted. Since acceleration is a *rate of change* of velocity, the velocity *kept* changing at that rate—in this case, kept increasing—as long as the force continued to act. When the bullet lost contact with the gun, the force became zero. The acceleration was then zero, which means that the velocity thereafter remained the same—the bullet going in a straight line with constant speed.

FIGURE 6.9.
For motion in a circular orbit, the velocity must keep the same size while continually changing direction. The central force accomplishes this, by always acting perpendicular to the direction of motion.

A different example illustrates the nature of velocity as a vector quantity. Consider a planet moving in a circular orbit around the Sun, with constant speed. Its direction of motion continually changes as it moves around the circle, so its velocity is always changing. This continual acceleration is caused by the Sun's gravitational force on the planet; if this force ceased to exist, the planet would continue moving in a straight line into the distance. Thus no force is needed to keep a planet moving, but a force *is* needed to make its path curved.

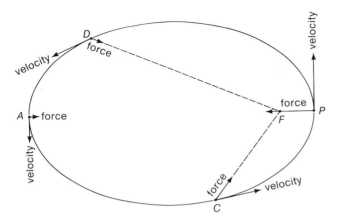

FIGURE 6.10.
Motion in an elliptical orbit. The force is always directed
toward the focus. At *P*, a large force changes the direction
of a large velocity; at *A*, a small force changes the direction
of a small velocity. At *C*, the force acts partly to change
the direction of the body's motion and partly to speed it
up; at *D*, the force changes the direction of motion while
slowing the body down.

When a planet moves in an eccentric orbit, the circumstances are more
complicated, but the correspondence between the forces and the motion
is still easy enough to see in a general way. The force is always directed
toward the Sun. At any point in the orbit, it can be resolved into a component of force perpendicular to the orbit, which makes the orbit curve,
and a component along the orbit, which slows the planet down when it
is moving away from the Sun and speeds it up when it is moving toward
the Sun.

This simple, quantitative argument may make sense of motion in an
elongated orbit, but it does not of course prove that the orbit is an ellipse.
That proof involves a number of complicated mathematical steps that
require the use of calculus. With such mathematical reasoning, however,
it can be shown that Kepler's laws are the direct result of the laws of
motion, if the forces are prescribed by the law of gravitation.

The mathematical approach to motion leads, further, to more general
conclusions than the laws of Kepler. Kepler's first law said that the orbit
of each planet is a closed ellipse. The more general analysis shows that a
different orbital shape is also possible: an open curve of infinite length,
called a hyperbola. An object in such an orbit would not, of course, be
called a planet; it would approach from infinitely far away, pass once by
the Sun, and recede forever into the distance. Thus, in general, all permanent members of the Solar System have elliptical orbits, while a chance
interloper passing only once through would have a hyperbolic orbit.

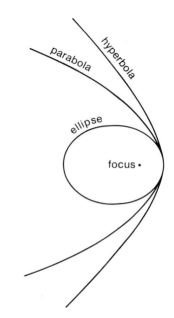

FIGURE 6.11.
Elliptical, parabolic, and hyperbolic orbits, all
with the same focus and the same point of
closest approach. The parabola separates
the class of ellipses from the class of
hyperbolas.

Theoretical dynamics does not modify Kepler's second law, but it does clarify its basis. Unlike the first and third laws, the law of areas does not apply solely to an inverse-square force such as gravitation; it turns out to be true for any force that is always directed toward the same center.

To Kepler's third law, dynamics adds a dependence on the masses of the gravitating bodies. This Kepler could not see, because all the planetary orbits are governed by the same dominant mass: the Sun. In the more general case, two bodies orbit around each other. Each body moves in a similar orbit, and each moves by an amount that is in inverse proportion to its mass. Thus the Sun does move in response to the force exerted on it by a planet; but because its mass is so great, its motion is very small. Kepler's third law, in its generalized form, applies to the relative orbit—that is, the orbit of either body relative to the other. It states that the square of the orbital period is proportional to the cube of the semi-major axis of the relative orbit *and* inversely proportional to the sum of the two masses.

Kepler's third law is our chief tool for determining masses in the universe. The orbit of a satellite around a planet tells us the mass of the planet. (Strictly speaking, what we find is the sum of the masses of the planet and the satellite, but the planet is so much the larger that this is practically the same as just the mass of the planet itself.) Similarly, the orbits of double stars around each other tell us the masses of the stars; and even for the Milky Way and for other galaxies, it is the size and the period of orbits that allow us to determine the total masses.

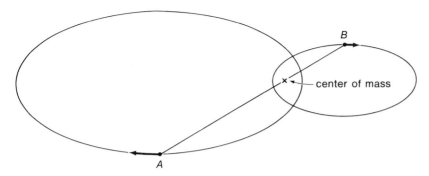

FIGURE 6.12.
When two bodies revolve around each other, the two orbits are ellipses
of the same shape, but the relative size of the orbits depends on the relative
mass of the bodies. In this example, body *B* has twice the mass of body *A*.

The motion of two bodies around each other, as described by Kepler and explained by Newton, is the simplest problem of celestial mechanics. When several bodies exert forces on each other at the same time, the problem is much more complicated—indeed, it has no neat solution at all. Fortunately, however, many situations in the real universe are well approximated by the two-body problem. In the Solar System, for instance, the mass of the Sun is so much greater than the mass of any other body that each planet moves almost as if the only force on it were that of the Sun. It is then a good approximation to describe the orbit by the simple rules of Kepler. The more exact truth, however, must take account of the forces that all the other planets exert. Because their effect is small, though, each planet can be considered to follow an elliptical orbit that is slightly perturbed by the other planets. These perturbations not only make small changes over short times, but also produce gradual changes in the size, shape, and orientation of the orbit. For very long periods of time, however, the tendency is for each orbit to shift its characteristics back and forth only by small amounts; and the orbits of the planets have probably been much the same for billions of years.

GRAVITY ON THE EARTH'S SURFACE

The most familiar effect of gravitation is the downward force that the Earth exerts on ordinary bodies on its surface. According to the law of gravitation, this force is proportional to the mass of the Earth and to the mass of the body that we are considering, and it depends on distance from the center of the Earth. This last factor is nearly the same everywhere on the Earth's surface, however. As we saw in Chapter 3, there are gravity differences that depend on latitude, on our rotating, slightly flattened planet; but the largest differences are less than 1 per cent.

The weight of a body is caused by this gravitational force. Because of the proportionality, we can use the weights of bodies as a measure of their masses. One must remember, however, that the weight of a body depends specifically on its gravitational interaction with the Earth; in a different gravitational field, it would have a different weight.

To understand the motion of falling bodies, we combine the law of gravitation with the second law of motion, to get a surprisingly simple result: all bodies fall with the same acceleration. To see this, we simply note that the Earth's gravitational force on a body is proportional to the mass of the body, whereas the acceleration that a given force produces on the body is inversely proportional to that same mass. As a result, the mass of the body cancels out of the calculation, and all bodies accelerate the same. To put it a little differently, a body of greater mass feels a stronger gravitational force, but the force is stronger in just the right proportion to produce the same acceleration. We talk loosely of the force of gravity, but it is more accurate to refer to the acceleration of gravity.

In everyday experience the picture is often confused by the resistance that air exerts on falling bodies. The effects of this retarding force depend on the body's density and on its size. It is air resistance that makes a feather float slowly to Earth, instead of falling like a rock. In the absence of air they would both fall the same; and, indeed, a common demonstration in physics courses is the rapid fall of a feather in an evacuated tube.

The fall of an object such as a stone also illustrates the combined operation of Newton's second and third laws. Since the stone and the Earth exert equal forces on each other, why does the stone fall to the Earth, instead of the Earth rising to meet the stone? The two forces are indeed equal; but the difference is that the Earth, with its tremendously greater mass, is accelerated upward by an amount so tiny that it is completely negligible.

Interestingly, we can also take an astronomical view of falling bodies near the Earth's surface. Except for the effects of air resistance, a body following a trajectory above the Earth's surface is actually in an elliptical orbit that has the center of the Earth at one focus. Because the end of such a long, narrow ellipse is extremely close in shape to a parabola, we normally describe the trajectory of a thrown body more simply as a parabola. Strictly speaking, however, its path is part of an ellipse. The body is temporarily in orbit, just like an Earth satellite; the difference is that this orbit intersects the solid surface of the Earth, and the motion of the body stops there.

The equal acceleration of gravity, for bodies large or small, combines both of the properties of mass: inertia and gravitation. This actually poses a philosophic puzzle. Why should a single quantity, mass, govern two such different areas of behavior? Or, to put it differently, why should the quantity that is called mass in the second law of motion always be the same as the quantity that is called mass in the law of gravitation? It is only this identity of gravitational mass with inertial mass that makes the acceleration of all falling bodies the same.

For many years, physicists regarded the identity of the two kinds of mass as a strange and puzzling coincidence. Only in this century has a likely resolution been suggested for the paradox. Einstein's general theory of relativity takes the identity of gravitational and inertial mass as a starting point; it characterizes gravitation as an accleration that is a natural property of the space around a body and the way in which we describe that space. (We shall see general relativity again in Chapter 23, where it serves as a way of understanding the large-scale properties of the universe.) The concepts are complicated; but to the physicist who follows them through, they offer a satisfying explanation of two properties that would otherwise have no causal relation to each other. Logical connections of this sort are the bones of physics.

THE ORBITS OF SPACECRAFT

A space vehicle follows an orbit just like that of a planet or a satellite, subject to the same rules of motion and gravitation. The only difference is that we provide the vehicle with an initial motive power to set it in the right orbit, and in some cases we change it later on from one orbit to another. For most of the vehicle's travel, however, no motive power is used—or needed—since orbital motion, as we have seen, persists without any force other than gravitation.

Rocket propulsion is a good illustration of Newton's third law. If the rocket needs to be pushed, so as to change its orbit, it must push against something else in turn. In empty space there is nothing to push against except the rocket's own exhaust gas, and this is indeed how the rocket gets its propulsive force. As the fuel burns, it heats and expands, driving the exhaust gas out the back. The rocket pushes on the exhaust gas, and by Newton's third law the gas pushes equally on the rocket. This is the source of its acceleration.

Rocket fuel thus plays a double role: it provides the power, and it is also the mass that is driven out the back. The speed that the rocket eventually attains will depend on the total mass of gas that is exhausted and on the speed with which it is ejected. Unfortunately, chemical burning can produce only a limited exhaust speed, and a rocket must move many times this fast if it is to leave the Earth. It can do this only by ejecting a very large amount of exhaust mass. The fuel must therefore have many times the mass of the "payload" that is put into the desired orbit. At the time of launch, nearly all the mass of a rocket is fuel.

The exact path that a space vehicle follows has the same complexities as that of a planet, acted on by the simultaneous gravitational force of all the bodies of the universe. Again, however, we can understand its approximate motion by using the same simple two-body approximations. At any given time, the force on the spacecraft is usually dominated by that of a single body, around which it can be considered to be following a Keplerian orbit.

Take, for example, a spacecraft that is to be sent to Mars. The first step is to raise it above the Earth's atmosphere and give it a horizontal velocity that is appropriate for a nearly circular orbit around the Earth. The motors are then turned off; the spacecraft will be able to continue going around the Earth for an indefinite length of time in this "parking orbit," while the necessary engineering checks and tests are made.

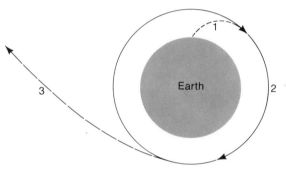

FIGURE 6.13a.
The beginning of a space flight to Mars: (1) launch into a circular orbit, (2) the parking orbit, (3) departure in the Earth-Mars transfer orbit.

The next step is to turn the motors on again, just long enough to accelerate the spacecraft into an orbit that is an open hyperbola around the Earth, so that it will escape from the Earth. This is not just *any* escape orbit, however; it is carefully chosen so that when the spacecraft is far from the Earth it will have just the right velocity with respect to the Sun. Hereafter the spacecraft will be very little affected by the Earth; the chief gravitational force on it will be that of the Sun. Hence we can consider it to be in an orbit around the Sun, which is determined by this initial velocity. With a suitable choice of velocity, this can be an orbit that reaches that of Mars. Once the spacecraft begins moving along this orbit, it is on its way to Mars—or, at least, to a crossing point with the orbit of Mars. The final condition is that this whole process be carried out at the right time, so that Mars reaches the point of crossing at the same time that the spacecraft does.

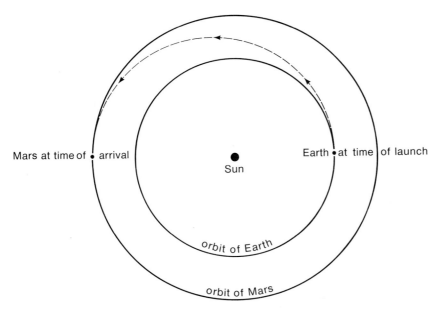

FIGURE 6.13b.
A transfer orbit from the Earth to Mars is an ellipse with the Sun at one of its
foci. The orbit is tangent to the orbit of the Earth at the launch point and tangent
to the orbit of Mars at the arrival point.

During its months of travel to Mars, the spacecraft is not driven; it
simply coasts along in its orbit. Its position and velocity are carefully
observed, however, because the initial aiming is rarely perfect. Usually
a brief amount of power is applied, eventually, in a midcourse correction
to the orbit. Except for this, however, we simply rely on the laws of motion; the destination of the spacecraft is just as sure as the eternal circling
of the planets.

REVIEW QUESTIONS

1. What is an ellipse? What is its relation to a circle? to a hyperbola? to a
   parabola?

2. State the laws of planetary motion, as found by Kepler. In what ways does
   dynamical theory generalize them?

3. State Newton's laws of motion, and give examples that clearly illustrate each
   law.

4. Kepler's second law predicts that a planet will move fastest when it is at its closest point to the Sun and slowest when it is at its farthest point from the Sun. Explain, in terms of forces and accelerations, why this should be so.

5. Draw the path of a spacecraft that goes from the Earth to Venus in an orbit whose aphelion is at the Earth's orbit and whose perihelion is at the orbit of Venus. Show the orbits of the Earth, Venus, and the spacecraft. How could you calculate how long the trip would take?

7 | THE MOTIONS OF
THE PLANETS

The nine planets circle the Sun in a simple, orderly way. Their orbits are nearly round and nearly in the same plane, and they all move in the same direction. In accordance with Kepler's third law, the inner planets go around their orbits much faster than those that are farther from the Sun. Mercury, the innermost planet, takes only three months to go around, but Pluto, the outermost, takes a thousand times as long, almost 250 years.

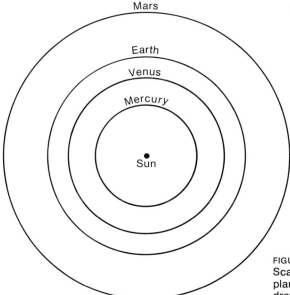

FIGURE 7.1.
Scale drawing of the orbits of the innermost four planets. For the outer five, see the smaller-scale drawing in Figure 7.2.

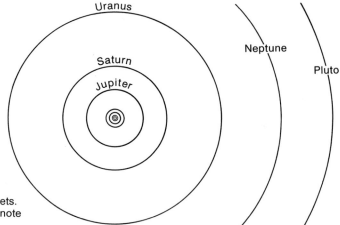

FIGURE 7.2.
Scale drawing of the orbits of the planets.
See Figure 7.1 for the innermost four; note
the difference in scale.

APPARENT MOTIONS AND POSITIONS

As we saw in the preceding chapter, this picture is simple enough when
we refer all motions to the Sun, as if we were looking at the Solar System
from an outside point. The motions of the planets are different, however,
when we look at them from our moving platform on the Earth. First of
all, for everyday purposes the most meaningful question about the position of a planet is how far its direction differs from the direction of the
Sun. If the planet is very close to the direction of the Sun, it rises and sets
with the Sun and also moves across the bright daytime sky close to the
Sun. In the nighttime darkness, it is below the horizon, just like the Sun;
so we do not see it at all. By contrast, if a planet is opposite in direction
to the Sun, it rises when the Sun sets and is in the sky all night, setting
again when the Sun rises.

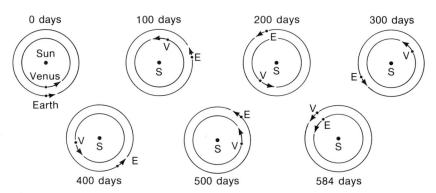

FIGURE 7.3.
The synodic period of Venus is the length of time that Venus takes to gain one full revolution on the
Earth, as both planets move around the Sun. In the synodic period of 584 days, the Earth has gone a
little more than 1½ times around the Sun, and Venus a little more than 2½ times.

The apparent positions of a planet, relative to the Sun, are called its *aspects.* For each planet, they repeat in regular sequence, over a length of time called the *synodic period* of the planet. The synodic period of one planet with respect to another can be defined as the time in which the inner planet gains one full revolution on the outer one. The formula for it is

$$\frac{1}{P_{\text{syn}}} = \frac{1}{P_{\text{inner}}} - \frac{1}{P_{\text{outer}}},$$

or equivalently

$$P_{\text{syn}} = \frac{P_{\text{outer}} \times P_{\text{inner}}}{P_{\text{outer}} - P_{\text{inner}}},$$

where P_{outer} and P_{inner} are the orbital (sidereal) periods of the two planets. If the two periods are close to each other in length, the synodic period can be much longer than either sidereal period.

The aspects that a planet goes through, as seen from the Earth, depend on whether the planet is closer to the Sun than we are or whether it is farther out. Mercury and Venus, which are closer to the Sun, are called *inferior planets.* Because their orbits are contained completely within the

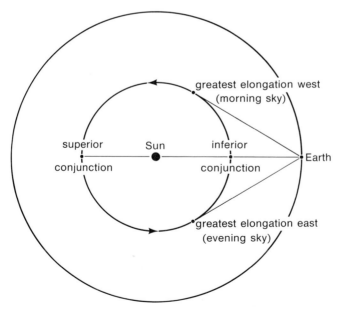

FIGURE 7.4.
Aspects of an inferior planet (Mercury or Venus). To show the relative positions more clearly, the Earth is held fixed in this diagram.

Earth's orbit, they never appear to be very far away from the Sun. In the course of its synodic period, an inferior planet goes through two *conjunctions,* occasions when it is in line with the Sun. At *superior conjunction* the planet is on the far side of the Sun; at *inferior conjunction* it is between the Earth and the Sun. Some time before and after inferior conjunction are the times of *greatest elongation,* when the planet is farthest from the Sun in the sky. For Mercury, this maximum angular separation averages only 22°; and as a result Mercury is never better than an elusive object in the evening or dawn twilight. Venus, on the other hand, is 46° from the Sun at greatest elongation, and is often conspicuous for months at a time in the evening or morning sky.

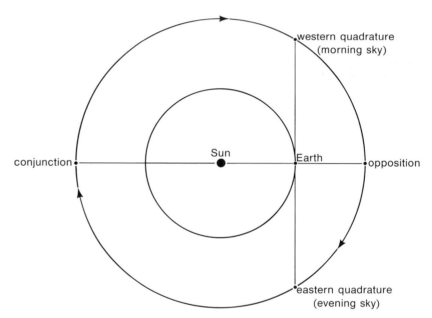

FIGURE 7.5.
Aspects of a superior planet, with the Earth held fixed, for clarity. Because we go around the Sun faster, the other planet appears to move backward.

A planet that is farther from the Sun than we are is called a *superior planet.* Because its orbit is completely outside that of the Earth, its aspects are quite different from those of an inferior planet. A superior planet is in conjunction with the Sun once in each synodic period, at the time when it is on the far side of the Sun from us. Half a synodic period later, however, we see it in *opposition;* at that time it is opposite the Sun in the sky and is thus visible all night long. Before and after opposition, there are times when the planet is 90° from the Sun. These are called the *quadratures;* they correspond to the times when an observer on the other planet would see the Earth at greatest elongation.

FIGURE 7.6.
Near the time of inferior conjunction, when Venus is almost between the Earth and the Sun, it appears as a narrow crescent. This photograph is taken with a large telescope, but during this period the crescent is large enough to be distinguished even with ordinary binoculars. (Hale Observatories.)

The appearance of a planet in the telescope depends on its aspect; in this respect inferior and superior planets again behave very differently. An inferior planet goes through all phases, just like the Moon. It is clear from the relative positions of Sun, planet, and Earth that we see the illuminated side of the planet near superior conjunction, whereas its dark side is turned toward us around inferior conjunction. For this reason the planet does not look brightest at inferior conjunction, even though it is closest to us at that time. Instead, its greatest brilliancy occurs at the point between inferior conjunction and greatest elongation where the closeness of the planet to us and the size of its illuminated crescent combine in the most favorable way possible.

By contrast, the geometry of the Sun, the Earth, and a superior planet shows that we always see the planet illuminated nearly full-face; at most, it can appear somewhat gibbous. We see that gibbous effect most strongly at quadrature; even then it is small, except for Mars, the closest superior planet. Greatest brilliancy, of course, occurs at opposition, since the planet is then closest to us as well as seen full-face. Here again, Mars shows the strongest effect, since it is so much closer at opposition than at conjunction.

CHARTING THE PLANETS

Although the aspects of a planet relative to the Sun govern the times of night at which we can see it, they do not tell where we will find the planet against the background pattern of the stars. This kind of motion is more complicated; the combination of the planet's orbital motion with our own leads to a series of loops. Throughout most of its synodic period a planet appears to move eastward among the stars; but when the Earth and the other planet are passing closest to each other, the planet appears to loop backward for a while, while the inner of the two bodies overtakes the outer one. This phenomenon of *retrograde* motion is easy enough to see in a diagram of the two orbits. The diagram also makes it clear why the retrograde loop is centered at opposition (or, for an inferior planet, at inferior conjunction).

If we saw the orbit of the other planet exactly edge-on, its direct and retrograde motion would take it back and forth along the same line. In fact, it will usually follow a narrow loop on the celestial sphere, because the orbit of each planet is slightly inclined to ours. Also, the orbit of each planet has some small eccentricity, so that it does not travel at a uniform rate. (Neither do we, of course.) Thus the apparent motion of each planet, as carefully observed, depends on all the details of its orbit and ours.

The orbit of each planet requires six numbers to describe it; they are called the *orbital elements.* The first two give the size and the shape of the

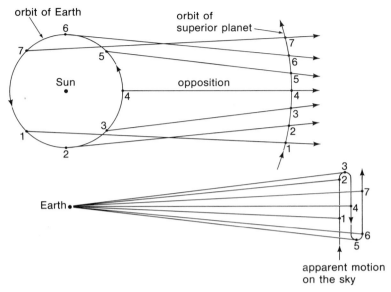

FIGURE 7.8.
A planet's "position" in the sky is really its direction as seen from the
Earth. This diagram shows, above, the real orbital positions, and below,
how these motions produce retrograde motion for an interval of time
around opposition.

FIGURE 7.9.
The apparent path of Mars among the stars in 1971.
The planet was at opposition in the middle of the retrograde loop.

orbit; as we saw in the previous chapter, these are the *semi-major axis*
and the *eccentricity.* Next, the orientation of the orbital plane must be
specified. This requires two angles: how much is the orbit tilted, and in
which direction? The amount of tilt is called the *inclination;* it is the
angle between the plane of the planet's orbit and the plane of the Earth's
orbit (the ecliptic plane). The second angle tells the direction in which
the tilt leans; what it actually states is the direction along which the two
orbital planes intersect—specifically, the direction from the Sun (which
lies in both orbital planes) to the *ascending node,* the point where the
planet crosses from the southern to the northern side of the Earth's orbital plane. The fifth number is also an angle: within the orbital plane,

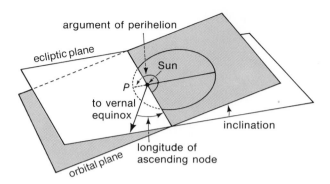

argument of perihelion

ecliptic plane

Sun

P

to vernal
equinox

inclination

longitude of
ascending node

orbital plane

FIGURE 7.10.
The angular elements of a planet's orbit.

where is the perihelion point with respect to the ascending node? This angle is called the *argument of perihelion*. Finally, a sixth number tells where the planet is in its orbit; this is specified by giving a *time of perihelion passage*. The occasions when the planet passes the perihelion point of its orbit are separated in time by the period of the planet's orbit; so it is enough to specify any one of them. The period can of course be calculated from the semi-major axis, by means of Kepler's third law; but it is often given along with the six orbital elements, just as a matter of convenience.

With the orbital elements well-established by observation, astronomers can calculate the positions of the planets at any time. Tables of right ascension and declination of each planet, on each day of the year, are published annually in volumes such as the *American Ephemeris and Nautical Almanac*. (Several other countries have almanac offices that produce similar volumes.) The almanac calculations include, of course, corrections for the perturbations of the planets on each other, so that the planetary positions are given with a high precision.

Such calculations can also be carried back through the centuries, to reveal where a planet was at any moment in the past. Thus, for example, historians of astronomy have calculated where the planets really were each time that Copernicus observed them, in order to check on the accuracy of his observations. Similarly, when planetary observations have been recorded in ancient times, it is sometimes possible for astronomers to use them to establish the true date of the historical era when the observations were made.

The planets swing regularly and predictably around their orbits, and on the surface of one of the nine of them, impersonal molecules have joined themselves together into organisms complex enough to exist as living creatures. All the parts of this picture work according to straightforward rules of physics and chemistry. How strange it is, then, that some people should continue to believe that the positions of the planets have an effect on our individual lives!

The belief in astrology is as old as astronomy itself. Many ancient cultures identified the bright, moving planets with the gods themselves; and indeed our own names for the planets are names of Roman gods. (This association has become a firm literary tradition, so that names for newly discovered planets, satellites, and even asteroids are normally sought in the pages of Roman and Greek mythology.) It was quite natural, in ancient times, to associate the influences of the individual gods with their positions in the sky, particularly at the crucial moment of birth. What is hard to understand, however, is the persistence of such a belief in the twentieth century. Its basis is emotional, not rational. The desire to know the future has always been a powerful force, and in a simplistic view it can override the rationality of the universe. In addition, the role of science in the modern world is a disquieting one, and many of the results of modern technology have created fear, resistance, and distrust. No doubt these antiscience feelings contribute to the prevalence of astrology and the other fads of pseudoscience and antiscience. The truth, however, was stated by Shakespeare nearly four centuries ago: "The fault, dear Brutus, is not in our stars but in ourselves." Far outside the human scene, the stars and planets move their eternal way, demonstrating the invariable laws of physics for any person who wishes to understand them.

REVIEW QUESTIONS

1. Describe the aspects of an inferior planet and those of a superior planet. Explain the differences.

2. Explain why we see Venus go through all the phases of the Moon, whereas Mars shows only phases around full.

3. Why do Venus and Mars have the longest synodic periods?

4. Draw a diagram similar to Fig. 7.8 for Venus and Earth, and show how the same sort of retrograde motion results.

5. List the six elements of an orbit, and explain what each one specifies.

8 | THE COPERNICAN REVOLUTION

Science, as we know it, is a phenomenon of the modern era; its whole development goes back only a few centuries. Indeed, the present is the only period, and ours the only culture, in which science has flourished; and seeing why this is so offers us another dimension in understanding science.

Science is a body of knowledge and understanding, but it is also a way of looking at things and, even more generally, an attitude about the world. This attitude is so characteristic of modern times and comes so naturally to us that it is hard to recognize that the attitude itself is an important step toward establishing science.

Many of the basic premises of science become more evident when we study a period in which they were still in question. Such a period was the Copernican revolution, when modern astronomy began, and with it modern science. That century-and-a-half is sometimes presented as a dawning of light upon a benighted world, but such a simplistic view misses the whole essence of the Copernican revolution. It was a revolution not so much of fact and understanding as of attitude. And it was a slow revolution. Copernicus, in placing the Sun at the center of the Solar System, took only the first step. The real revolution in man's view of the universe was still to come, for the world of Copernicus was nearly as far from Newtonian mechanics as the medieval monk was from the modern laboratory. In this chapter we shall trace the background of those times, and the development from Copernicus to Newton; and each step will illuminate a part of the premises on which modern science is built.

ANCIENT AND MEDIEVAL VIEWS OF ASTRONOMY

The roots of the Copernican revolution go back two thousand years, to the philosophers of the ancient Greek world. These lovers of wisdom devoted themselves to all knowledge, but their particular concern was with questions of truth, right, reason, and reality—questions that have remained at the heart of philosophy ever since. In particular, the systems of thought developed by Plato and Aristotle have repeatedly influenced later thinkers.

Not all the Greek philosophers were generalists, however; some directed their attention in particular to the workings of the heavens. These ancient astronomers used all their knowledge of geometry to construct schemes for representing the apparent motions of the planets in the sky. The most successful of them was Ptolemy of Alexandria, who around A.D. 150 put together a set of circular motions that gave a fairly good representation. Like nearly all ancient world views, the Ptolemaic system had the Earth at the center. To represent the apparent retrograde motions of the planets, it made use of a device that by that time was well-known: the *epicycle*. This is a circle on a circle; the center of the epicycle moves smoothly along a larger circle called the *deferent*. The net result is a looping path that gives retrograde motion in the innermost part of each loop. (We moderns would say that the epicycle merely adjusts the

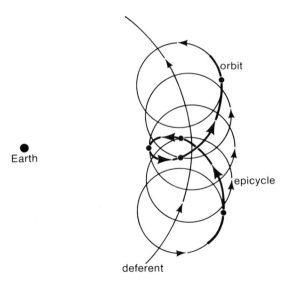

FIGURE 8.1.
The Ptolemaic system used a deferent and an epicycle to represent the retrograde motion of a planet in the sky. This figure gives the Ptolemaic representation of the same orbit that is shown in Figure 7.8.

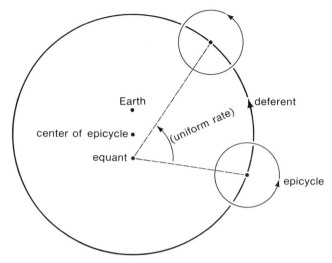

FIGURE 8.2.
Where we would describe the motion of a planet as
nonuniform, according to Kepler's second law, the Ptolemaic
system took the motion to be uniform with respect to the
"equant" point, and it placed the center of the deferent at
a point removed from the Earth.

planet's apparent orbit for what is really the motion of the Earth.) In
addition, the Ptolemaic system used two other geometric devices, both
of which help it to represent what we would call the eccentricities of
elliptical orbits. One was to make the deferent *eccentric*, that is, with its
center displaced from the center of the Earth. The second was to assume
that the motion of the planet was uniform not with respect to the center
but with respect to a displaced point called the *equant*.

Ptolemy was the last great astronomer of ancient times. After the
collapse of classical civilization, his works were among the Greek manu-
scripts that were preserved, in Arabic translation, in the Moslem culture
that dominated the eastern and southern shores of the Mediterranean.
The West, meanwhile, had been overrun by successive migrations of
barbarian tribes, to whom a smattering of culture was eventually brought
by the monks who converted them to Christianity. Along with their
theology, the Christianizers brought a few secondhand accounts of ancient learning. It was clear that the learning of the ancient Greeks was
far superior to any contemporary knowledge—but most of it had been
lost. Imagine, then, the intellectual excitement of finding that an untold
amount of Greek learning was preserved in the hands of the Arabs!
This is what happened in the twelfth century. At the points of contact
between the Christian and Arab worlds—principally Spain and Sicily—
many scholars labored at translating a large number of Greek treasures
from their Arabic versions into the Latin that all medieval scholars used.

The new learning was exciting, but it was also bewildering. The writers of eight centuries of Greek culture had not all agreed with each other, and some of their Arabic commentators had widened the range of opinion even further. But more important, nothing must conflict with the body of religious doctrine that had been enunciated by the fathers of the church. Somehow the eye of faith had now to survey a wider horizon, but it had to discern a truth that would not contradict the articles of that faith. Out of this task, and the intellectual ferment that it created, were born the first modern universities.

Of the ancients whose writings were rediscovered, by far the most impressive was Aristotle. He had written brilliantly in nearly every area of philosophy, and he acquired such prestige among medievals that he became known as The Philospher. The task remained, however, of fitting the philosophy of Aristotle into an intellectual world that was dominated by the church. This synthesis was carried out by Thomas Aquinas, whose brilliant writings are still official teachings of the Catholic Church.

Aristotle had argued that every effect has a cause, which must in turn have its cause. But the chain cannot go on forever; there must be a First Cause. Aquinas used this as a proof of the existence of God. Similarly, Aristotle considered motion and argued that everything that moves must be moved by something else—but again the chain must end with an unmoved mover, which Aquinas identified with God.

The triumph of Aquinas was to have an unfortunate effect on science. The theology was that of Aquinas, but some of its aura rubbed off onto Aristotle. The trouble is that whereas Aristotle was among the greatest of all thinkers within the realms of logic, metaphysics, ethics, and the like, his physics was terrible. Much of it was wrong in fact, and it was misconceived in method. Although he was a careful observer of plants and animals, Aristotle had not observed how inert bodies really move. Even more, his principles of physics were often based on metaphysical arguments; a heavy body fell, for instance, because its nature included heaviness, and this caused it to seek its natural abode in the heaviness of the Earth.

Such arguments seemed very appropriate to scholars whose highest aim was to understand everything within the framework of their religious faith. When it was a matter of doctrine that a sacrament administered by a priest was valid not because he was a good man (many were not) but because the spirit of God was in the sacrament, it was easy enough to imagine that downward motion was also a property that inhered in a physical body just because of what it was.

This overriding tendency toward qualitative rather than quantitative arguments left astronomy in a strange position. Aristotle's work had included an astronomical system, made up of spheres surrounding the Earth, each sphere moved by the one outside it. An account of it had

FIGURE 8.3
A sixteenth-century representation of the Earth-centered universe of the middle ages.

been translated from Arabic during the twelfth century, but the Ptolemaic system was translated shortly afterward and clearly gave a more accurate description of the motions of the Sun, Moon, and planets. This posed a problem, because the epicycles of the Ptolemaic system did not fit at all into the Aristotelian cosmology, in which all circles had to be centered on the Earth. Moreover, astronomy could not be ignored, or be dismissed as an imperfect manifestation of an ideal world. The heavenly bodies had to be followed in order to keep the calendar, and in particular to fix the date of Easter, and they could be best followed by the rules of Ptolemy rather than those of Aristotle and his followers. Medieval scholars resolved this paradox by declaring that the principles of Aristotelian physics represented the real truth, whereas the Ptolemaic system was just a set of mathematical devices that happened to fit the observations in a convenient way but had nothing to do with the real nature of the world.

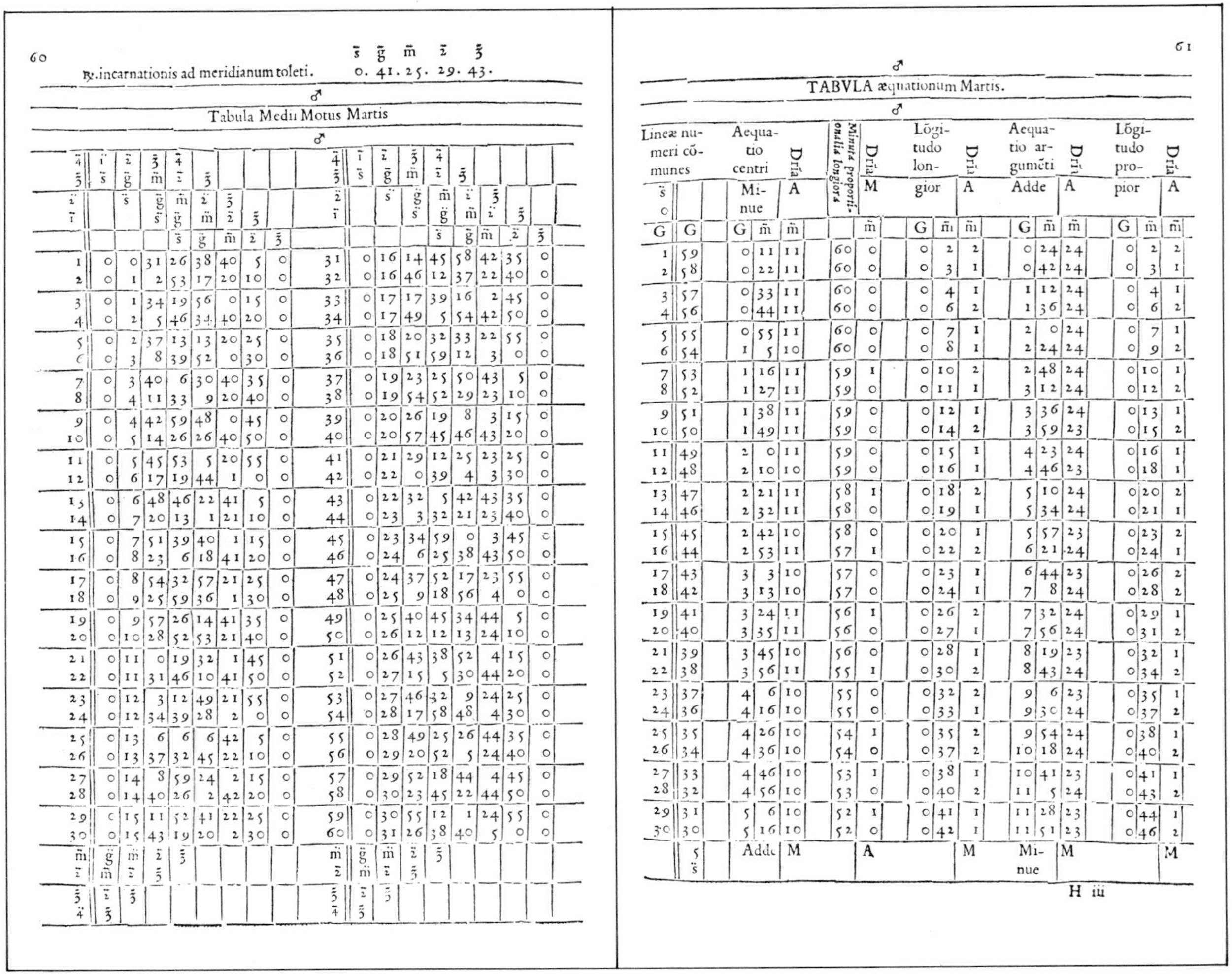

60

℞. incarnationis ad meridianum toleti. s g m 2 3 — 0. 41. 25. 29. 43.

♂

Tabula Medii Motus Martis

♂

			s	g	m	2	3	
1	0	0	31	26	38	40	5	0
2	0	1	2	53	17	20	10	0
3	0	1	34	19	56	0	15	0
4	0	2	5	46	34	40	20	0
5	0	2	37	13	13	20	25	0
6	0	3	8	39	52	0	30	0
7	0	3	40	6	30	40	35	0
8	0	4	11	33	9	20	40	0
9	0	4	42	59	48	0	45	0
10	0	5	14	26	26	40	50	0
11	0	5	45	53	5	20	55	0
12	0	6	17	19	44	1	0	0
13	0	6	48	46	22	41	5	0
14	0	7	20	13	1	21	10	0
15	0	7	51	39	40	1	15	0
16	0	8	23	6	18	41	20	0
17	0	8	54	32	57	21	25	0
18	0	9	25	59	36	1	30	0
19	0	9	57	26	14	41	35	0
20	0	10	28	52	53	21	40	0
21	0	11	0	19	32	1	45	0
22	0	11	31	46	10	41	50	0
23	0	12	3	12	49	21	55	0
24	0	12	34	39	28	2	0	0
25	0	13	6	6	6	42	5	0
26	0	13	37	32	45	22	10	0
27	0	14	8	59	24	2	15	0
28	0	14	40	26	2	42	20	0
29	0	15	11	52	41	22	25	0
30	0	15	43	19	20	2	30	0
31	0	16	14	45	58	42	35	0
32	0	16	46	12	37	22	40	0
33	0	17	17	39	16	2	45	0
34	0	17	49	5	54	42	50	0
35	0	18	20	32	33	22	55	0
36	0	18	51	59	12	3	0	0
37	0	19	23	25	50	43	5	0
38	0	19	54	52	29	23	10	0
39	0	20	26	19	8	3	15	0
40	0	20	57	45	46	43	20	0
41	0	21	29	12	25	23	25	0
42	0	22	0	39	4	3	30	0
43	0	22	32	5	42	43	35	0
44	0	23	3	32	21	23	40	0
45	0	23	34	59	0	3	45	0
46	0	24	6	25	38	43	50	0
47	0	24	37	52	17	23	55	0
48	0	25	9	18	56	4	0	0
49	0	25	40	45	34	44	5	0
50	0	26	12	12	13	24	10	0
51	0	26	43	38	52	4	15	0
52	0	27	15	5	30	44	20	0
53	0	27	46	32	9	24	25	0
54	0	28	17	58	48	4	30	0
55	0	28	49	25	26	44	35	0
56	0	29	20	52	5	24	40	0
57	0	29	52	18	44	4	45	0
58	0	30	23	45	22	44	50	0
59	0	30	55	12	1	24	55	0
60	0	31	26	38	40	5	0	0

61

♂

TABVLA æquationum Martis.

♂

Lineæ numeri cõmunes		Aequatio centri Minue		Dria A	Minuta proportionalia longiora	Dria M	Lõgitudo longior		Dria A	Aequatio argumẽti Adde		Dria A	Lõgitudo propior		Dria A
s	o	G	m	m		m	G	m	m	G	m	m	G	m	m
1	59	0	11	11	60	0	0	2	2	0	24	24	0	2	2
2	58	0	22	11	60	0	0	3	1	0	42	24	0	3	1
3	57	0	33	11	60	0	0	4	1	1	12	24	0	4	1
4	56	0	44	11	60	0	0	6	2	1	36	24	0	6	2
5	55	0	55	11	60	0	0	7	1	2	0	24	0	7	1
6	54	1	5	10	60	0	0	8	1	2	24	24	0	9	2
7	53	1	16	11	59	1	0	10	2	2	48	24	0	10	1
8	52	1	27	11	59	0	0	11	1	3	12	24	0	12	2
9	51	1	38	11	59	0	0	12	1	3	36	24	0	13	1
10	50	1	49	11	59	0	0	14	2	3	59	23	0	15	2
11	49	2	0	11	59	0	0	15	1	4	23	24	0	16	1
12	48	2	10	10	59	0	0	16	1	4	46	23	0	18	1
13	47	2	21	11	58	1	0	18	2	5	10	24	0	20	2
14	46	2	32	11	58	0	0	19	1	5	34	24	0	21	1
15	45	2	42	10	58	0	0	20	1	5	57	23	0	23	2
16	44	2	53	11	57	1	0	22	2	6	21	24	0	24	1
17	43	3	3	10	57	0	0	23	1	6	44	23	0	26	2
18	42	3	13	10	57	0	0	24	1	7	8	24	0	28	2
19	41	3	24	11	56	1	0	26	2	7	32	24	0	29	1
20	40	3	35	11	56	0	0	27	1	7	56	24	0	31	2
21	39	3	45	10	56	0	0	28	1	8	19	23	0	32	1
22	38	3	56	11	55	1	0	30	2	8	43	24	0	34	2
23	37	4	6	10	55	0	0	32	2	9	6	23	0	35	1
24	36	4	16	10	55	0	0	33	1	9	30	24	0	37	2
25	35	4	26	10	54	1	0	35	2	9	54	24	0	38	1
26	34	4	36	10	54	0	0	37	2	10	18	24	0	40	2
27	33	4	46	10	53	1	0	38	1	10	41	23	0	41	1
28	32	4	56	10	53	0	0	40	2	11	5	24	0	43	2
29	31	5	6	10	52	1	0	41	1	11	28	23	0	44	1
30	30	5	16	10	52	0	0	42	1	11	51	23	0	46	2
	s	Adde		M		A			M	Minue		M			M

H iii

FIGURE 8.4.
A page from a printed version of the Alphonsine Tables. These tables, produced in the thirteenth century by the astronomers of King Alfonso the Wise of Castile, were used for many centuries to calculate planetary positions according to the Ptolemaic system. (Courtesy of Bancroft Library, University of California, Berkeley.)

This was not a completely absurd position to take. Even today it is common for scientists to describe a process by empirical formulas that have no deeper basis than their mere ability to fit the observations. Moreover, the scholars of the thirteenth century were thus preserving an Aristotelian world picture that seemed as powerful and consistent to them as our scientific picture seems to us today.

Yet their physics was wrong; and worse, they had created a cleavage between physics and mathematics that would hobble the progress of astronomy until the two were finally rejoined four centuries later. The Copernican revolution began, in fact, with the substitution of one numerical scheme of representation for another; and it was only in the final stages that the new geometric rules acquired a physical basis.

COPERNICUS AND HIS WORK

Nicholas Copernicus was born in 1473. After an education at the University of Cracow, he was sent for further study in Italy. We do not know much about the detailed development of his thought, but we do know that in the course of his studies he learned Greek and was thus able to read some of the ancient writers who would otherwise have been inaccessible to him. And he was certainly exposed to the spirit of inquiry of the Italian renaissance. His studies completed, Copernicus returned to his homeland and spent the remainder of his life as a canon of the cathedral at Frauenburg, in East Prussia—that is, as a member of the administrative staff of the bishopric. (He is generally regarded as a Pole, although a few German voices have claimed him.)

The real life-task of Copernicus, however, was to work out a new system for describing the motions of the planets. His system differs from that of Ptolemy, as everyone knows, by placing the center at the Sun rather than the Earth. This change was not a pure stroke of inspiration on his part, however; he was aware of ancient astronomers who had preferred that orientation. He was also well acquainted with the work of medieval philosophers and astronomers who had criticized the Ptolemaic system. Some of these had already suggested that a daily rotation of the Earth could explain the daily rotation of the celestial sphere. Others had pointed out inaccuracies in the Ptolemaic tables, and it had even been noticed that the motion of each planet around the Earth seemed to mimic, in some significant way, that of the Sun.

But the chief motivation of Copernicus was not to make the planetary system more accurate; it was to simplify it. In understanding this, however, we must be careful not to impose our modern prejudice and to say—incorrectly—that by putting the Sun in the center he eliminated the need for a system of epicycles. Copernicus did indeed eliminate the large

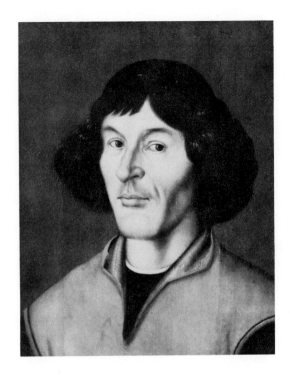

FIGURE 8.5.
Nicholas Copernicus (1473–1543) published, in
the year of his death, the book that began
modern astronomy. (Courtesy of Muzeum w
Toruniu.)

epicycle of each planet, but he replaced it by a smaller one. What actually had offended him most in the Ptolemaic system was the equant,
the geometric device that made nonuniform motion look uniform. Copernicus himself considered his greatest achievement to be the elimina-
tion of the equant and the restoration of the Platonic ideal of uniform
circular motion. The actual motion of the planets still had to be correctly
represented, however, and for this purpose Copernicus quite unconcern-
edly reintroduced a small epicycle for each planet.

Nor was Copernicus overly concerned with accuracy. His system represented planetary motions only a little better than that of Ptolemy.
What probably made more difference was the fact that he presented
more convenient methods of computation, and that one of his followers
used his work to calculate new and up-to-date planetary tables. Thus the
work of Copernicus was read and used by many people who did not
necessarily believe in its underlying truth.

The question of physical truth, indeed, was still a sore point and a
confused one. Copernicus never faced it directly, although he did devote
some of his attention to countering physical objections, such as the argument that falling bodies would not behave properly on a spinning Earth.
It may have been concern with attitudes toward the question of physical
truth that led him to withhold publication of his book, the first copy of

which was brought to him only in the last weeks of his life. And in the book itself, the question was further obscured by a preface inserted into the text by the clergyman who saw the manuscript through printing. It stated flatly that the Copernican system was only a computational device and did not represent physical reality.

The trouble, of course, was that the Copernican system was in gross contradiction with Aristotelian cosmology, which insisted that everything must be centered on the Earth. The traditional way out was to continue the medieval fiction that astronomical systems were only devices to "save the appearances"—the standard phrase used to describe the process of empirical fitting of observations. The alternative was to face the issue squarely and, in accepting the Copernican system, to reject Aristotelian cosmology in spite of all its entrenched blessing by the theology of Aquinas.

This radical step had become much more possible, because the ways in which the world had changed in the three centuries between Aquinas and Copernicus had led to a complete change in outlook, which did more to make modern science possible than could any particular discoveries. In a single phrase, the God-centered outlook of the middle ages had been replaced by the man-centered outlook of the renaissance. The change had flowed over every aspect of human activity. The feudal system was gone, with its attitude that every man knew his ordained place. The political development of national status had brought with it a strong merchant class, whose ideal of "rugged individualism" is still a catchword. The arts had flowered and had gloried in Man's nature, achievements, and worldly aspirations. The great explorations had replaced the narrow medieval world by a whole Earth. And last of all, the dominance of the church over thought had been shattered by the Protestant Reformation, and doctrinal questions were now a matter of heated debate.

Thus the philosophical changes of science came into a world that was ready for them. The 150-year road from Copernicus to Newton would require many difficult steps, but the ground was laid. In a deep sense the publication of *De Revolutionibus,* in the year 1543, was one of the watersheds of human history. Copernicus, in his quaint, half-medieval language, began the road that has led to the industrial revolution and the atomic bomb.

The seed began its growth only slowly, however. A few hundred copies of *De Revolutionibus* were read, discussed, and sometimes commented on. One important change soon appeared. Copernicus had never bothered to change the outermost sphere, to which the stars were attached; his was still a closed universe. But since it was the Earth that turns, there was no need for a stellar sphere that could rotate once a day; and soon accounts of the Copernican system began to dispense with this limited

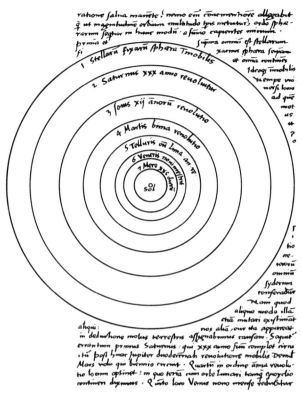

FIGURE 8.6.
In this diagram, in his 1543 book *De Revolutionibus Orbium Coelestium,* Copernicus placed the Sun in the center of the universe and placed the Earth among the other planets. Note, however, the outermost sphere, on which the stars were supposed to lie. (New York Public Library.)

sphere and to scatter the stars through space instead. Although this was a direct and natural step, one of its extensions led to one of the most bizarre disasters in intellectual history, when a priest named Giordano Bruno extended the stellar universe into theology. He reasoned that an omnipotent God would have created life like ours on planets surrounding the other stars too. Our drama of sin and salvation must have been reenacted on more worlds than we know of. This was too much for the Church authorities; when Bruno imprudently returned to Italy in 1600, he was seized by the Inquisition, tried, convicted, and burned at the stake for heresy. It is sometimes said, incorrectly, that Bruno was executed for supporting the Copernican system. His actual views, however, were certainly burnable heresies under the severe standards of the Inquisition.

None of the churches welcomed Copernican ideas. At first, however, the Catholic Church showed a casual tolerance toward them; after all, *De Revolutionibus* had been written by a servant of the Church and dedicated to the Pope. It was the Protestants who first showed indignation, perhaps because their return to the simple verities of the Bible had led them into a kind of fundamentalism. Martin Luther ridiculed Copernicus as a fool who wanted to turn the world upside down, and cited scripture to make his point. John Calvin, himself a burner of heretics, insisted on scripture. Eventually, however, it was the Catholic Church in which repression seated itself. Reeling under the blows of Protestantism, the Church tightened its lines, defining its doctrines more categorically and setting the Roman Inquisition to watch over them. Galileo was to feel the weight of its hand.

TYCHO AND THE WORLD OF OBSERVATION

The world view of Copernicus was far from modern science in many ways; for the immediate development of astronomy, the most important of these was an indifference to accurate observational detail. This indifference reflected a widely held philosophical prejudice that went all the way back to the ideas of Plato, who held that the world of the senses was only an imperfect reflection of an underlying ideal world. The true philosopher should therefore discover the truth by pure reason and not waste his time looking at the stars. This metaphysical view rang resonant chords in medieval theology: how could we even aspire to perfect knowledge in this imperfect world? Finally, in astronomy itself there would be no virtue in great accuracy, if it had nothing to do with the real physical world but would merely "save the appearances" a shade better—but here, perhaps, is the crucial point at which a change was to take place.

In spite of equivocal prefaces, most readers considered *De Revolutionibus* to present an actual view of the universe. If such a change could be made, might not others be necessary? And if so, how better to chart the way than to look carefully at the motions of the planets?

The man who lived out this idea was Tycho Brahe, a Danish nobleman who was encouraged and partly supported by his king. He made the best observing equipment that he could design—telescopes were not yet available—and used it as accurately as he could. At his death he left behind a 20-year series of observations of unparalleled accuracy. No existing planetary system would succeed in fitting the observations of Tycho—although much work would be expended before this was recognized. The time was ripe for a new system.

FIGURE 8.7.
Even without the use of a telescope, Tycho made observations of high accuracy, by using instruments that relied on large, carefully inscribed arcs of circles. This picture is taken from one of his own books. (Courtesy of Bancroft Library, University of California, Berkeley.)

KEPLER AND THE WORLD OF MATHEMATICS

The next chapter in the Copernican saga begins with a great stroke of luck. In his old age Tycho had quarreled with the new king and with others under him, and he had moved to the protection of the Emperor Rudolph of Bohemia. Settled in Prague, the aged Tycho sought an assistant to help him carry on his work. He hired a young Central European who had already attracted notice by an imaginative book on the role of mathematics in the structure of the planetary system.

Of all the intellectuals in the Copernican firmament, Johann Kepler is certainly the strangest. He was a mathematical mystic. Like the school of Pythagoras in ancient times, he believed that number, geometry, and harmony formed the basis of the world. His first book had attempted to show why the planets Mercury, Venus, Earth, Mars, Jupiter, and Saturn were the only ones that existed (the others had not yet been discovered), and why they are placed as they are. His explanation involved a nesting of six concentric spheres, separated by the five "regular polyhedra." These latter have been known since ancient times as the only possible solid figures that can be made up of flat faces and straight edges, all of them perfectly symmetric. (The cube is one of these solids.) If the nesting was arranged in the proper order, then the six spheres had radii that were proportional to the sizes of the six planetary orbits. It later turned out that these numbers did not represent the planets accurately enough—but the story reveals Kepler's attitude toward the universe.

Tycho lived just long enough for Kepler to be thoroughly impressed with the quality of Tycho's observations. At the death of Tycho the young Kepler, with his mystical belief in the role of mathematical figures, was left in possession of the only set of observations that could allow him to select the truth of the universe from out of the many flights on which his fancy took him.

The work that he did was anything but fanciful, however; it was plain hard work. For years Kepler labored over the orbits of the Earth and Mars, adjusting eccentrics and equants until he had a better representation than ever before. But then he found that some of Tycho's observations disagreed with his orbits by 8 minutes of arc—a tiny amount, but more than he believed Tycho should err by. He thereupon threw out the whole representation and started over again.

This was a great moment in the history of science. With Kepler's insistence on the exactness of the universe—and his eventual success in pursuing that idea—observational science came into its own. Only then did the contributions of Tycho bear fruit.

Returning to the problem anew, Kepler studied the variation of speed of Mars in its orbit. Trying various geometrical devices, he hit upon his law of areas (now called his second law). He then reexamined the shape of the orbit and was finally forced to abandon the 2,000-year-old ideal

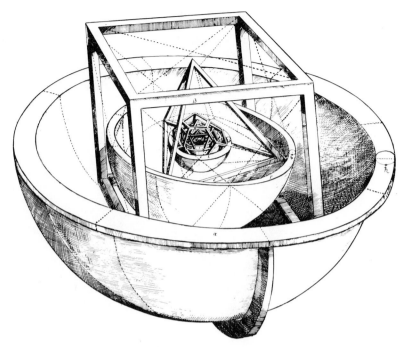

FIGURE 8.8.
Kepler's bizarre idea of the sizes of the planetary orbits. He believed that they
were determined by the way in which polyhedra and spheres could be nested
together.

of the perfect circle. The shape had to be some kind of oval—and his
published account, in 1609, describes his excitement when he found that
it was actually an ellipse!

This still left Kepler with the problem of the sizes of the planetary
orbits. Ten years later he discovered his third law, relating the cube of
the size of each orbit to the square of its period. Again he was ecstatic
to find the harmony of the universe.

Thus Kepler established a new role for mathematics in astronomy.
It was not only a means of calculating; it was actually a part of the answer. Even this answer, however, was not yet complete; there was still
no understanding of why the planets move as they do. What was lacking
was a new physics, and its development fills the remainder of the Copernican story.

GALILEO AND THE TRIUMPH OF COPERNICANISM

While Kepler was replacing Copernican epicycles by a better mathematical picture, the whole Sun-centered picture was under attack in

FIGURE 8.9.
Galileo. The painter seems to have caught his
questioning attitude toward the universe. (Yerkes
Observatory.)

another arena. The issue was over the conflict with the old Earth-centered view of medieval times. The established forces of religion held to the Earth-centered view, whether because of the metaphysical subtleties of Aquinas or because of a literal interpretation of passages in the Bible. As fate would have it, the astronomer who produced the irrefutable proofs of the Sun-centered system was the one who came into direct and disastrous conflict with the Church.

Galileo Galilei was a firm believer in the Copernican system even before he began his own observations. Hearing that a Dutch spectacle-maker had found a way to put lenses together so as to make distant objects look closer, Galileo quickly made his own telescope, in 1609, and directed it to the sky. Crude as his telescope was, observation after observation either discredited the Aristotelian and Ptolemaic views or else directly supported the Copernican picture. Mountains on the Moon and spots on the Sun showed that the heavenly bodies were not perfect, as they were supposed to be. Likewise the protuberances on Saturn, which his telescope was not sharp enough to show as rings. Around Jupiter, Galileo found four satellites, thus demolishing the anti-Copernican argument that a moving Earth could not carry the Moon with it.

Galileo's crucial observation, however, was the phases of Venus, which his telescope showed to be a full set, just like those of the Moon. This showed that Venus must sometimes be on the far side of the Sun, because otherwise we could not see its full face illuminated (see Figure 8.1).

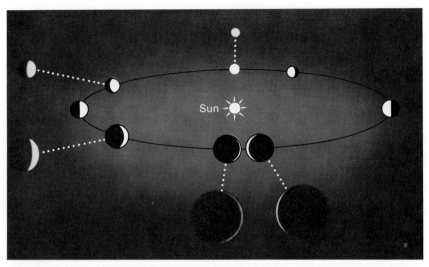

FIGURE 8.10.
The phases of Venus, as observed from the Earth. The photographs (courtesy
of Lowell Observatory) show how Venus actually looks through the telescope at
the positions that are marked in the diagram. Because Venus goes around the
Sun, we see a full set of phases.

Such a behavior is perfectly proper in the Copernican system; but it is
impossible in the Ptolemaic system, where Venus always remains between the Sun and the Earth and can therefore show only crescent
phases.

Although he occasionally showed some caution, as when he described
his observations of Venus in "hidden" language, Galileo was generally
a militant advocate of the Copernican system. It is likely that he mis-
judged his political support; in any case, he could not escape the notice
of the Inquisition. He was ordered in 1616 to cease supporting the Copernican system, and in the same year *De Revolutionibus* was placed, for the
first time, on the Index of Prohibited Books.

Galileo kept silent for more than a decade, but then in 1632 he published his famous *Dialogue Concerning the Two Great World Systems*,
a "debate" that goes through all the arguments but leaves no doubt
about where the truth lies. The reaction of the Inquisition was immediate; Galileo was brought to trial, convicted, and forced to recant. On
his knees he swore, before the Inquisition, that the idea that the Earth
goes around the Sun is false and heretical.

Legend has it that when Galileo rose unsteadily to his feet he mut-
tered, "Eppur si muove" (but it *does* move). This is a story that ought to
be true even if it is not. Galileo had lost the battle but won the war. The
Inquisition had silenced Galileo, but no educated person could doubt,
ever again, that the Earth goes around the Sun.

THE BIRTH OF MODERN DYNAMICS

Galileo was sentenced, at age 69, to "perpetual imprisonment." This consisted, in fact, of confinement to his own home, where he completed and published his work on dynamics. This work was so important that the eventual effect of the Inquisition, in removing Galileo from the astronomical arena, was to further the progress of science.

Galileo, in fact, plays a double role in the Copernican story. Not only was he the founder of telescopic observation and the champion of Copernicanism, he was also one of the founders of modern dynamics. In the over-all development of modern science, this was of greater importance than his more spectacular telescopic discoveries.

The scientific coin has two sides, knowledge and understanding. With Kepler's laws the motions of the planets were accurately described, but one side of the coin was still blank. There was no understanding whatever of why the planets move as they do. Kepler himself speculated at great length about what drives the planets in their orbits, but he always missed the crucial point.

The great stumbling-block was inertia—not as a resistance to motion, which was obvious, but as a *persistence* of motion. Again the doctrine that had to be overcome was Aristotelian; Aristotle had stated that everything that is moved must be moved *by* something else. If a body continued to move, it did so because some agent kept it moving. This is an insidious idea that comes to all of us naturally, since our experience continually reinforces it. Repeatedly, we see moving bodies stop, apparently of their own accord. It is an act of insight to recognize that what stops the body is friction, and it is an inspired abstraction to imagine that in the absence of friction a body would move forever.

(So important is the law of inertia that Newton later chose to state it separately, as his first law of motion, despite the fact—already discussed in Chapter 6—that it is completely contained in the second law, as a special case. The law of inertia was an idea that had to be understood first, before one could appreciate Newton's second law. Even today, it clarifies the study of motion to take Newton's first law as a simpler first step.)

Even worse for medieval science, the Aristotelian law of motion was intimately related to other aspects of Aristotle's philosophy; this immediate causality of motion was just one manifestation of his whole theory of causality. Since one of the greatest strengths of Aristotle's philosophical system was its interrelatedness and tight internal consistency (note how it shares this one characteristic with modern science), it was very difficult to change the analysis of motion without disturbing the whole picture.

Yet moving bodies had posed problem that perplexed many medieval philosophers. After the arrow leaves the bow, what *is* the agent that keeps the arrow moving? Explications were devised that were more ingenious than convincing, but gradually the idea began to grow that a moving body carried within it a virtue that was its own agent of motion. This virtue acquired the name *impetus*. The impetus theory had the seed of the idea that a modern physicist would call conservation of momentum, but it was still acceptable Aristotelian physics. The virtue called impetus served as the agent of motion, and inherent virtues were often cited as causes—as, for instance, in Aristotle's explanation of gravity. The impetus theory was one of the seeds of modern dynamics.

The other great problem was the motion of falling bodies. Here Aristotle had explained the fall as due to a heaviness (Latin: *gravitas*) within the body, which caused it to seek its natural abode at the center of the Earth. The more heaviness, the faster a body would fall; hence a ten-pound weight should fall ten times as fast as a one-pound weight. This is, of course, quite contrary to fact, but it might seem to accord with common sense, because all heavy bodies fall so fast that differences are hard to see, whereas a feather or a leaf is so slowed by air resistance that the slowness of its fall is obvious.

The behavior of falling bodies became, in the hands of Galileo, the key to further developments in dynamics. By observing balls as they rolled down inclined planes, he was able to observe motions that were slow enough to time carefully, and he was able to observe the uniform acceleration produced by a gravitational force. (The story of dropping weights from the Leaning Tower of Pisa is another Galileo legend that is probably not true.)

It would be an exaggeration to say that Galileo knew all the facts about motion that were used by Newton, but he did almost as much for the description of falling bodies as Kepler did for planetary orbits. Since the behavior of falling bodies is so much simpler than that of planetary orbits, it was through their properties that the laws of motion were to become clear. Finally the gap between astronomy and physics, opened in the middle ages by the contradiction between Ptolemy and Aristotle, was to be closed.

NEWTON AND MODERN MECHANICS

Kepler had changed the spheres of Copernicus into the elliptical orbits of the planets, and Galileo had made sense of the behavior of moving bodies, but the joining of the pieces into a rational system came only half a century later. Newton's achievement was double: not only did he state the simple laws that govern motion, but he then applied them, to show that the motions of the planets are an expression of the same rules. In another sense, too, his contribution was double: alongside his new physics he developed the new mathematics that was needed to apply his physical rules.

Isaac Newton was born in 1642, a year after the death of Galileo and a century after the publication of *De Revolutionibus*. In his college days he was intrigued by the problems of motion; and when Cambridge University was recessed during the plague of 1665, he sat in his country home and pondered these problems. Everyone knows the story of how Newton saw an apple fall to the ground and wondered, could the same force that draws the apple to the Earth also hold the Moon in its orbit? The curve of the orbit, as it deviates from a straight line, is a kind of falling; and what one first needs to know is the amount of acceleration that corresponds to this continual curving away from a straight line. Finally, of course, one needs to know how the force of the Earth varies with distance, and thus the law of gravitation enters.

Even beyond the straightforward problems, however, the route from the apple to the Moon was not an easy one. When Newton first made the calculation, the numbers did not agree. The trouble was that he needed to use the radius of the Earth, which was not accurately known. Five years later, however, the first modern measurement of the Earth's radius was made, and the Moon fell into step with the apple. A more serious difficulty lay in the application of gravitation itself. Newton had assumed that all the distances should be measured from the center of the Earth, but could he prove it? This proof, that a spherical body attracts as if all its mass were concentrated at its center, taxed even the inventor of the calculus.

In taking his first step, however, Newton had struck the final blow to the old system. With the fall of Newton's apple, the Aristotelian distinction between the heavens and the Earth was gone forever. What acted on Earthly bodies acted on heavenly bodies too.

The new system was far from complete, however. To understand the acceleration of the Moon is one thing, but this is far short of the detailed statements in Kepler's laws. This was Newton's remaining task, to show that Kepler's laws were an inevitable mathematical consequence of his laws of motion and law of gravitation.

The publication of Newton's *Principia,* in 1687, marks the end of the Copernican revolution. In a century and a half, man's view of the heavens had changed from a mysterious set of Earth-centered spheres to a space in which planets moved according to the laws of falling bodies; and Newton's contemporary Halley was already speculating that if the brightest star was a body similar to the Sun, it must be more than a million million miles away.

Such was the birth of modern science. Other branches of science began their development during the same period; but nowhere except in astronomy, and the related physics, was it so clearly shown what basic

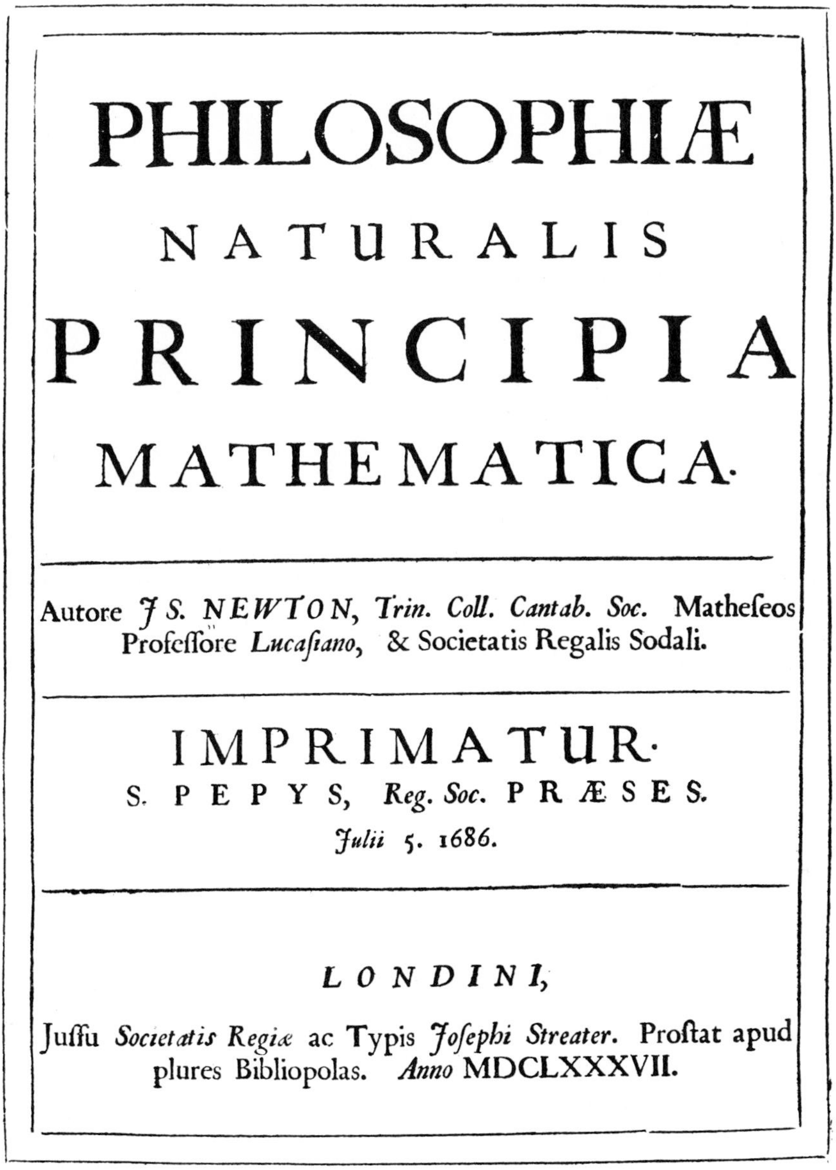

PHILOSOPHIÆ

NATURALIS

PRINCIPIA

MATHEMATICA.

Autore *JS. NEWTON*, *Trin. Coll. Cantab. Soc.* Mathefeos Profeſſore *Lucaſiano*, & Societatis Regalis Sodali.

IMPRIMATUR·

S. PEPYS, *Reg. Soc.* PRÆSES.

Julii 5. 1686.

LONDINI,

Juſſu *Societatis Regiæ* ac Typis *Joſephi Streater.* Proſtat apud plures Bibliopolas. *Anno* MDCLXXXVII.

FIGURE 8.11.
Like other learned works of the time, Newton's *Principia* was written in Latin. The full title means "Mathematical Principles of Natural Philosophy." This book included the laws of motion, the law of gravitation, and many of their applications. (Courtesy of Bancroft Library, University of California, Berkeley.)

intellectual steps were needed, and how the birth of science was an indissoluble part of a basic change in human outlook.

REVIEW QUESTIONS

1. Each of the five great figures of the Copernican revolution, Copernicus, Tycho, Kepler, Galileo, and Newton, played a very different role in its story. Describe and contrast the role of each.

2. Of the five, Galileo played the most varied part. What were three different aspects of his role?

3. Choose one of these five figures, and argue that his contributions were the most important of all.

4. Imagine that you are a traditional scholar in 1543. Give all the arguments that you can *against* the Copernican system.

5. In what ways was the Copernican revolution a part of a general change in attitudes? To what extent do you believe it was a consequence of this change, and to what extent a cause?

9 | THE PLANETS

It is a long way from Man to the depths of the universe, both in distance and in concept. Not only the Earth but the whole Solar System is just a speck in the celestial picture. Nevertheless we, as rational creatures contemplating the whole scene, must wonder at our local place in it. The first step toward pursuing this question was to examine the Earth, the body whose surface chemistry and physics have produced the biological chain that leads to our existence. To follow our own setting further, we shall now study the other bodies of the Solar System, probing their nature and seeking to understand our role among them.

As we saw in the preceding chapter, astronomy began with the study of the planets, the recognition that the Earth is one among them, and the understanding of their motion, by means of simple physical rules. The basic premise—that the heavens work by knowable, earthly physical laws—built a bridge to the universe; and while some astronomers calculated the clockwork of planetary motions to the nth decimal place, others probed the nature of the stars. One area languished, however: the study of the planets as physical bodies. Whereas the hot gas of the stars behaves in very simple ways, the cold solid (or liquid) material of the planets shows infinite complexity; and it is very difficult to discover the nature of these substances by means of the simple sort of observations that we are able to make. Furthermore, our ability to see fine detail is limited by the Earth's atmosphere rather than by the capability of large telescopes. Thus the telescopes of 1950 gave views of Mars that were hardly better than those of 1850. In recent years, however, knowledge of the planets has taken a great leap forward, as a result of the development of modern technology. Much of the new information has come from the space program, but a large part has also come from ground-based applications of new techniques of infrared and radio observation. Before 1950, planetary astronomy was static; today it is moving rapidly forward.

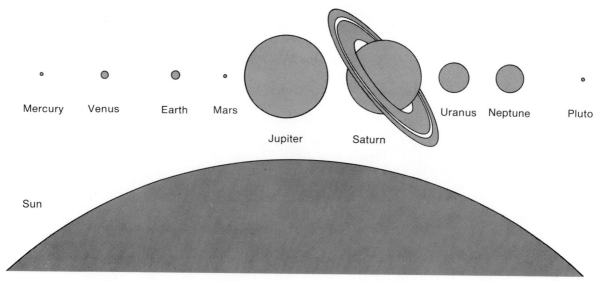

FIGURE 9.1.

The sizes of the planets, compared with the Sun. In looking at this drawing one should remember, however, that the volume of a body is proportional to the cube of its radius, so that in volume even Jupiter is only 1/1000 as large as the Sun.

The planets are very much smaller than the Sun, and in a real sense they are subordinate to the Sun and dependent on it. Not only does the Sun's gravitational force control the way that the planets move, but its radiation also largely determines, according to the distance of each planet from the Sun, the temperatures of their surfaces. In a gross way, in fact, the characteristics of each planet are determined simply by its distance from the Sun and its size. In both these respects, eight of the nine planets divide clearly into two groups. (Pluto, the outermost planet, is an exception that we shall take up separately later.) The inner four, which are called the terrestrial planets, are small, are made mainly of heavy elements, have moderate or thin atmospheres or none at all, and have few or no satellites. The next four planets, which are much farther from the Sun, are called the Jovian planets. They are large, are composed mostly of hydrogen and helium, have dense, deep atmospheres, and have numerous satellites.

MERCURY

Mercury, the planet closest to the Sun, is the smallest of the terrestrial planets. Its radius is less than half the Earth's radius, and its mass is just over a twentieth of the Earth's mass. From the mass and the radius we can calculate the mean density, which is close to that of the Earth. Mercury probably has a rocky mantle and an iron core, like the Earth.

FIGURE 9.2.
This montage of photographs taken by the Mariner 10 spacecraft shows almost half of one side of Mercury. Note how similar it looks to the Moon. (National Aeronautics and Space Administration.)

FIGURE 9.3.
This region of Mercury contains a tremendous
structure of concentric rings (from the left
edge nearly to the center), which must have
been formed by a large, early impact.
(National Aeronautics and Space
Administration.)

FIGURE 9.4.
A small part of the surface of Mercury is shown
in this Mariner 10 photograph. The finely
spattered appearance of this region is unlike
the surface of the Moon. The large crater,
with its unmarked floor, must have formed
after the spattering took place. (National
Aeronautics and Space Administration.)

Because of its low gravity (about 4/10 that of the Earth) and its high surface temperature, Mercury is unable to hold any atmosphere. Lacking both protection and erosion, its bare, rocky surface is cratered just like that of the Moon. The high temperature of Mercury's surface results from its closeness to the Sun and from the large amount of sunlight that it consequently receives.

Mercury rotates on its axis in a way that surprised astronomers very much when, in 1965, they discovered its true rotation period, which is exactly two-thirds of its orbital period around the Sun. Just like our Moon in its orbit around the Earth, Mercury has a rotation that is locked by tidal forces to its rotation period. Unlike the Moon, however, Mercury does not keep the same face toward its parent body. Instead, its rotation is in *resonance* with its orbital period; in every two revolutions around the Sun, Mercury completes exactly three turns on its axis.

Like the synchronism of the Moon's rotation, the resonance of Mercury's rotation was produced long ago by tidal friction, which acted on a faster rotation and gradually slowed it down. And, like the Moon, Mercury has a permanent tidal bulge that keeps it aligned to the direction of the Sun in its own particular way. With Mercury, however, the rotation was "trapped" in this 3-to-2 ratio to the period of revolution, and will not slow down any further. The reason for the difference is the large eccentricity of Mercury's orbit, about 2/10. Because tidal forces depend so strongly on distance, it turns out that the Sun's tidal force on Mercury at perihelion is more than three times as strong as the tidal force at aphelion. Hence it is only in a brief interval around perihelion that tidal forces do their real work. But in this part of its orbit Mercury is swinging around the Sun at much more than its average rate, according to Kepler's second law. As a result, during the time when tidal forces matter the most, Mercury's orbital speed temporarily matches its rotational speed, and during this time the planet keeps almost the same face turned toward the Sun. Then at the next perihelion, because of the resonance of rotational and orbital periods, Mercury has the exact *opposite* face turned toward the Sun. The tidal effect will again be just the same, however, because tidal forces act identically on the front and back sides of a body. As a result, the 3-to-2 resonance locks the rotation of Mercury just as effectively as the 1-to-1 resonance locks the rotation of the Moon.

The rotation of Mercury is not measured by directly watching it turn. Even with the highest magnification that telescopes on Earth can produce, one can rarely glimpse any features of its surface at all. Instead, the rotation measurement is made by radar. A carefully timed train of radio waves is beamed at Mercury, reflects off the planet, and is detected again by the same radar telescope after its round trip, which takes 15 or 20 minutes. Because of the rotation of Mercury, the wave comes back slightly changed. The fact that one edge of the planet is rotating toward us and the other side away from us changes the shape of the reflected

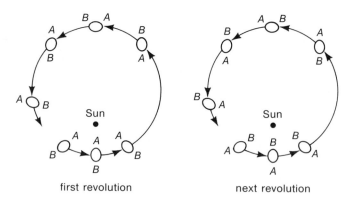

first revolution   next revolution

FIGURE 9.5.
The rotation of Mercury in two successive revolutions around
the Sun. In alternate revolutions, alternate ends of the tidal
bulge face the Sun at perihelion. (The sizes of Mercury and
its tidal bulge are greatly exaggerated here.)

waves in a way that can be measured, and from this the speed of rotation
can be calculated. (This is actually an example of the Doppler effect of
velocity on wavelength, which will be explained in Chapter 12.)

Before the radar measurements succeeded in measuring the correct
rotation period of Mercury, astronomers had convinced themselves that
Mercury rotated once per revolution period, keeping the same face
toward the Sun. Their conclusion was based on a series of difficult telescopic observations of features on the surface of Mercury, which seemed
to repeat the same orientation with respect to the Sun. The observations
were right, but the interpretation was wrong. There were indeed observations of the same side of Mercury turned repeatedly toward the Sun,
but they actually came from alternate revolutions. On the revolutions
in between, when Mercury was turned the opposite way, the observations
did not fit and were unconsciously left out. Synchronous rotation was
a plausible answer, and it was human nature to stop there!

VENUS

The second planet, Venus, is only a little smaller than the Earth, having
95 per cent of our radius, and 4/5 of our mass. The mean density is
almost the same, suggesting a similar internal structure.

There the resemblance ends. Venus has a hot surface and a dense
atmosphere, a hundred times as massive as the atmosphere of the Earth.
This dense Venus atmosphere consists largely of carbon dioxide. The
minor constituents include a small amount of water vapor, and there

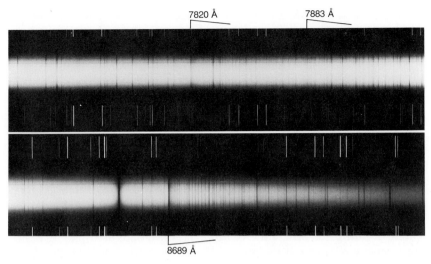

FIGURE 9.6.
The spectrum of Venus is that of reflected sunlight, except that the carbon dioxide in the atmosphere of Venus adds extra bands of dark lines; three bands are marked here. (Lick Observatory.)

are traces of a few exotic substances, such as hydrochloric acid. There is probably some nitrogen also, but its amount has not yet been measured.

Much of the analysis of planetary atmospheres can be done from the Earth's surface, by study of their spectra. The light that we receive from a planet is sunlight, reflected off the surface of the planet. Thus a bare planet should have the same spectrum as the Sun, when we spread the light out according to wavelength and analyze it in detail. If the planet has an atmosphere, however, we see light that has passed down through the atmosphere, been reflected at the surface, and passed back up through the atmosphere. As the light passes through, the gases of the atmosphere absorb certain wavelengths, and the lack of these wavelengths leaves dark lines in the spectrum, in addition to the dark lines that were present in the original solar spectrum. Some gases are much easier to detect in this way than others, however; it depends on whether their characteristic absorption lines are in the part of the spectrum that we can easily study. The problem here is the Earth's atmosphere; as we saw in Chapter 2, it blocks off some parts of the spectrum completely. Thus it has been possible to detect hydrochloric acid in the atmosphere of Venus at the level of one part in a million; but nitrogen, which produces no lines in the observable part of the spectrum, could make up one part in ten and still escape spectroscopic detection.

Space probes have improved our knowledge of the atmosphere of Venus. Five Soviet craft have floated down through the atmosphere, sending radio data as they went; and three American vehicles have made close passages, during which radio signals from the spacecraft to the Earth passed through various levels of Venus's atmosphere. Since only the top levels of the cloudy atmosphere can be seen from the Earth, these spacecraft observations have given us our first direct information on the deeper layers.

Beneath the dense atmosphere of Venus lies a hot surface, whose temperature is between 600° and 700° on the absolute scale, or about 700° Fahrenheit. (The absolute scale measures temperature, in centigrade degrees, from a zero that is the lowest temperature possible. On the absolute scale, whose usual symbol is K, water freezes at 273° K and boils at 373° K.) The surface temperature of Venus is measured by means of its radio radiation, which is a direct measure of temperature. Unlike visible light, radio waves pass freely through the cloud layer; so our radio observations come from the solid surface below the atmosphere. For infrared observations, however, the clouds are opaque; hence the intensity of infrared radiation tells us the temperature of the cloud layer, which is only 220° K.

These different ways of looking at Venus show how astronomy plays on the variety of observations that can be made in different regions of the electromagnetic spectrum. The visible light of Venus is reflected sunlight, but many of the gases of the Venus atmosphere leave their characteristic imprint on its spectrum. The infrared radiation that we receive is emitted by the clouds and tells us their temperature. The radio radiation comes from the surface and passes unimpeded through the clouds; this is how we measure the temperature of the hot surface of Venus.

This high temperature was a surprise when it was first discovered, but with a little thought it is easy to understand. The carbon dioxide in the atmosphere of Venus produces a strong greenhouse effect, blocking most of the wavelength region in which a body of planetary temperature gets rid of the heat that it is constantly soaking up from the Sun. Thus energy gets in easily and escapes only with difficulty, and the result is a much higher temperature than Venus would otherwise have. The small amount of carbon dioxide in the Earth's atmosphere (aided by water vapor) warms us only a few degrees, but on Venus the difference is hundreds of degrees.

Like the Earth's lower atmosphere, the atmosphere of Venus is warm below and cool above. The warm air rises, cooling as it expands, while the cool air sinks and warms. On Venus this mixing layer, or troposphere, is very deep, however. Radar measurements show the position of the surface, which is about 60 kilometers (40 miles) below the cloud layer. (By contrast, the Earth's troposphere has a depth of only about 9 km, or 6 miles.)

The composition of the clouds is not known at all. Many suggestions have been made, ranging from water to such exotic things as sulfuric acid; but none of them is completely convincing. The answer may have to wait for studies by future space probes.

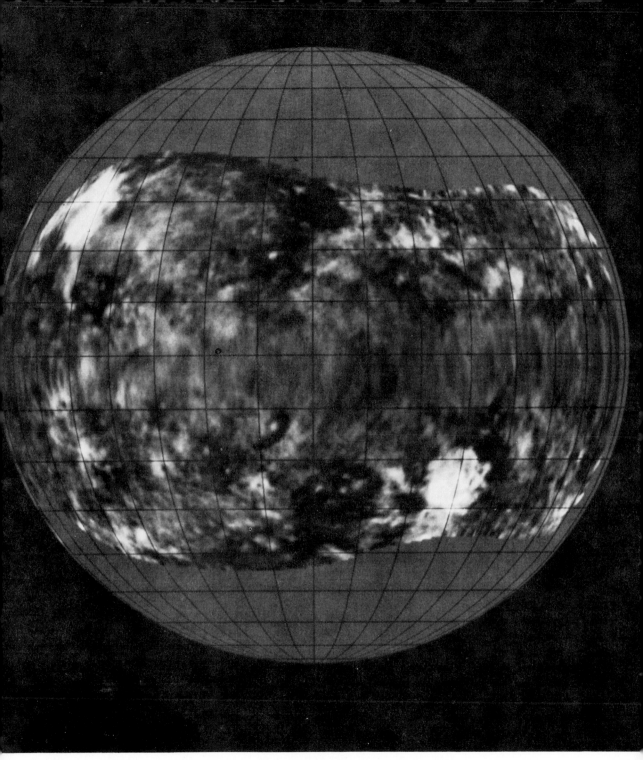

FIGURE 9.7.
A radar map of part of the surface of Venus. Unlike visible light, these radio waves penetrate the atmosphere of Venus and are reflected from the solid surface of the planet. (Courtesy of Dr. R. M. Goldstein, Jet Propulsion Laboratory, California Institute of Technology.)

FIGURE 9.8.
This Mariner 10 photograph of Venus shows the pattern of its cloud layer, high in the atmosphere. The principal features center on the region that directly faces the Sun, but they are drawn into trailing swirls by the rapid rotation of the planet's upper atmosphere. (National Aeronautics and Space Administration.)

We know even less about the surface of Venus. Radar studies have produced a crude map of high spots and low spots, as well as spots that reflect in different ways. The map that is emerging shows a rough surface, like that of the Earth. Since we would expect erosion to be rapid in the dense atmosphere of Venus, the roughness probably indicates that Venus has an actively changing surface, perhaps like that of the Earth. (This idea is confirmed by pictures sent by two Soviet spacecraft in November 1975. The pictures show rocks on the surface of Venus.)

Another contribution of radar has been to measure the rotation of Venus, which is both a surprise and an enigma. Its direction is retrograde (that is, opposite to the direction of orbital motion), and its period is 243 days, or about 10 per cent longer than its orbital period of 225 days. After the example of Mercury, it was natural to look for a tidal force with which this rotation might be in resonance. The Sun does not work at

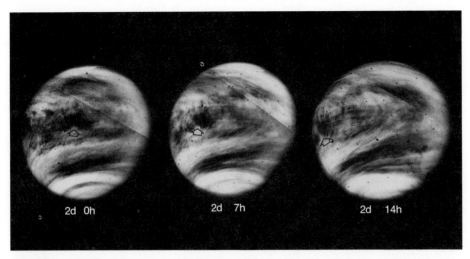

FIGURE 9.9.
This sequence of Mariner 10 photographs shows the rapid rotation of the upper
atmosphere of Venus. Note the motion of the dark patch that is marked by an arrow in
each picture. (National Aeronautics and Space Administration.)

all. If we work out all the relative turnings correctly, we find that a given
point on Venus faces the Sun every 117 days, a period that does not fit
with anything. The only resonance of the rotation of Venus is that, within
the uncertainties of observation, it turns on its axis exactly four times,
with respect to the *Earth*, in every synodic period. At every inferior conjunction Venus has the same face toward the Earth; and as the planet
passes closest to us, its rate and direction of rotation are such that it keeps
this face turned toward us almost perfectly, even while it is swinging by.
This coincidence has led some astronomers to suggest that the rotation
of Venus is locked by tidal forces that are due to the Earth. Such an explanation is difficult to accept, however, since the largest tidal force that
the Earth exerts on Venus is less than a ten-thousandth that of the Sun.

Because the rotation of Venus is so slow, its surface is heated very
unevenly, as the sunlight filters down through the atmosphere. Yet the
radio measurements of temperature show that the dark side of Venus is
nearly as hot as the side that is turned toward the Sun. Winds must blow
around Venus, to keep the temperature even. At the upper cloud level,
which we can observe directly, there are very strong winds, in fact; we
can see the cloud patterns drifting around Venus with an apparent rotation period of only 4 days. Such observations are made with difficulty
from the Earth's surface, but the Mariner 10 fly-by, in 1974, produced
cloud pictures that show the motion very clearly. It appears that as the
surface of Venus slowly rotates, the resulting temperature differences set
up winds that go around much faster.

MARS

Throughout the history of planetary astronomy, Mars has received more attention than any other planet, not because it is the closest or the largest—it is neither—but because it is the easiest to observe. Unlike cloudy Venus, whose dark side is turned toward us when the planet is closest, Mars has an atmosphere that is usually clear, and when Mars is closest to us we see its sunlit side.

FIGURE 9.10.
This sequence of photographs of Mars was taken when the planet was at different distances from the Earth. Note how much more detail can be seen when Mars is near opposition, at its closest to us (extreme right). (Lowell Observatory.)

Mars is a small planet, with only a little more than half the Earth's radius and about a tenth of our mass. Its mean density is lower than that of the Earth, Venus, or Mercury, but not as low as the density of the Moon. The density suggests that if Mars has a central core of iron at all, it is much smaller than that of the Earth.

Mars has a thin atmosphere, only about a hundredth as much as the Earth. It is nearly all carbon dioxide. A small amount of nitrogen is not out of the question; but, as usual, nitrogen is difficult to detect spectroscopically. Water vapor, by contrast, can be detected spectroscopically in quite small amounts; and small amounts of water vapor are indeed present in the Martian atmosphere.

Clouds are rare on Mars, and we usually have an unimpeded view of its surface. Mars is small and distant, however; even under the best observing conditions, examining Mars with an earthbound telescope is like trying to study this page with the unaided eye from 30 feet away. Under these circumstances only crude and vague maps can be made. A few astronomers even claimed to see the infamous "canals," while most other astronomers discreetly disbelieved them. In short, observations made from the Earth's surface were simply unable to produce any clear idea of the nature of the Martian surface.

Thus it was a revolutionary step forward when the Mariner 9 spacecraft, put into orbit around Mars, was able to take thousands of close-

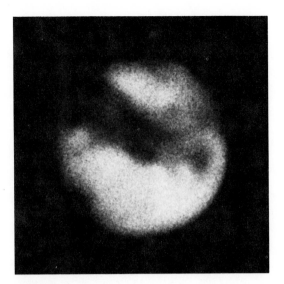 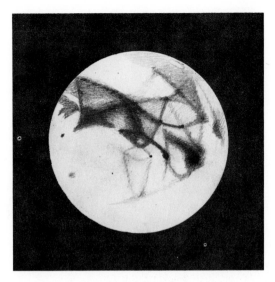

FIGURE 9.11.
When Mars is observed from the surface of the Earth, little detail is visible, and the boundary between observation and imagination is poorly defined. At the left is a photograph of Mars, and at the right is a drawing that an observer made of details that he believed he could see during moments of superior seeing. Photographs taken by spacecraft show that the "canals" in the righthand picture are completely imaginary. (Lick Observatory.)

up pictures in 1971–72. No other step in the space program has ever produced such a dramatic change in an area of knowledge. Small contributions had been made by the earlier Mariner fly-bys—establishing the low atmosphere density beyond any doubt and discovering numerous craters—but the detailed Mariner 9 pictures have told a completely new story.

More than anything else, the surface of Mars is windblown. Even though the air is thin, Martian winds can and do raise large quantities of dust. The first Mariner 9 photographs were made during a planet-wide dust storm that covered everything to a depth of more than 20 kilometers (12 miles). As the dust gradually settled, the mountain tops began to show through, and then the lower elevations, until at last, months later, everything was clear.

When the last of the dust had settled, more was clear than just the Martian atmosphere. The Mariner pictures showed that many of the large-scale geographic features that we see on Mars are merely regions in which the winds have deposited lighter or darker-colored dust. In particular, the long-recognized seasonal changes are not changes in vegetation, as some fanciful interpreters had suggested, but rather a movement of dust by winds that change with the seasons. The larger, permanent light and dark areas are regions in which rocks of different sorts predominate.

FIGURE 9.12.
This map of the surface of Mars was drawn by hand, from examination of photographs taken by Mariner 9. Note the numerous craters, and the four large extinct volcanoes near the center. (Lowell Observatory.)

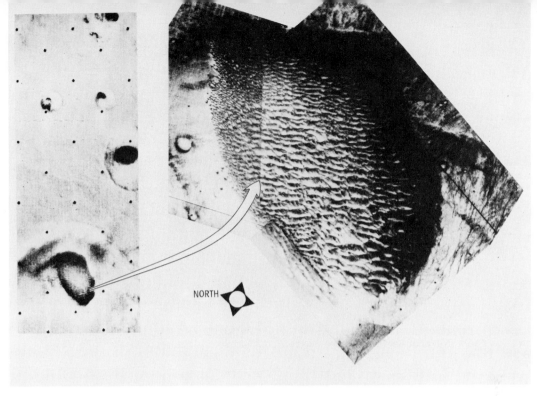

NORTH

FIGURE 9.13.
This small dark spot on Mars turns out, when seen under higher magnification, to be a range of gigantic, windblown sand dunes. (National Aeronautics and Space Administration photograph by Mariner 9.)

FIGURE 9.14.
This Mariner 9 photograph of Mars shows many small craters. The black spot is the shadow of the Martian satellite Phobos; its narrow dimension is 21 kilometers (13 miles). (National Aeronautics and Space Administration.)

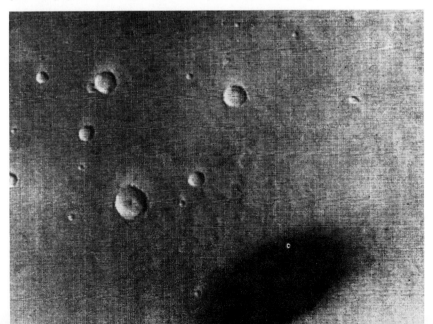

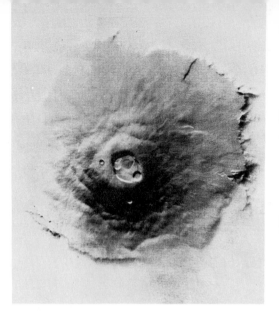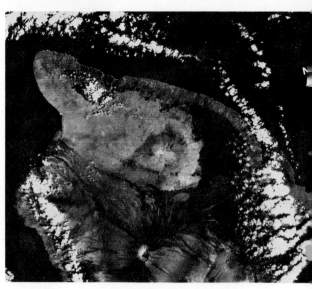

FIGURE 9.15.
The largest "volcanic shields" on Mars and on the Earth. Left, Nix Olympica on Mars; right, the island of Hawaii on Earth. (National Aeronautics and Space Administration, [right, reproduction from USGS/ EROS Data Center.])

On a smaller scale, Mars shows interesting geologic features of various kinds. There are many craters, as might be expected from the poor protection that the thin atmosphere gives. Many of the craters are noticeably eroded, however, and this is also to be expected of the Martian atmosphere. Mars has also changed some of its surface from within, at least during some eras in its past. Its highest mountains are giant extinct volcanoes. The largest of these, visible on pre-space-age maps as a small black spot called Nix Olympica, is 20 kilometers (12 miles) high, and flows of lava from its central vent have covered an area about 600 kilometers (over 300 miles) in diameter. By comparison, the largest similar structure on the Earth, the island of Hawaii, is less than a third as large.

More meteorological than geological are the polar caps of Mars, visible from the Earth as white spots that grow and shrink with the seasons. Mars has an inclined axis of rotation, just like the Earth; hence it has a similar sequence of seasons, as each hemisphere alternately receives a greater share of the sunlight. The Martian seasons differ from those of the Earth in one important way, however. The orbit of Mars is much more eccentric than that of the Earth, and as a result the whole planet receives almost 50 per cent more sunlight at perihelion than at aphelion. These orbital variations of sunlight happen to coincide with the southern-hemisphere seasons, but in the northern hemisphere they go opposite to the ordinary seasonal changes and greatly reduce the seasonal variations in temperature. One important consequence is that whereas the south polar cap disappears completely at the height of the southern

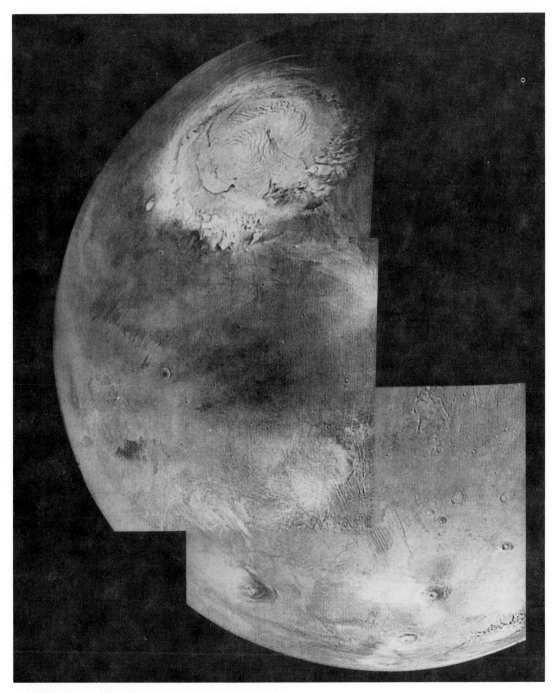

FIGURE 9.16.
Near the top is the north polar cap of Mars. Its successive terraced edges give an impression of how much frozen material it contains. Note also Nix Olympica, near the lower edge. (National Aeronautics and Space Administration.)

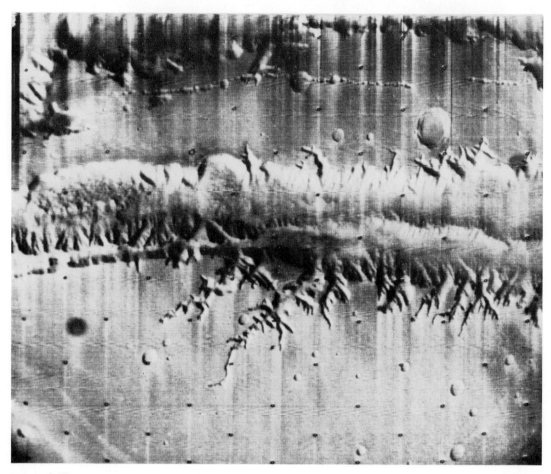

FIGURE 9.17.
A section of the long canyon that winds more than 5,000 kilometers (3,000 miles) around the equator of Mars. (National Aeronautics and Space Administration photograph by Mariner 9.)

FIGURE 9.18.
Mariner 9 took this photograph of Mars' satellite Phobos, showing its irregular shape and its craters.
(National Aeronautics and Space Administration.)

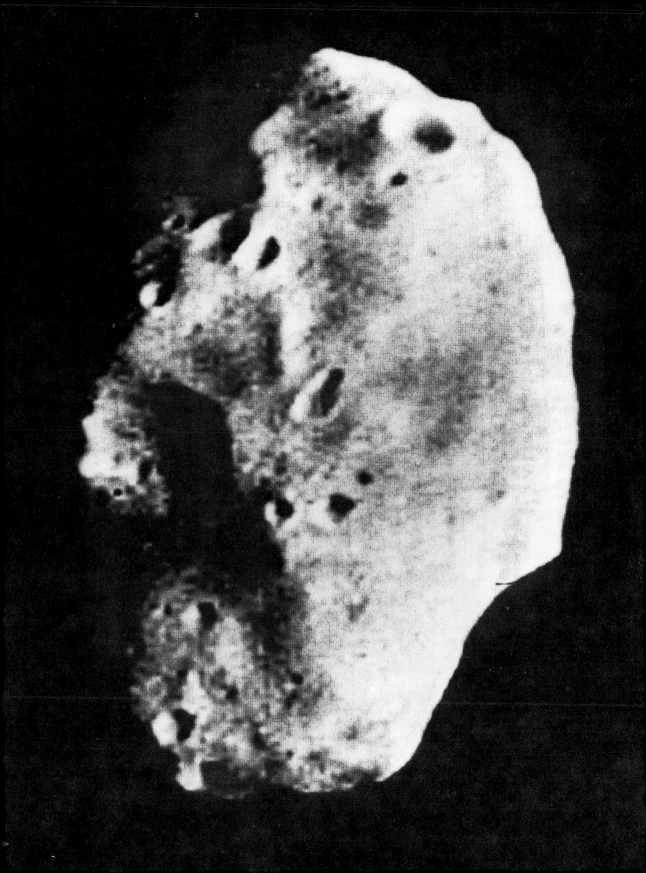

summer, the north polar cap of Mars has a central region that never melts. The Mariner photographs show this as a thick, terraced covering of frozen material. Spectroscopic studies have shown that the white polar caps that grow and shrink are a frost of solid carbon dioxide. The thick, permanent caps might possibly be frozen water, however. Its composition is extremely important to an understanding of the nature and history of the Martian atmosphere, because there is more material in the thick north polar cap than in the whole atmosphere of Mars.

Of all the surface features on Mars, the most intriguing are the winding canyons and stream beds. Although there is no liquid water on Mars, and very little water vapor in the atmosphere, these features look just like the meandering beds of rivers, with smaller streams merging into larger ones. Their existence suggests that the atmosphere and climate of Mars have not always been what they are today. It may well be that Mars has had eras in the past when the material of the polar cap vaporized and formed a much denser atmosphere than Mars has at present. If the solid polar cap does indeed consist of frozen water, its amount would be ample to produce rain and heavy surface flows. Some astronomers believe, in fact, that the polar cap is in only a precarious balance and that Mars goes periodically through times in which a warming trend melts the polar cap and quickly turns the Martian atmosphere from a thin, dry one to a dense, wet one.

The temperatures of the Martian surface are generally lower than those on the Earth, simply because Mars is farther from the Sun and receives less sunlight. But the warmest places have temperatures, in the daytime at least, that are pleasantly warm by Earth standards. The daily temperature range is large, as one would expect in a desert climate.

The Martian day is very similar to ours, since the rotation period of Mars is 24½ hours. The seasons are also similar, as we have seen, because the axis of rotation is tilted 25° from the perpendicular to its orbit. The year is of course much longer, nearly twice as long as ours.

Instead of a Moon like ours, Mars has two tiny satellites, each only a few kilometers in radius. They are very difficult to see from the Earth, partly because they are only a few thousand kilometers away from Mars, and they were discovered only in 1877. Phobos and Deimos, as these two satellites are called, are irregular chunks of rock; and Mariner photographs have shown that their surfaces are cratered, just as one would expect.

JUPITER

The Jovian planets are very different from the terrestrial planets in almost every respect: size, atmosphere, composition, and structure; distance from the Sun and surface temperature; satellites. The best ob-

FIGURE 9.19.
Jupiter as seen from the Earth. The cloud belts are easy to see, as is the Great
Red Spot, an atmospheric disturbance that has persisted for more than a century.
One of Jupiter's large satellites is just off the edge of the planet's disc, and its
shadow is on the disc. See also Plates 2 and 3. (Hale Observatories.)

served of these giant planets is Jupiter, which is the largest, the nearest
to us, and the first to have been studied by spacecraft fly-bys.

Jupiter has 11 times the Earth's radius and over 300 times our mass.
(This is still only 1/1000 of the mass of the Sun, however.) The ratio
of mass to volume gives a mean density of only 1.3 times the density of
water, less than a quarter of the density of the Earth. Clearly Jupiter
must have a very different composition and structure from ours.

All that we can see of Jupiter is the top of a dense layer of clouds,
high in its atmosphere. Yet from observations of this layer, and of the
gravitational field and magnetic field of the planet, we can deduce a
great deal about the structure below the clouds and throughout the
interior.

The spectrum of Jupiter reveals immediately that its atmosphere is
very different from those of the terrestrial planets. The spectrum has
strong "bands" of absorption lines due to methane and ammonia. The
chemical formulas for these two compounds, CH_4 and NH_3, respectively,
suggest that hydrogen must be present in great quantity. We fail to see
hydrogen strongly in the spectrum merely because hydrogen has no
strong absorption lines in the part of the spectrum that we ordinarily

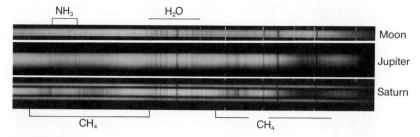

FIGURE 9.20.
A part of the spectra of Jupiter and Saturn, compared with that of the Moon. The sunlight reflected from all three bodies has of course passed through the Earth's atmosphere in reaching our telescope; the water-vapor (H_2O) band is produced by this passage through our own atmosphere. In the spectra of Jupiter and Saturn we see, in addition, bands of ammonia (NH_3) and methane (CH_4), impressed on the underlying solar spectrum. (Lick Observatory.)

observe; the weak lines that we do observe indicate, however, that hydrogen makes up most of the atmosphere.

Jupiter rotates rapidly. Features in its clouds persist long enough for us to see them carried around in a rotation period, which is just under ten hours. We can be sure that this is the real rotation period of the planet and not just an upper-atmosphere current, such as Venus has, because the magnetic field, which must be anchored in the planet itself, goes around with the same period. For a body as large as Jupiter, this is an extremely rapid rotation. Unlike the Earth, whose flattening has to be measured carefully to be detected at all, Jupiter is quite noticeably flattened. The rotation also sets up strong winds parallel to the equator of Jupiter, and the clouds are marked by a series of prominent dark belts, also parallel to the equator. There are no seasons on Jupiter, because its equator is almost in the same plane as its orbit, and local features in its cloud belts persist for long times. The most prominent of these, the Great Red Spot, has been observed for more than a hundred years.

Infrared measurements of Jupiter show that the cloud layer is very cold, about 140° K. This is so far below the freezing point of water that Jupiter's clouds cannot be made of water droplets. The most likely candidate is ammonia; at the atmospheric pressure that exists at the cloud layer, the temperature is just appropriate for the condensation of ammonia.

The radiation from Jupiter has been measured over a large part of the infrared spectrum, and these measurements combine to give a very surprising result: Jupiter is radiating away more heat than it receives from the Sun. This means that Jupiter must have an internal source of heat. This is not to say that the planet is generating new energy all the time, like a star; it may still be cooling down from an original warmer state. What the outward heat flow does mean, however, is that the interior of Jupiter is warmer than the cloud layer that we see. It is this temperature difference that keeps the atmosphere vertically stirred, as warm gases rise up from below and cool gases sink down from above. The cooling of the rising gas causes the condensation of the clouds in the upper atmosphere.

But the most bizarre result of the temperature gradient is that Jupiter has no solid surface. In fact, it has no surface at all. Deeper layers are denser and denser, until the material is liquid rather than gas, and the transition is completely gradual; there is no level that one could call a surface, with liquid below and gas above.

Most of the interior of Jupiter must be liquid—a warm hydrogen-helium soup. The pressure deep inside this giant planet is tremendous, however, and at pressures of many million pounds per square inch, hydrogen behaves differently. At these pressures the hydrogen atoms no longer pair into electrically neutral molecules; instead they move about singly, and their charged electrons are also able to move about. At the pressures and temperatures that exist throughout a large central core in Jupiter, the hydrogen is a liquid metal—a state that it cannot take at low pressures. It is probably this rapidly rotating core of metallic hydrogen that is responsible for Jupiter's strong magnetic field.

We detect the magnetic field of Jupiter indirectly, by radio observations. Just like that of the Earth, the Jovian magnetic field holds high-energy charged particles trapped in its lines of force, and these particles move rapidly back and forth. As they gyrate around the magnetic field lines, the charged particles emit radio waves, and thus we detect their existence.

In addition to the radio radiation from these particles, which is high-frequency and steady, Jupiter also emits bursts of low-frequency waves, like the radio static from a thunderstorm. These bursts reach us only when Io, the innermost large satellite of Jupiter, is in certain positions. Although we do not understand the mechanism, it is clear that the passage of Io disturbs something in the magnetic field surrounding Jupiter. Somehow, currents in this disturbed region produce the radio bursts.

FIGURE 9.21.
In these three photographs of Jupiter, one can see how the positions of the "Galilean" satellites change from night to night. Note that one does not always see all four of them; a satellite is invisible when it is in front of or behind Jupiter, or in the planet's shadow. (Yerkes Observatory.) See also Figure 9.22.

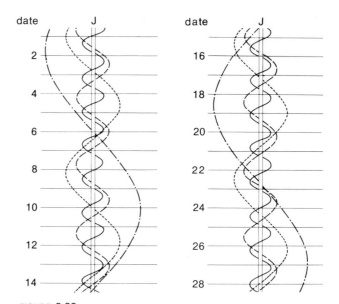

FIGURE 9.22.
The positions of the four Galilean satellites during February 1971, shown graphically. The double line below the "J" indicates the diameter of Jupiter. Note how, in accordance with Kepler's third law, the satellites with larger orbits move more slowly. (Adapted from *American Ephemeris and Nautical Almanac.*)

Jupiter has a rich family of satellites. Fourteen are known, and it is likely that some small Jovian satellites have so far escaped detection. Four of the satellites are quite large. They are easy to see, even in a small telescope; and in fact, Galileo discovered them as soon as he pointed his telescope at Jupiter. The larger two of these are about as large as the planet Mercury, while the next two are about the size of our Moon. They are very different physically from Mercury or the Moon, however, as their low densities show. (Their radii can be measured directly, from their apparent sizes, and their masses can be calculated from their gravitational effects on each other.) The densities of Io and Europa seem rather too low for rock, and those of Ganymede and Callisto are far too low. Yet their densities are too high for them to consist of the Jovian mixture, which is mostly hydrogen and helium. Most likely they are intermediate in composition, with large quantities of ice, ammonia, and methane, mixed with compounds of the heavier elements—in a phrase, frozen mud and ice. None of Jupiter's satellites has any appreciable atmosphere; apparently they were unable to hold onto their hydrogen and helium, and their interaction with the charged particles in the Jovian magnetosphere makes it impossible for them to hold any atmosphere now.

The other Jovian satellites are much smaller bodies, with radii ranging from 10 to 100 kilometers. One of them is Jupiter's closest satellite, less than two Jupiter radii from its surface. The other nine have much larger orbits. They are sometimes called the irregular satellites, because they are so far from Jupiter that their orbits are perturbed strongly and erratically by the Sun. All have orbits that are eccentric and somewhat inclined. Four in one group have similar orbits that go around in the same direction as the inner satellites. The four in the outer group are twice as far from Jupiter and go around in the opposite direction. (The newly found fourteenth satellite no doubt belongs to one of these groups.)

It is possible that the irregular satellites did not begin their lives in association with Jupiter. Dynamical studies have shown that it is possible for Jupiter, with the aid of solar perturbations, to capture passing bodies; and the most likely capture orbits are very similar to those of the two groups of irregular satellites.

SATURN

Saturn is the second of the Jovian planets, both in distance from the Sun and in size. Its size is nearly as large as that of Jupiter—more than 9 Earth radii—but it falls far short of Jupiter in mass. Still, it is nearly 100 times as massive as the Earth. Saturn has the lowest mean density of all the planets, only about 7/10 of the density of water. Its composition and structure must be nearly the same as those of Jupiter, though; the difference is that the smaller mass and gravity of Saturn do not compress its hydrogen-helium interior as much as that of Jupiter. Saturn also radiates more heat than it receives from the Sun; so, like Jupiter, it must have a warm, liquid interior. Its innermost core is probably also metallic, and the radio radiation that would go with a magnetic field has been weakly detected from Saturn. The planet rotates rapidly, with a period just over 10 hours, and is even more flattened than Jupiter.

The atmosphere of Saturn is similar to that of Jupiter, and the small differences are probably explained by the fact that Saturn is farther from the Sun and therefore has a cooler atmosphere. Ammonia is much less prominent in its spectrum than methane; at the temperature of the top of Saturn's atmosphere, it is likely that most of the ammonia has condensed out. The cloud belts are much less prominent than those of Jupiter; this difference tends to support the idea that the clouds on both planets are ammonia.

Saturn also has a rich family of satellites. The largest of them, Titan, is similar to the largest satellites of Jupiter. It differs in one important respect, however; it has an atmosphere. Methane is seen in its spectrum, and some other characteristics suggest that a considerable quantity of hydrogen is present too. It is not easy to understand why Titan has an

FIGURE 9.23.
This photograph of Saturn clearly shows the largest gap in the rings. Notice also the cloud belts in the planet's atmosphere. (Hale Observatories.)

atmosphere, whereas the large satellites of Jupiter do not; perhaps this indicates that the magnetic field of Saturn is not strong, after all.

The other satellites of Saturn are much smaller than Titan, but they are larger than the small satellites of Jupiter. Because of the distance of Saturn, it may be that smaller satellites exist but have not yet been detected.

The masses of several of the Saturnian satellites have been calculated, from their gravitational perturbations on each other's orbits, and approximate values of their radii can be calculated from the amount of sunlight that they reflect. Their densities turn out to be low; so they are probably made of icy material.

Saturn's unique glory is its system of rings. These consist of a broad, thin sheet of small bodies, all in the equatorial plane of Saturn, each body going around in its own circular orbit. When we see the rings edge on, they disappear almost completely; from their faintness at that time, it is estimated that the rings are only one or two kilometers thick (about a mile). The thinness is an indication of how closely neighboring bodies in the rings stay together; from this we can calculate that, unlike other bodies in the Solar System, when chunks of Saturn's rings collide they merely jostle each other gently. The rings reflect sunlight well; they look about as bright as the surface of Saturn itself. Yet their region cannot be densely filled, for stars can be seen through them. The whole picture adds up to a swarm of icy bodies ranging from snowball size perhaps up to iceberg size. Their spectrum suggests a large component of frozen water, but radar reflections are more like those from rock. The bodies

FIGURE 9.24.
The changing appearance of Saturn as the planet and its ring system move around Saturn's orbit.
(See Figure 9.23.) Note how the rings disappear when seen edge-on. (Lowell Observatory.)

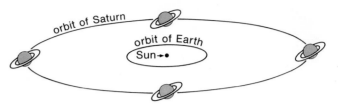

FIGURE 9.25.
Because the ring system of Saturn keeps the same spatial orientation as Saturn moves around in its 29-year orbit, we alternately see the rings from above and below and, in between, edge-on.

that make up the rings are probably ice-covered rock, a composition much like that of the satellites of the outer Solar System.

The rings extend from just outside the atmosphere of Saturn out to a little more than twice the radius of the planet. They are far from uniform, however. The zones of different brightness are often referred to as separate rings, although only between the outermost two (called rings A and B) is there a distinct gap. This division is at a distance from Saturn where a body in the rings would have close to half the period of the closest large satellite of Saturn. In this resonant situation, perturbations by the satellite would repeat again and again in the same way, and the orbit of the body would soon change. Since no ring particles can remain there long, there is a gap in the rings.

Saturn's equator, and the ring system along with it, is inclined 27° to the plane of Saturn's orbit. The Sun is of course in the orbital plane; and since we are so close to the Sun, we are practically in the orbital plane of Saturn. As Saturn moves around the Sun in its 29½-year period, the planet keeps its equator oriented the same in space, and as a result we see the rings alternately from the north side and from the south side. At times in between, about every 15 years, we look at the ring system edge-on.

Why Saturn should have rings is a puzzle that is still unsolved. One thing is clear, however: the rings occupy a region in which the tidal force of Saturn is too strong for a satellite ever to have formed. Closer than about 2½ times the radius of a planet (a distance called the Roche limit), the stretching by the planet's tidal force is stronger than the gravitational cohesion of a body that is so close to the planet. Thus any large satellite that ever came that close to Saturn would have been pulled apart by the tidal force. What is even more likely is that material within the Roche limit was simply never able to collect into any sizeable bodies and remained as the separate swarm that we see today. Although the total mass of the rings is difficult to estimate, it is probably about as large as that of a medium-sized satellite. As plausible as all this may be, however, it does not explain in any way why Saturn should be the only planet to have rings.

URANUS AND NEPTUNE

The five bright planets, Mercury, Venus, Mars, Jupiter, and Saturn, are conspicuous and have been known since man first directed his attention to the sky. The outermost planets are much fainter, however, and were discovered only in recent centuries. Uranus is bright enough barely to see with the naked eye, but it was never identified as a planet until 1781, when Sir William Herschel came across it accidentally with his telescope and recognized it as a disc rather than a starlike point. From its motion, its orbit was calculated; and astronomers soon found that Uranus had been charted a number of times previously as a star, without anyone noticing, among the thousands of faint stars, that this one object never appeared in the same place twice.

The advantage of these inadvertent observations was that, shortly after its discovery, positions of Uranus were available for more than a whole 84-year revolution around the Sun, and a really accurate orbit could be calculated. As time went on, this orbit was improved by the inclusion of new observations; but somehow the orbit never fitted all the observations in a satisfactory way, even when the perturbations due to Jupiter and Saturn were carefully allowed for. The indication became stronger and stronger that some additional, unknown body was affecting the motion of Uranus.

As the decades of the nineteenth century passed, the science of celestial mechanics progressed to the point where some astronomers could undertake the task of calculating, from the motion of Uranus, where the unknown body must be. Two astronomers independently solved the

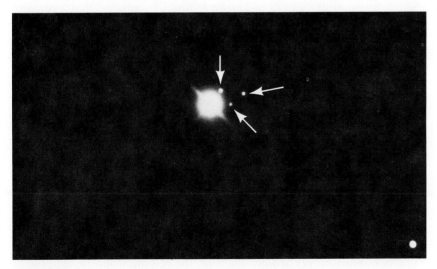

FIGURE 9.26.
Because Uranus is so far away, its satellites seem almost lost in the glare of the planet itself. Three satellites are marked in this picture. (Lick Observatory.)

problem. A young Englishman, John Couch Adams, calculated the correct answer, but he was unable to convince the members of the scientific "establishment" that it would be worth their trouble to search the indicated area for a new planet. Meanwhile a Frenchman, Urbain Leverrier, also arrived at the correct solution. He wrote to the Berlin Observatory, where there was a larger telescope than any in France. They quickly found Neptune, close to the indicated position. This triumph of scientific theory took place in 1846, when three centuries had elapsed since Copernicus. Half of this time had gone into the quest for basic understanding, which was complete with Newton's *Principia* in 1687. In the following century and a half, mathematical astronomers learned to apply Newton's simple rules to situations that were more and more complex, and the discovery of Neptune was the culmination of this development of dynamical theory.

Because they are so far from us, Uranus and Neptune are difficult to observe, and we still know little about them. They are giant planets, compared with the Earth, although they are not nearly as large as Jupiter or Saturn. Each has about four times the radius of the Earth and about fifteen times the mass. The radii of Uranus and Neptune are difficult to measure and poorly known, however. According to the present data, Uranus has a slightly larger radius than Neptune, whereas the mass of Neptune is larger than that of Uranus. Both planets have densities like that of Jupiter. But because they are less massive and therefore less compressed by strong gravity, they must have a higher fraction of medium-weight and heavy elements in their makeup.

Like the other Jovian planets, Uranus and Neptune have dense atmospheres. Both planets show strong absorption lines of methane in their spectra, along with enough hydrogen features to suggest that hydrogen is the principal constituent of their atmospheres. Ammonia is absent from their spectra, no doubt because both planets are too cold for ammonia to remain in a gaseous form. Uranus and Neptune are very cold, indeed, because they are so far from the Sun and receive so little sunlight. At the level of their atmospheres that we see, the absolute temperature is about 60° K on Uranus and somewhat cooler on Neptune. Neither planet shows any distinct atmospheric features, at least none that we can observe at such a great distance.

Uranus and Neptune rotate moderately fast; the period is about 11 hours for Uranus and 16 hours for Neptune. Both of these rotations are measured spectroscopically, by means of the small shifts that velocity produces in spectral lines (Chapter 12).

The rotation of Uranus is very unusual in one respect: its equatorial plane is almost *perpendicular* to the plane of its orbit. Its five known satellites also move in this plane. Since the equatorial plane remains fixed as Uranus moves around the Sun, the region around each pole of the planet spends about 42 years in the sunlight and 42 years in darkness

FIGURE 9.27.
Neptune's large satellite, Triton, shows easily in this photograph. (Lick Observatory.)

during each revolution around the Sun. The climate on Uranus must be exceedingly strange.

Neptune has a more normal rotation, with an inclination of 29°. Its satellites are unusual, however. One of them, Triton, is about the size of our Moon and moves in a circular orbit close to Neptune, but its direction is retrograde. Neptune's second satellite is small and moves in a large, eccentric orbit. Triton is large enough and cold enough to hold an atmosphere, but none has ever been detected. It may simply be too cold; perhaps all the potential atmosphere of Triton is frozen out as a solid layer on its surface.

PLUTO

The ninth planet, Pluto, was discovered in 1930 as the result of a long, intensive search of photographs that collectively covered a large part of the sky. Percival Lowell, who set up the search, was responsible for the discovery of Pluto, even though the actual event took place many years after his death.

January 23, 1930   January 29, 1930

FIGURE 9.28.
Percival Lowell's search for the ninth planet culminated, 14 years after his death, in this pair of photographs, on which C. W. Tombaugh found the small image that had moved. Further intensive observations showed it to be a planet. The name Pluto commemorates Lowell's initials. (Lowell Observatory.)

On the average, Pluto is almost 40 times as far from the Sun as the Earth is, and at that distance it takes nearly 250 years to go around its orbit. Pluto has the most eccentric orbit of all planets, however, and for many years around perihelion (which occurs next in 1989) it is actually closer to the Sun than is Neptune. Pluto also has the most highly inclined orbit (17°). Although in a plan view of the Solar System it would appear that the orbit of Pluto intersects that of Neptune, the orbit of the outer planet is actually tilted in such a way that the two orbits nowhere come close together. At perihelion, Pluto is above the plane of Neptune's orbit by more than a quarter of the distance of either of them from the Sun.

In fact, the tilt of Pluto's orbit and the timing of its motion are arranged in a way that gives the largest possible margin of safety against close encounters between Pluto and Neptune. Not only is Pluto's perihelion point as far as possible from its nodes, which are the places where its orbit crosses the plane of Neptune's orbit, but the speeds of the two planets are synchronized in such a way that when Pluto passes perihelion, Neptune is always far away from the corresponding point in its orbit. The average orbital period of Pluto is exactly 1½ times that of Neptune; and as a result the two planets repeat their relative positions almost exactly, every 495 years, during which time Neptune has circled the Sun three times and Pluto twice. Neptune always overtakes Pluto when the outer planet is at aphelion; and when Pluto is at perihelion, Neptune is always 60° away in its orbit.

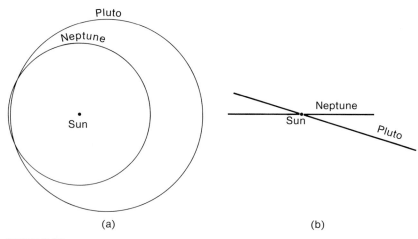

FIGURE 9.29.
The orbits of Neptune and Pluto (a) as seen from above the Solar System, and (b) as seen edge-on.

This uncanny synchronism is, of course, no accident. It is a dynamic phenomenon called a stable resonance. What this means is that if Pluto were ever to drift out of lockstep with Neptune, the changed positions would change the forces, but in just such a way that Pluto would be brought right back into its present relationship with Neptune. Furthermore, if Pluto had, in the distant past, been in a nonresonant orbit that was at all similar to its present orbit, the forces of Neptune would have changed Pluto's orbit until it was trapped in this resonance, to remain so forever after.

The resonance of the Pluto-Neptune system is a recent discovery, a result of the new-found ability to carry out long-term, accurate orbital calculations on high-speed electronic computers. The phenomenon was discovered by "observing" the motions of Neptune and Pluto in a calculation that simulated their motions for 4½ million years. Although there are small periodic variations, the motions of the two planets stayed so closely in resonance during this length of time that they seem likely to have been in this relationship since the time of their formation.

As the most distant of the planets, Pluto is the most difficult to observe. Its radius would be easy to observe if it were as large as Uranus or Neptune, but in fact it is as small as Mars or Mercury, and its diameter as seen from the Earth is a small fraction of a second of arc. Our best chance of measuring the size of Pluto is to watch it pass in front of a star, and to see for how long a time it covers the star; unfortunately, such events are rare, and none has yet been successfully observed.

The mass of Pluto is also very poorly known. Since it has no known satellites and no space probe has ever passed near it, Pluto's mass can be calculated only from its gravitational influence on Neptune. This gives a very weakly determined result, since Neptune is a hundred times as massive and Pluto consequently affects it very little. All in all, it appears that Pluto is probably intermediate between Mars and Mercury, both in radius and in mass.

The resemblance ends there, however. As little as we know about Pluto, we can tell that it is extremely cold—about 50° K—and has little or no atmosphere. Even though Pluto's gravitation could hold an atmosphere, its lack of atmosphere is not surprising; most of the gases that might contribute to an atmosphere would be frozen out at the frigid temperature of Pluto's surface.

Only one physical property of Pluto has been measured in a definite way: its rotation period. The brightness of Pluto varies by 10 per cent with a very regular period of 6.4 days; this must be caused by the rotation of a surface that has brighter and darker places on it. Unfortunately, there is no way of telling from this fluctuation which way the axis of rotation is oriented—or, for that matter, whether the rotation is direct or retrograde.

Pluto lies in the outer part of the Solar System, which we think of as the realm of the Jovian planets, but it is very different from them. Its singular nature has led to the suggestion that Pluto is not properly a planet at all, but instead began its life as a satellite of Neptune and later escaped. In this picture, Pluto would have been ejected by a close encounter with Neptune's large satellite, Triton. Its orbit originally would have intersected that of Neptune, but would have been shifted gradually to its present orbit by perturbations. Once it reached its current resonant relationship with Neptune, it was trapped permanently in its present orbit. Whether this hypothesis is true, or whether Pluto was simply born as a small planet, no one knows.

In the past 200 years, the number of known planets in the Solar System has increased from 6 to 9. Will others be discovered in the future? Obviously, we cannot flatly predict that such discoveries will or will not occur, but some limits can be set on the possibilities. There are three ways in which new planets can be discovered: by accidental notice, as happened with Uranus; by studying perturbations of the motions of known bodies, by means of which Neptune was found; or by searching for slowly moving bodies, as was done for Pluto.

Accidents are, of course, impossible to predict, but there are some things that we are sure will not happen. A discovery like that of Uranus will never be made again. If a body the size of Uranus were much farther away than Pluto, in a single observation it would be indistinguishable from a star. If it were less remote than that, it would have been detected long ago by its perturbations on Neptune. Even more likely, it would have been detected by searches like the one that discovered Pluto. A planet like Uranus or Neptune, even at twice the distance of Pluto, would look much brighter than Pluto, which was difficult to find only because it is so small. Thus if undetected planets exist, they could be either small, Pluto-sized bodies appreciably farther than Pluto, or else Jovian-sized planets many times as far as Pluto. Such a body would be discovered only if someone noticed its night-to-night motion, among all the millions of equally faint but stationary stars. Considering the type of observations that astronomers make, an accidental discovery of this type would be very unlikely.

THE FAMILY OF PLANETS

Among the planets and satellites are bodies with a wide range of properties, from the bare, baked rock of Mercury to the hydrogen-helium soup of Jupiter. All were made at the same time, out of the same batch of material, yet they are so different. It is, of course, the task of the astronomer to seek out the simple underlying regularities that have led to such a diversity of results. What do the planets have in common, and why do they differ as they do?

As we saw when discussing the composition of the Earth, the material of the universe is mostly hydrogen and helium, and most of the remainder consists of carbon, nitrogen, oxygen, and neon. Material that is not already in stars—interstellar material, as we call it—is nearly all gas; and nearly all the atoms move about as individuals. Combination of atoms with each other, into molecules and then possibly into solid material, can take place best when the gas is dense, so that atoms encounter each other frequently, and when the temperature is low, so that they do not collide so violently as to break up molecules. Such were the conditions when the Sun formed, and along with it the planets and the other solid bodies of the Solar System.

The crucial questions are how the condensation and the cooling proceeded in each part of the Solar System, and how strongly the gravitation of each body was able to hold its parts together. As we saw previously, different atoms tend to combine chemically in different ways; each such combination has its own sensitivity to temperature and its own ability to escape from a gravitational field. Hydrogen atoms combine into molecules of H_2. These hydrogen molecules remain in a gaseous state at all except the very lowest temperatures, and they are so light that they escape from all but the most massive bodies. Helium behaves almost the same, except for the fact that its atoms remain separate and do not join into molecules. Thus it is not surprising that among the bodies of the Solar System, only the four large planets seem to have retained a nearly normal abundance of hydrogen and helium. Everywhere else,

TABLE 9.1. CHARACTERISTICS OF THE PLANETS.

ORBITS

	Semimajor axis	Period	Eccentricity	Inclination	Synodic period (in days)
Mercury	0.39 a.u.	88 days	0.21	7°	116
Venus	0.72	225 days	0.01	3°	584
Earth	1.00	365 days	0.02	–	–
Mars	1.52	687 days	0.09	2°	780
Jupiter	5.20	11.9 years	0.05	1°	399
Saturn	9.54	29.4 years	0.06	2°	378
Uranus	19.1	83.6 years	0.04	1°	370
Neptune	30.0	164.5 years	0.01	2°	368
Pluto	39.3	246.2 years	0.25	17°	367

PHYSICAL DATA

	Radius[a]	Mass[a]	Mean density (g/cm³)	Flattening	Rotational period	Inclination of axis	Number of satellites
Mercury	0.38	0.06	5.4	almost 0	59 days	?	0
Venus	0.95	0.82	5.3	almost 0	243 days	?	0
Earth	(1.00)	(1.00)	5.5	1/298	24 hours	23°	1
Mars	0.53	0.11	3.9	1/200	24½ hours	25°	2
Jupiter	11.2	318	1.3	1/15	10 hours	3°	14
Saturn	9.4	95	0.7	1/10	10½ hours	27°	10
Uranus	4.1	14.6	1.2	1/14	11 hours	98°[b]	5
Neptune	3.9	17.2	1.7	1/50	16 hours	29°	2
Pluto	0.4?	0.02?	?	almost 0	6½ days	?	0

[a] Here the radius and mass of the Earth are taken as the unit of measurement.
[b] Inclination 98° means retrograde rotation inclined at 82° (i.e., 180°–98°).

these two elements have been lost, or else they were never able to condense in the first place.

The next most abundant group of elements, carbon, nitrogen, oxygen, and neon, have fared very differently in the inner and the outer parts of the Solar System; and it is clear that the difference has been temperature. Carbon, nitrogen, and oxygen all combine readily with hydrogen; and the molecules that result from such combination remain gaseous at the temperature of the inner planets, but freeze out at the colder temperatures that prevail in the outer parts of the Solar System. The satellites of the Jovian planets, with their low densities and highly reflecting surfaces, seem to be made largely of ices of these molecules. Since many of these bodies have far too weak a gravity to hold on to such molecules in gaseous form, they must have formed cold and stayed cold. It is interesting to note, in this connection, that the inner two of Jupiter's large satellites have higher densities than the outer two; were they once heated by Jupiter itself, so that much of their icy material evaporated and was lost? It will be interesting, eventually, to study the relative abundances of methane (CH_4), ammonia (NH_3), and water ice (H_2O) in the bodies of the outer Solar System. Each of these substances changes from gas to liquid at a different temperature, and their relative amounts should reveal much about the mode of formation and history of the bodies that contain them.

Neon is an interesting contrast. In the stars it is almost as abundant as carbon, nitrogen, or oxygen; but neon forms no molecules, and it remains gaseous down to very low temperatures. It is difficult to detect spectroscopically, so that we cannot measure its abundance on the Jovian planets and their satellites. We can predict, however, that it is nowhere an important constituent. Jupiter and Saturn should have held their neon, but there it is far outweighed by hydrogen and helium. In the satellites, where neon might otherwise be an important part, it presumably has been lost, because it did not freeze onto the smaller bodies out of which they were made.

The bodies of the inner Solar System seem to have been formed at temperatures at which the compounds of carbon, nitrogen, and oxygen were unable to condense. They include much less than a "normal" proportion of these elements. As was discussed in Chapter 3, the condensation of these bodies was based on chemical binding of less abundant elements, such as silicon, aluminum, and magnesium, which form solids that can endure higher temperatures. The inner planets and their satellites are thus made of heavy-element residues, with nearly all the light elements lost. The difference between the terrestrial and the Jovian planets is simply a consequence of closeness to the Sun, which maintained a higher temperature in the region in which the terrestrial planets were forming.

TABLE 9.2. SATELLITES OF THE PLANETS

Name	Mean distance (in planet radii)	Orbital period (in days)	Radius (in km)
Earth			
Moon	60	27	1,740
Mars			
Phobos	2.8	0.3	12
Deimos	6.9	1.3	6
Jupiter[a]			
V (Amalthea)	2.5	0.5	90
Io	5.9	1.8	1,800
Europa	9.4	3.6	1,500
Ganymede	15	7.2	2,600
Callisto	26	17	2,500
VI	161	251	70
VII	164	260	20
X	164	260	10
XIII	173	282	10?
XII	297	625	10
XI	317	699	10
VIII	329	739	20
IX	331	758	10
Saturn			
Janus	2.6	0.7	100
Mimas	3.1	0.9	200
Enceladus	4.0	1.4	300
Tethys	4.9	1.9	500
Dione	6.3	2.7	600
Rhea	8.8	4.5	800
Titan	20	16	2,900
Hyperion	25	21	200
Iapetus	59	79	600
Phoebe	215	550	100
Uranus			
Miranda	5.0	1.4	300
Ariel	7.4	2.5	700
Umbriel	10	4.1	500
Titania	17	8.7	900
Oberon	23	13	800
Neptune			
Triton	14	5.9	2,500
Nereid	225	360	200

[a]A fourteenth satellite of Jupiter was discovered in late 1975, but its orbit has not yet been calculated.

The great difference in size between the terrestrial and the Jovian planets is also caused largely by this difference in chemical selection of the original material. The heavy elements that formed the terrestrial planets make up only a small fraction of the material of the universe; the light elements, which were lost, were most of the material with which the process started. Approximately the same amount of original Solar-System material went into forming the Earth as went into Jupiter; but the Earth ended up with only a three-hundredth as much mass, because it was able to hold only the small fraction that was made up of the heavy elements and their chemical compounds.

When we come to the details of planetary characteristics, however, we are able to explain much less. A large part of the problem is that our knowledge of their compositions is still very crude. We can analyze atmospheres by studying the spectrum of light that has passed through them; but no such study can be made of solid surfaces, because solids simply do not produce the sharp, distinctive spectral features by which we study gases so easily. For major steps forward in understanding how the planets came to be what they are, we may have to wait for samples of their material to be brought back by spacecraft in the distant future.

For the terrestrial planets, however, we have enough observational facts at least to begin an attempt to understand them. They have much in common, but there are also great constrasts between them. Most striking are the differences in atmospheres. These may prove very difficult to understand, however, because the atmospheres of the terrestrial planets are such superficial things. Unlike the atmospheres of the Jovian planets, which are the gaseous outer surface of the planet itself, the atmospheres of the terrestrial planets are like the frosting on a cake. They are superficial layers of a quite different composition, which overlie the surface of a solid planet. Furthermore, the solid material of each planet contains, bound into its various substances, far more of the atmospheric gases than are in the atmosphere itself. In the Earth's crust, for instance, carbonate rocks such as limestone contain thousands of times more carbon dioxide than the tiny quantity that exists in free form in the atmosphere. Thus it is most likely that the atmospheres of the terrestrial planets are an outgrowth of their surfaces. Their origin is to be sought not in the origin of the terrestrial planets, but in the later chemical behavior of the surfaces.

Mercury is an exception, in a sense. Because of its low gravity and high temperature, it is unable to hold gases, and anything that has come out of its surface has been lost. Thus it has remained without an atmosphere.

The difference between Venus and the Earth is puzzling. The two planets have approximately equal amounts of atmospheric material, but with compositions that differ in a strong but simple way. On Venus the dominant atmospheric gas is carbon dioxide, which lies in a layer a hundred times as massive as all the Earth's air. Of the Earth's "atmosphere," however, the majority of the material is the water of the oceans; it happens to be liquid rather than gas, at the temperature of our surface, but it is the sort of material out of which atmospheres are made. The essential difference between the atmospheres of the Earth and Venus seems to be the relative amounts of carbon and oxygen. On the Earth, the carbon is not in the atmosphere but is bound up in the rocks of the surface, whereas on Venus it is in the atmosphere as carbon dioxide. Furthermore, the carbon on Venus seems to have tied up all the oxygen as carbon dioxide, whereas on the Earth there was an excess of oxygen, which combined with the ever-present hydrogen, as water. The water of the oceans then dissolved the carbon dioxide of the atmosphere and deposited it as solid carbonate rock.

Aside from defining the problem in this general way, however, we can do little toward understanding the differences between the Earth and Venus. The study of the chemical behavior of the Earth's surface layers, geochemistry, is a whole difficult science in itself; and it has made only limited progress, in spite of the extensive data that are available. Since we know nothing about the surface of Venus, we can only speculate about its behavior.

Similar difficulties also limit our understanding of Mars. Its gravity could hold much more atmosphere than Mars has at present. Its atmosphere is very thin; there is presumably far more material tied up in carbonate and oxide rocks, and perhaps even in ice within the cold soil. Why so little gas covers the surface of Mars, we can only speculate.

We know even less about the chemical details of the Jovian planets. At the temperatures of their atmospheres, the heavier elements are frozen out, and are to be found only in the deeper layers, to which our sight is unable to penetrate—and which may be equally inaccessible even to future exploring spacecraft. Understanding of these bodies may have to come indirectly, by study of the satellites that were made along with them.

Finally, there are some things that we will never understand, because they are largely the results of accident. Although in principle the behavior and the motion of each of the parts of the Solar System are predictable, there are always situations in which the result is so sensitive to the exact starting conditions that we cannot hope to follow them in detail. The everyday metaphor, "That's the way the ball bounces," expresses it perfectly. The bounce of a football is absolutely predictable, *provided* you know exactly how it is going to strike the ground. The trouble is that a small error in the orientation of the ball leads to a large change in the direction of the resulting bounce, and after the next bounce the difference is even greater, so that after only a few bounces we can only guess. Physical science is full of situations like this, where we know the rules fully but can never follow all the details of the picture. In the Solar System there are many such details. Why do the Earth and Mars rotate so similarly, while Venus is so different? Why does Jupiter have four large satellites, rather than two or five? Why does only Saturn have rings? Yet along with these mysterious details, we do have a good understanding of the fundamental processes that have made the Solar System what it is.

REVIEW QUESTIONS

1. List the planets, in order away from the Sun. Distinguish between the terrestrial group and the Jovian group.

2. What are the major physical differences between the terrestrial and the Jovian planets? Explain how they are simple results of the differences in size and in distance from the Sun.

3. The bodies of the inner Solar System are rocky; those of the outer Solar System are icy. What connection does this difference have with the chemical behavior of atoms and the way in which the Solar System probably formed?

4. Discuss resonant rotations and the way in which tidal forces may cause them. Is there some parallel between these processes and the behavior of the orbit of Pluto?

5. Contrast the surfaces of Mars, the Earth, and the Moon.

6. Describe the internal structure of Jupiter, and explain how it is deduced.

7. Compare the magnetic fields of the various planets (insofar as they are known). How are these probably related to the internal structures of the planets?

8. How were Uranus, Neptune, and Pluto discovered? Why did the method used for Neptune not work for Pluto?

10 | THE OTHER BODIES OF THE SOLAR SYSTEM

The planets are not the only bodies that move around the Sun; there are also many smaller bodies, each traveling in its own orbit. Although these make up only a tiny fraction of the total mass of the Solar System— a total much less than the mass of the smallest planet—their existence and nature offer further clues to the way in which the Solar System has developed.

THE ASTEROIDS

In the wide gap between the orbits of Mars and Jupiter, a large number of small bodies move their separate ways, each in its own orbit around the Sun. These are called the asteroids. The largest of them has a radius of almost 800 kilometers (500 miles), and their sizes range downward to tiny chunks of rock far too small for us ever to observe directly.

The first asteroids were discovered, accidentally, at the beginning of the nineteenth century. A few dozen discoveries accumulated over the ensuing decades; but the real flood came with the introduction of photography into astronomy, near the end of the nineteenth century. Photographs cover so much of the sky, and reach such faint objects, that dozens of asteroids can be discovered in a single night by any astronomer who cares to take the trouble. From a few examinations of this sort, it has been estimated that 50,000 to 100,000 asteroids are observable. Astronomers no longer make any attempt to chart them all; only the brighter

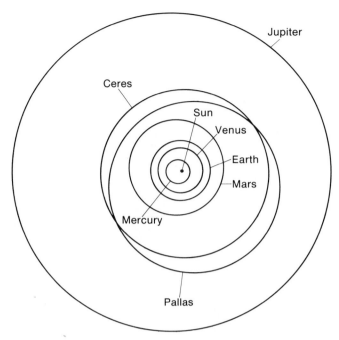

FIGURE 10.1.
The orbits of the two first-discovered asteroids, Ceres and Pallas,
shown with the orbits of the terrestrial planets and Jupiter.

asteroids are catalogued and tracked systematically, along with a few
fainter ones that have orbits of unusual interest. The first asteroids were
given mythological names, like the planets and satellites, but the length
of the modern list taxes even the richest imagination, and names of
almost any sort have been used. For more formal cataloguing, numbers
are also used; so an asteroid may be 4 Vesta, 433 Eros, 1685 Toro, etc.
About 1,800 asteroids have been catalogued; orbital calculations are
kept up-to-date on most of them, but a few asteroids have been lost as
a result of inadequate calculations or observations.

Each asteroid moves, of course, according to Kepler's laws, following
an elliptical orbit around the Sun. The asteroids affect each other very
little, because they are bodies of such small mass; but their orbits are
perturbed by the attractions of the planets, especially massive Jupiter,
In extreme cases the orbit can be strongly changed, or even controlled,
by the perturbations, so that there are gaps in the distribution of the
asteroids, just like the gap in the rings of Saturn. The clearest break in
the asteroid distribution is at the distance from the Sun where the orbital
period would be just half that of Jupiter. An asteroid in such an orbit
would experience the same force, in the same direction, again and again

FIGURE 10.2.
Two asteroids left their trails on this photograph of a field of stars. The telescope
was guided, as usual, to follow the stars, as the Earth rotated; but the asteroids
moved relative to the stars during the exposure. (Yerkes Observatory.)

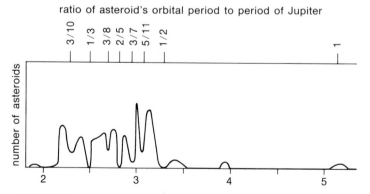

FIGURE 10.3.
The distribution of orbit sizes for the known asteroids, showing the
"Kirkwood gaps." The group with periods equal to that of Jupiter
is the Trojans.

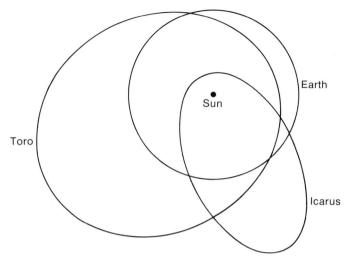

FIGURE 10.4.
The orbits of the asteroids Icarus and Toro. Both are unusual
in passing inside the orbit of the Earth.

every second revolution; and the resonant repetition of these forces
would shift the orbit into a quite different one.

Jupiter has also captured some asteroids and locked them into orbits
like its own. These bodies (called the Trojan asteroids, because the first
ones found were given names taken from the story of the Trojan war)
travel around the Sun ahead of or behind Jupiter, staying 60 degrees
away from Jupiter on the average. This is a stable configuration; as a
Trojan asteroid drifts away from the average point, the perturbations
of Jupiter tend to bring it back.

Most asteroids, however, move in orbits that are influenced relatively
little by the planets and change only slowly. Their eccentricities and
inclinations tend to be larger than those of planetary orbits; eccentricities
of 0.1 to 0.2 and inclinations of 10 degrees are common. Nearly all the
asteroids remain securely between the orbits of Mars and Jupiter, but
a few have orbits that cross those of the planets. One extreme is 944
Hidalgo, which at aphelion, its farthest distance from the Sun, is beyond
Saturn. At the other extreme, 1566 Icarus comes closer to the Sun at
perihelion than even Mercury.

Icarus is one of a small group of asteroids that cross the Earth's orbit.
They can of course be conspicuous—to sharp-eyed astronomical photog-
raphers, at least—because they occasionally come so close to us and move
by us so fast. At least one member of this group, 1685 Toro, seems to
have been "captured" by the Earth; its orbit, although of longer period,
appears to be in resonance with ours.

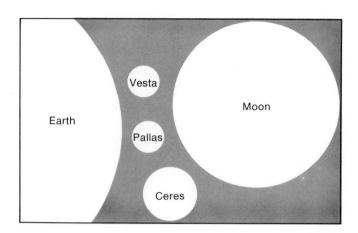

FIGURE 10.5.
The sizes of the three largest asteroids,
relative to the Earth and the Moon.

Because the asteroids are such small bodies, it is difficult for astronomers to measure their sizes. Even the largest asteroids look hardly
bigger than points of light. Only recently have their diameters begun to
be measured with any accuracy. The technique that is used depends on
sensitive measurements of infrared radiation, which tells us how much
energy the asteroid emits. This must exactly balance the amount of
energy that the asteroid receives from the Sun, which of course depends
on the size of the asteroid. The exact calculation also involves the reflec-
tivity of the asteroid, and this can also be found from comparing the
infrared brightness of the asteroid with its brightness in visible light.

The reflectivities tell us something about the material of which asteroids are made. Asteroids are poor reflectors and must be made of rock,
without any bright covering of ice. Apparently, when the bodies of the
Solar System formed, even in the asteroid zone the temperatures were
not low enough for the light elements to be held as ices. It begins to
appear, however, that there may be composition differences within the
asteroid zone. Some asteroids are relatively light in color, and others
are very dark. This distinction corresponds in an intriguing way with the
division between the two main classes of meteorites, which, as we shall
see below, are asteroids that have struck the Earth. We do not know
yet what the differences mean, but it may well be that asteroids, as observed through the telescope, and meteorites, as analyzed in the laboratory, will combine to reveal some of the deepest secrets that lie hidden
in the past history of the Solar System.

Careful study of the asteroids also reveals their shapes. Only the
largest dozen or two of the asteroids are round; the smaller ones have
irregular shapes. We do not see the shape directly, but the brightnesses
of most asteroids fluctuate as they tumble end-over-end in their rotation.

To understand why small asteroids are not round, we should first ask why larger bodies, like the Earth, *are* round. The answer is the force of gravitation, which tends to draw the parts of a body as close together as it can. The closest-together shape is a perfect sphere, and a body will adopt this shape unless the stiffness and strength of its material prevent it from doing so. Thus the question is whether the gravitational forces of the body are strong enough to crush its material. Rock is not difficult to crush; the operators of gravel quarries do it all the time. Even bodies of modest size have large enough gravitational forces to squeeze themselves into roundness. The dividing line comes at a radius of about 100 kilometers; smaller bodies can, and do, maintain irregular shapes.

The origin of the asteroids is a mystery. The planets are spaced fairly regularly, except for the large gap between Mars and Jupiter. The pattern of spacings suggests that there should be another planet there; instead we have the asteroids. For some reason the material in this region did not condense into a single planet, but instead remained as a swarm of bodies. There is the romantic possibility, of course, that a planet originally existed there but was destroyed, leaving the asteroids as fragments. This seems very unlikely, however; among other difficulties, the total mass of the asteroids is much less than that of even a small planet. As we shall see, however, the compositions of meteorites give evidence that at least some asteroids have indeed broken up.

COMETS

In many ways, the comets are the strangest bodies of the Solar System. When a comet swings close to the Sun, it becomes much brighter, swells in apparent size, and sprouts a flowing tail that may extend for millions of miles. To add to their strangeness, most comets appear unexpectedly from the outer reaches of the Solar System, swing once around the Sun, and return to the cold distances from which they came.

Strange as comets may seem, their mystery lessens when we study them more carefully. From their motions we can calculate their orbits, and this reveals their place in the Solar System. The average comet has an orbit that takes it very far from the Sun, many times the distance of Pluto. Its period in this large orbit is thousands of years, and it spends most of its time in the far outer part of the orbit. We can see it only during its brief swing through the inner parts of the Solar System. Since we observe several comets every year, there must be a large number that are now drifting slowly around the outer parts of their orbits and will appear in future years. Furthermore, we can observe a comet only if its orbit passes through our tiny central region of the Solar System. There is every reason to believe that there are many other comets, far from the

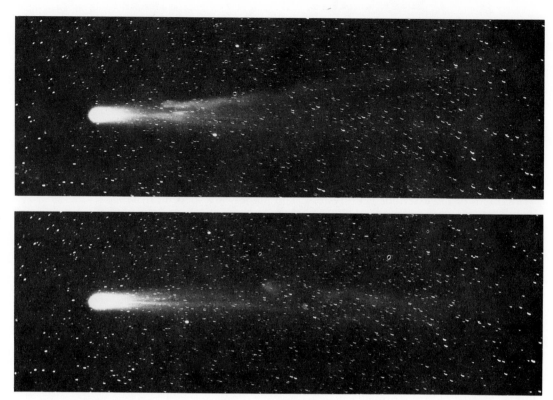

FIGURE 10.6.
Two views of Halley's comet, on successive nights in June 1910. Note the rapid changes in the tail. (Lick Observatory.)

Sun, whose orbits are not so elongated and narrow, so that they never come close enough to the Sun for us to observe them. The total number of comets must be very large indeed.

In spite of their very large average distances from the Sun, comets are certainly members of the Solar System rather than interlopers from outside. This we can tell from their orbits, as we observe them passing through our inner part of the Solar System. As we saw in Chapter 6, a body that is permanently bound to the Sun will follow an orbit whose shape is an ellipse, while a one-time passerby will have a hyperbolic orbit. Even from the small arc of a comet orbit that we observe, we can determine the orbit well enough to show that it is an ellipse rather than a hyperbola. In many cases the ellipse is a very elongated one, which we cannot distinguish from a parabola; but this is quite reasonable, because elongated ellipses do in fact resemble parabolas very closely. Some hyperbolas also resemble parabolas closely; the crucial point, however, is that a hyperbolic orbit differs from a parabola by an amount that depends on the velocity with which the body approaches the Solar System.

FIGURE 10.7.
The French artist Daumier enjoyed poking fun at the savants of his time.
(Courtesy of the Museum of Fine Arts, Boston.)

Any object with a velocity that remotely resembles those of the stars would pass through the Solar System on an orbit that was unmistakably hyperbolic in shape—and we do not observe any such comet orbits.

The material of comets, insofar as we can analyze it, seems to be appropriate for their position in the Solar System. When a comet brightens, in the inner part of its orbit, we see in its spectrum molecules containing carbon, nitrogen, oxygen, and hydrogen. In addition, if a comet passes so close to the Sun that it is heated past the vaporization temperature of metals, its spectrum usually shows lines of iron and other metals. It thus appears that comets are made of the mixture of "dirty ices" that characterizes all bodies that formed far from the Sun. In the outer parts of the comet's orbit, its material remains frozen; but as it approaches the Sun and its surface warms, the ices begin to vaporize, and the solid nucleus is then surrounded by a cloud of gas, along with particles of dust that were frozen into the ice. The gravitation of the tiny nucleus is far too weak to hold on to this surrounding "coma," however, and the material streams out to form the tail.

The tail does not stream behind the comet, however, as it would if the comet were moving through air. In the empty space of the Solar System, material released from the comet would move right along with it, except that something acts to blow it away. The agent that blows comet tails is

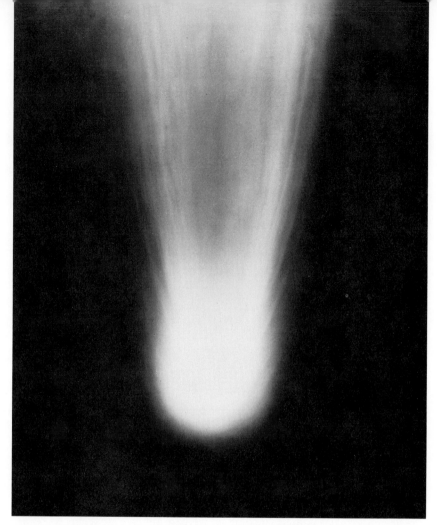

FIGURE 10.8.
The head of Halley's comet, photographed on a large scale near the time of its closest approach to the Earth in 1910. (Hale Observatories.)

the Sun. Its light exerts a gentle pressure on everything on which it falls, and the tenuous material released from the comet is gradually pushed away from it. In addition, a very sparse but rapid flow of gas continually streams outward from the Sun. This "solar wind" interacts with charged atoms and blows them harder than does the pressure of radiation. Thus many comets have two tails at once. The radiation-pressure tail curves gently away from the head of the comet, whereas the solar-wind-induced tail streams out faster, in a straighter line. Comet tails always point away from the Sun, since it is the Sun that blows them, in one way or the other. Thus when a comet has rounded the Sun and is moving away, its tail actually precedes it.

FIGURE 10.9.
Comet Mrkos, in 1957, showed a beautiful
example of a straight tail and a curved one.
(Hale Observatories.)

The solid nucleus of a comet is very small, probably only a few miles in diameter. If the nucleus were the only thing to see, we would never notice the comet at all. What makes the comet conspicuous is the gas and dust that are released when the comet approaches the Sun and is heated up. This material is lost to the comet forever, so that any comet whose orbit brings it close to the Sun is gradually eroded and depleted. The glory of each perihelion costs the comet something, and after a hundred passages, or perhaps as many as a thousand, the volatile material of the comet is used up. Its total lifetime as a bright comet is about a million years, only a fraction of the age of the Solar System.

How does it happen, then, that we still have comets? Why have they not used up all their material long ago? The answer must be that worn-out comets are continually replaced. Far from the Sun, out beyond the orbit of Pluto, is an almost inexhaustible reservoir of comets, each following its own orbit in the frigid outer reaches of the large realm that is dominated by the Sun's gravitation. Other stars pass us, and they perturb the orbits of the comets. Each time a star passes, among the perturbed

190

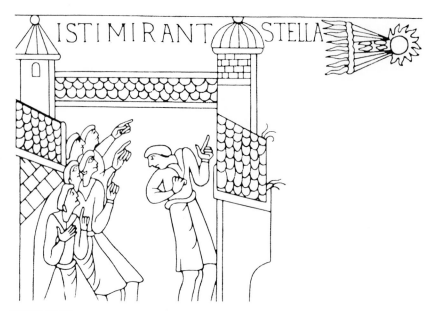

FIGURE 10.10.
In bygone times, comets were taken as signs of great events. When Halley's comet appeared in 1066, it was recorded in the Bayeux tapestry, which depicts the Norman conquest of England. (Yerkes Observatory.)

orbits there happen to be a few that accidentally swing close to the Sun, and these provide the new comets, transported from eternal cold storage to a brief life of glory.

Comets run additional hazards in their passage through the inner Solar System, besides slow evaporation. If a comet chances to pass close to one of the planets, its orbit is greatly changed. The comet may be speeded up, and thrown out of the Solar System forever. Or it may be slowed down, in which case the new orbit will not take it so far away from the Sun again, and its period will be much shorter. Several dozen of these short-period comets (often simply called "periodic") are now known. Only one of them, Halley's comet, is unusually bright. Its period is 76 years. Nearly all of this time is spent in invisibility, out beyond the orbit of Uranus. For a few months around each time of perihelion, however, it passes through our inner part of the Solar System, brightening tremendously. Every passage of Halley's comet has been noted, somewhere in historical records, since 240 B.C. In 1910 it passed unusually close to the Earth, and its tail spanned half of the sky. The passage of 1985–86 will be a much less favorable one. By a stroke of ill luck, the comet will be on the opposite side of the Sun from us when it passes perihelion. It will be a good comet, but not a spectacular one.

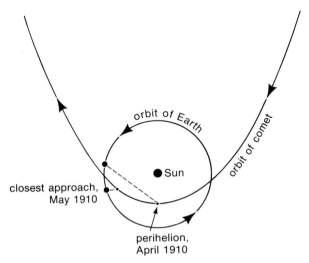

FIGURE 10.11.
The path of Halley's comet in 1910. The comet was
spectacularly visible from the time of perihelion because
of its close passage between the Earth and the Sun.

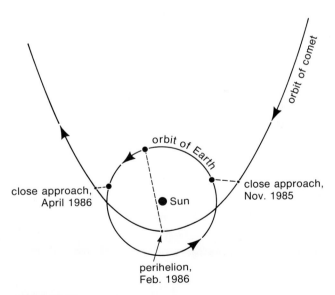

FIGURE 10.12.
The path of Halley's comet in 1985–86. At perihelion,
the comet will be on the far side of the Sun from us. It
will be relatively far from the Sun at the times when it
comes closest to us.

METEORS AND METEORITES

Anyone who watches a clear, dark sky will occasionally—perhaps a dozen times an hour—see a star-like spot streak across it. These "shooting stars," or meteors, as astronomers call them, are tiny bodies that happen to collide with the Earth. As soon as a meteor runs into even the thin upper atmosphere, it is slowed and heated by the drag of the air; in a rapid flash, it vaporizes and is gone.

Astronomers can measure the paths of meteors, by photographing them with special cameras that have rotating shutters in front of them. The meteor shows up as a streak across the picture, with breaks made by the shutter at known intervals of time, from which the speed of the meteor can be found. If two such cameras are aimed at the same region in the upper atmosphere and photograph the same meteor, its exact path can be calculated, showing just what direction it came from and how it slowed down when it encountered the atmosphere.

From the amount of light that the average meteor produces, we can tell that it is about the size of a grain of sand. It is its high speed—typically around 30 kilometers per second—that gives it so much energy. The average meteor burns at about 50 miles above us; so its light is considerable. Meteors slow down rapidly for their size, and it thus appears that they have a fluffy rather than a compact structure.

The origin of meteors is clear from their orbits, which can be calculated from the direction and speed at which they encountered the Earth. Although some meteors have quite individual orbits ("sporadic" meteors), others belong to well-defined streams, in which all the particles move in nearly the same orbit. These are called *shower* meteors, because many can be seen in a short time, at the time of year when the Earth crosses

FIGURE 10.13.
While this long-exposure photograph of a star field was being taken, a meteor flashed across the field of view, leaving its trail. (Kitt Peak National Observatory.)

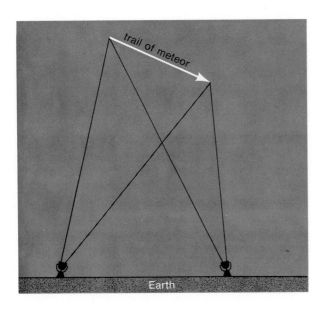

FIGURE 10.14.
If a meteor is photographed simultaneously
from two stations 20 or 30 miles apart,
its exact path can be calculated.

their orbit; if the stream of meteors is extremely dense, they may even
look like a shower of falling stars. The meteors of a shower appear to
diverge from a particular direction in the sky, in a perspective effect that
is just like the appearance of snowflakes in front of a rapidly moving car.
Most showers are named for the constellation from which they appear
to radiate: Perseids, Geminids, etc.

A meteor shower is a stream of debris from an old comet. The material
originally follows the same orbit as the comet, but perturbations gradually change the orbit of each particle. The first effect is for particles to
get ahead and behind, so that they diffuse completely around the orbit;
but eventually they spread out into a broader stream as well. If the orbit
of the meteor stream happens to cross that of the Earth, we see a meteor
shower at the time of year when we reach the crossing point. The particles of a young shower are concentrated around a single point that
moves around their orbit, and we may encounter an intense shower in
years when the meteor swarm happens to be crossing the Earth's orbit
just at the time when the Earth arrives at the intersection point. An old
stream, however, has particles spread evenly around the orbit, and we
encounter them every year.

The connection with comets is clear, because several showers have
orbits that are identical with those of periodic comets, some of which
still exist and some of which have disappeared in recent times. It is likely
that sporadic meteors are just a mixture of countless old showers that
are too thinly spread to be recognizable.

Ordinary meteors vaporize completely in the upper atmosphere, and
are gone. Occasionally, however, a larger body encounters the Earth.

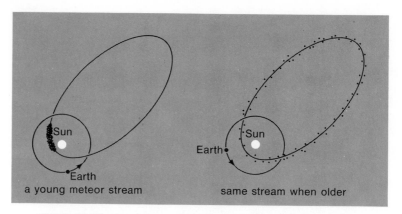

a young meteor stream   same stream when older

FIGURE 10.15.
A meteor stream consists of debris from a comet. As the stream
ages, material gradually spreads more widely around the
orbit. Each particle goes around with nearly the same orbital
period, but they pass a given point at different times.

Friction with the atmosphere heats its surface and melts off a part of
it, but some of the body survives the passage and strikes the Earth's
surface. The fall of such an object is spectacular, lighting up the whole
sky. The remnant, left lying on the Earth's surface, is called a *meteorite.*

Not all meteorites are easy to recognize once they are on the ground,
but thousands have been found, ranging from pebble-size up to many
tons. Some are stone, some are iron, and some are a mixture. Scientists
have studied the composition of meteorites with some interest, because,
except for a few Moon rocks, the meteorites give us our only chance to
examine material from outside the Earth.

Although a few meteorites may originate from the solid material in
the nuclei of old, dead comets, most meteorites are almost certainly small
asteroids whose orbits have been perturbed so far that they cross the
orbit of the Earth. Thus when we analyze meteorites, we are sampling
material from the asteroid belt.

One result comes clearly out of the analysis. When we determine the
ages of meteorites (by studies of radioactivity that will be described in
the following chapter), we find that the meteorites are 4.6 billion years
old. This must be the age of the Solar System itself—or at least the time
since the material of the asteroids and the planets solidified.

Otherwise, the meteorites tell a complicated story that is not at all
well-understood. The detailed sequence in which materials condensed
was a complex one, and many substances seem to have separated from
each other along the way. Particularly interesting are the iron meteorites.
Not only does the separation of iron from stone suggest a melting inside
a body of some size, where gravity is strong enough for the heavier iron

FIGURE 10.16.
This meteorite, one of the largest ever found, weighs 15 tons.
(American Museum of Natural History, New York.)

FIGURE 10.17.
If the surface of an iron meteorite is cut, polished, and etched with acid, these
characteristic patterns appear, indicating the regular, crystalline structure of the
metal. (American Museum of Natural History, New York.)

to sink to the center, but the iron of meteorites has a crystalline structure that many astronomers believe could have been formed only at the high pressures that exist inside a large body. The iron meteorites seem to be telling us that at least one large asteroid broke up, presumably by collision with another such body.

INTERPLANETARY DUST AND GAS

Between the planets is a scattering of dust and a flow of gas. The dust consists of finely divided solid material, each particle following its own orbit around the Sun. These orbits have rather small eccentricities and inclinations, as far as we can tell from the distribution of the dust. We see the dust by means of the small amount of sunlight that it scatters, which produces a very faint glow all around the ecliptic. It is called the *zodiacal light* (after the zodiac, the band on either side of the ecliptic within which the planets move).

The zodiacal light is brightest near the Sun; therefore we see it best when the Sun is just far enough below the horizon for the sky to be dark. Even then, its visibility depends on how sharply the ecliptic stands up from the horizon. This angle is always favorable in tropical latitudes, and in middle latitudes it is best after late-winter sunsets and before late-summer dawns. At such times the zodiacal light can be quite conspicuous; it has sometimes been called the "false dawn."

Immersed as we are in the material that is responsible for the zodiacal light, it is difficult for us to figure out its nature and distribution. Some of the scattered light comes from dust immediately around us and some comes from farther away, and the picture is further complicated by the tendency of the particles to scatter light more readily at glancing angles than in other directions. From the color of the zodiacal light, which is a little bluer than sunlight, we can tell that the particles are small—perhaps a little under a thousandth of an inch—but not as small as the particles in the Earth's atmosphere that make the scattered light of the sky so very blue. Spatially the particles seem to extend from the middle of the asteroid belt inward to the Sun, with their density increasing slowly inward.

Just like a larger body, a dust particle follows an orbit around the Sun. In empty space there is no resistance to its motion, which is controlled by gravitation in just the same way as that of a planetary-sized body would be. For a small particle, however, one additional force becomes important; this is the pressure of sunlight. In a way, this gentle force acts, in fact, as if it were a resistance. It affects small bodies much more than large ones, and it acts to slow a body down and to cause it slowly to spiral in toward the Sun. This dragging effect of radiation pressure is called the Poynting-Robertson effect, after the two scientists who first studied its properties.

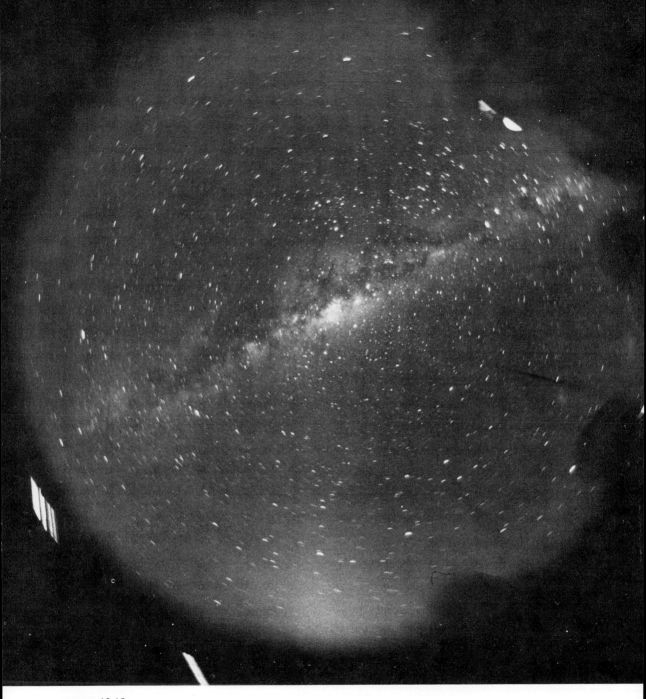

FIGURE 10.18.
This remarkable photograph was taken at the Cerro Tololo Interamerican Observatory, in Chile, with a "fish-eye" lens that shows the entire sky at once. Domes and flagpoles of the Observatory can be seen around the right edge. The Galactic center is in the middle of the picture, overhead, and the Milky Way stretches from horizon to horizon. Extending up from the horizon at the bottom is the zodiacal light. Faintly visible, on either side of the more prominent flagpole, are the Magellanic Clouds, our companion galaxies. (Courtesy of A. A. Hoag, Kitt Peak National Observatory.)

For small particles in the inner part of the Solar System, the Poynting-Robertson effect is the dominant process that controls the evolution of their orbits. At the Earth's distance from the Sun, even a particle as large as a tenth of an inch would be slowed down so effectively that it would fall into the Sun in only a million years, a tiny fraction of the age of the Solar System. The smaller particles that are responsible for the zodiacal light last an even shorter time. Not only is their distribution around the Sun controlled completely by the rate at which they drift inward, but, more importantly, the whole band of zodiacal-light particles must be continuously replaced, or else it would all be gone in a few tens of thousands of years.

We are not sure how the dust particles are replenished; there are two processes that seem capable of it. One is the erosion of comets, as the Sun evaporates their ices and leaves their dust sailing about in orbit; the other is the grinding up of asteroids when they occasionally but inevitably collide with each other. Either process might be sufficient in quantity to keep up the flow of short-lived dust that makes the zodiacal light. On the whole, the asteroids seem the more likely source, however, because they have a distribution that is flattened around the ecliptic plane, just like the zodiacal light.

In addition to the interplanetary dust, throughout the Solar System there also spreads a very thin gas. Unlike the dust, however, the gas flows away from the Sun. The gas comes, in fact, from the outer atmosphere of the Sun, which is continually overheated and evaporates away from the Sun in all directions in a flow called the "solar wind." (See Chapter 17 for a discussion of the Sun and its behavior.) It is a very high-speed wind—typically about 500 kilometers per second, or more than ten times the orbital speed of the Earth. But it has very little effect on bodies, because it is such a thin wind. A cubic inch of solar-wind gas contains about 50 atoms; compare this with the air around us, where the figure is about 10^{20}.

The particles in the solar wind are electrically charged, and they interact most strongly with other charged particles or with magnetic fields. Thus it is the solar wind that blows the straight tails out of comets, directly away from the Sun. The straight tails are made of whatever charged particles happen to be in the region around the comet's head. The solar wind sweeps these out and away, but has almost no effect on the uncharged gas and the dust. These are acted upon by the much less forceful pressure of sunlight, and they curve away from the comet in a slower and more graceful arc.

The solar wind also blows against the magnetic field that surrounds the Earth. Close to the Earth the field is strong, and the solar wind has no effect on it. But far away the field is much weaker, till it reaches a point where it can no longer hold off the pressure of the solar wind. This

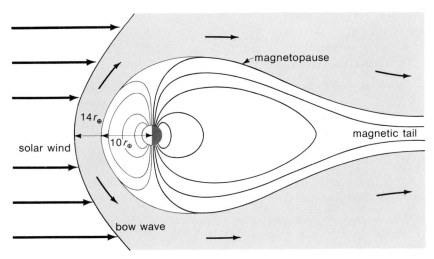

FIGURE 10.19.
As the solar wind blows past the Earth, it compresses the Earth's magnetosphere on the "bow" side, whereas a long magnetic tail stretches behind the Earth. Even outside the magnetosphere itself, the flow of the solar wind is disturbed by the Earth's presence. (Adapted from *New Horizons in Astronomy,* by J. C. Brandt and S. P. Maran. W. H. Freeman and Company. Copyright © 1972.)

balance point is the limit of the Earth's "magnetosphere." Actually, the magnetosphere is not a sphere at all. It is round only on the sunward side, where the solar wind washes around it and holds it back. On the rear side of the Earth, the magnetosphere extends back in a long tail, around which the solar wind streams on all sides as it rushes by the Earth.

The solar wind behaves similarly around Jupiter, which has a strong magnetic field. But it blows directly (but without effect) against the solid surface of the Moon, which has no magnetic field to hold off the charged wind. In some cases the solar wind is our most sensitive test for a magnetic field. It is perturbed slightly as it flows past Mercury, showing the presence of a tiny magnetic field there.

The material of the Solar System is divided into many diverse forms, from the sparse atoms of the solar wind to the giant masses of the planets, and from the baked rock of Mercury to the frozen ices of the comets. Yet all the forms developed in a rational way, each according to the conditions and processes that molded it. Seeing the unity in this diversity is the role of science.

REVIEW QUESTIONS

1. What distinguishes the asteroids from the planets? How do asteroids differ from satellites?

2. How many asteroids are known? How are they discovered? How many may remain undiscovered?

3. What do we learn from fluctuations in the brightnesses of asteroids? Why do they differ from planets in this respect?

4. Where do comets come from? Why is their arrival unpredictable?

5. What produces the light of a comet? How does it differ from the light of a planet or an asteroid?

6. Contrast the compositions of asteroids and comets. How do their differences accord with their places in the Solar System?

7. Define *meteor* and *meteorite*. What are the physical differences between them?

8. Why are meteorites valuable to study, and what do we learn from them?

9. Contrast the behavior of the interplanetary gas and the interplanetary dust. Why does each behave as it does?

11 | AGES, ORIGINS, AND LIFE

In contemplating the mysteries of the universe, we cannot fail to wonder at the problem that is closest to us personally, that is in a sense the greatest mystery of all: why are we here? In the great impersonal structure of physical systems, how is it possible that atoms have arranged themselves into the delicate balance that is life, let alone into the intangible that we call intelligence? And if we are here to ask this, who else may be somewhere "out there," asking the same questions, or perhaps asking other questions whose nature we cannot even conceive?

Problems like these can lead to the intricacies of philosophy, or to the speculations of science fiction. What we shall do in this chapter instead is to explore what astronomy, and its point of view, can contribute to the question of life in the universe. First we shall examine our own background: the age and origin of the Earth and of the Solar System as a whole. Next we shall ask, in a simple but basic way, what life is and what conditions are necessary for its genesis. Finally we shall examine the possibilities of life elsewhere in the universe and ask how we might possibly detect it.

THE AGE OF THE SOLAR SYSTEM

In a qualitative way, the history of the Earth's surface is known in great detail. Geologists have arranged the various layers of rock in their correct chronological order, and paleontologists have used fossils to establish correspondences that tie various geographical locations together in time.

What is lacking in both these approaches, however, is an absolute chronology, a length of time measured accurately in years. It is fortunate that nature does give us such a set of clocks, in the form of radioactive atoms. These allow us to determine the ages of various samples of rock.

Radioactive atoms have nuclei that are just a little bit unstable. Each nucleus lasts for a long time, but eventually it ejects one of its component parts and thus turns into a nucleus of an atom of a different kind. The breakup process happens to individual atoms at random; one can never tell when a particular nucleus will break up. What one can assert with certainty, however, is that the breakups will happen at a well-established average rate, which can be accurately measured in the laboratory for each type of radioactive atom. The rate is usually stated as a *half-life*, that is, the length of time in which half the atoms in any given sample can be expected to have broken up. There are several radioactive nuclei whose half-lives are comparable to the age of the Earth, and indeed, to the age of the universe itself. The amounts of these substances serve as excellent indicators of the ages of the oldest rocks.

The secret of the radioactive clock is that nuclei break up for internal reasons of their own; their decay rate is thus independent of external circumstances, such as the pressures or temperatures to which a rock is subjected. The decay would go on at the same rate, of course, even if the atoms were not bound up in a rock; the difference, however, is that inside the rock the products of the decay are trapped along with the radioactive atoms, so that from the relative amounts one can determine how far the process has gone since the rock was formed, and therefore how old the rock is. The radioactive atoms and their decay products exist in only small traces; so the measurement process is a very delicate piece of microchemistry. But it can be done with reasonable accuracy, and thus many rocks have been satisfactorily dated.

The rocks of the Earth's surface show a complete range of ages. The surface at each point has, in fact, formed, eroded, and reformed so many times that it is hard to find surface rocks that are really old. The oldest that have been dated so far go back considerably more than three billion years, but even those rocks are certainly not primeval. It seems impossible to find anywhere a piece of the Earth's original solid surface; everything has been remelted, eroded, or covered. It is elsewhere in the Solar System that we must look, to find out the age of the Earth. There are strong reasons for believing, as we shall see, that the entire Solar System had a single origin; so there are much cleaner places for us to look than the surface of the Earth, defaced as it is by the graffiti of the geologic ages.

Our best evidence comes from the asteroids. These relics of the sloppy creation of the Solar System include many bodies whose interiors have suffered no appreciable change—except for radioactive decay—since the Solar System was formed. Most asteroids circulate outside our reach, but a few have occasionally encountered the Earth, plunging down through the atmosphere and coming to rest on the Earth's surface as meteorites. Their mischance is our good fortune; here are samples from elsewhere in the Solar System that we can analyze in our laboratories. Such analyses have shown clearly that the asteroids were formed about 4.6 billion years ago. This, we presume, is the age of the Solar System.

As we saw in Chapter 5, rocks that have been brought back from the Moon's surface have a range of ages whose upper limit approaches the age of the asteroids but does not quite reach it. Apparently the surface of the Moon was reformed in its early days. With so much wear and tear going on in the Solar System, it is indeed fortunate that we have the meteorites as relics of the original formation of the asteroids.

THE ORIGIN OF THE SOLAR SYSTEM

What event, at that remote time, could have led to the formation of this hierarchy of bodies that surround the Sun? This is a question for which we still have no good answer, but it is nevertheless possible for us to note some significant facts and to outline a general picture. We can begin by looking at the Solar System and asking, "What are its over-all features and regularities?"

The first striking fact is that nearly all the mass is in the Sun. The other bodies make up, together, only about 1/750 of the mass of the Sun. But the distribution of rotation, which we shall find plays an important role in the story, is just the opposite; the rotational motion of the Solar System is nearly all in the outer planets.

To examine the problems of rotation, we need to be much more specific. There are many ways in which an amount of rotation can be measured, but the most fundamental among them is a quantity called angular momentum. This is defined as the mass of a body times its speed, in linear measure, times its distance from the center around which it moves. For a planet's orbital motion, this is simply the mass of the planet times its orbital speed times its distance from the Sun. For the rotation of a body, such as the Sun itself, the calculation is more complicated; one must add up the product of mass, speed, and distance from the axis for all the parts of the body. But for every rotation or revolution there is some specific amount of angular momentum.

The importance of angular momentum in physics is that it is one of the quantities whose total remains the same throughout any process; a physicist would say that angular momentum is *conserved*. If an interaction makes one body turn faster, some other body must turn more slowly in consequence; and by "turning" we mean here specifically the quantity that we have defined as angular momentum. The distribution of angular momentum in a system thus gives a clue to the processes that may have traded this quantity around. As we shall see, one of the persistent difficulties in devising an explanation for the origin of the Solar System—or of the Sun itself, for that matter—is to get the angular momentum to end up in the right place.

For the Solar System, the "right place" for the angular momentum is in the orbital motion of the four Jovian planets. Between them these orbits contain nearly all the angular momentum of the Solar System; only about a fiftieth of the total is in the rotation of the Sun itself.

Even from these rudimentary facts, we can begin considering possible modes of origin. First, did the planets originate in some natural way, in the same process that gave birth to the Sun, or are they the result of some great external catastrophe? The latter idea had great currency early in this century, in the form of the suggestion that another star had once passed so close to the Sun that its tidal forces ripped large chunks out of the Sun; these, or at least part of them, condensed to form the planets. One can calculate how often such close encounters between stars will happen. They do occur, but they are rare; there should be only a few thousand stars, among the 100 billion in our Galaxy, that have ever suffered such a close passage. If this were the way in which solar systems are made, they would be very rare indeed.

The tidal hypothesis, as it was called, suffers from some fatal difficulties, however. One of them is angular momentum. If another star passed so close to the Sun that it actually tore material out, its very closeness would insure that each bit of the detached material had an angular momentum much lower than that of the present outer planets. One of the factors in the product that defines angular momentum is the distance from the center, and at such a small distance there is no possible velocity that would give the material the necessary angular momentum without requiring that it move so fast as to escape from the Solar System completely. And to demolish the tidal hypothesis even more completely, one can show that material torn out of the hot Sun would not condense into anything at all; instead, its high temperature would cause it to expand almost explosively, and dissipate.

It is much more likely that the planets were born along with the Sun, as a by-product of the detailed way in which the Sun condensed out of a cloud of interstellar gas and dust. This idea has been considered, in a general way, ever since astronomers were first able to calculate anything about the dynamical processes that are involved in such a contraction. In the version of nearly two centuries ago, the so-called nebular hypothesis, the Sun was originally a much larger star than it is now. As it radiated its energy into space, the total balance of energy required that it contract. In this contraction, however, the Sun had to conserve angular momentum, so that as its size became smaller, its speed of rotation had to increase. Eventually the Sun's equator was spinning so fast that the centrifugal force of the rotation balanced gravity, and a rapidly rotating ring detached itself from the equator. With so much angular momentum left in this ring, the remaining Sun was able to turn at a more comfortable speed. It continued to contract, however, as a result of its continuing radiation, while the detached ring condensed into the outermost planet. This process repeated itself again and again, until the whole set of planets was formed.

Although the nebular hypothesis has a general plausibility, in detail it falls down, again because of angular momentum. In its picture, the great majority of the angular momentum would have to be left in the Sun, which thus remains spinning at a speed close to that of equatorial breakup. The actual Sun rotates less than a hundredth this fast; therefore any theory that wishes to trace the origin of the Solar System in this way must provide some additional mechanism of interaction, which can transfer angular momentum out from the Sun to the planetary rings. Furthermore, like the tidal hypothesis, the nebular hypothesis has the problem of explaining how hot gas from the Sun could have condensed to form solid bodies.

Even so, the nebular hypothesis marked a large step toward modern attitudes; it was an attempt to explain the origin of the Earth and other planets as the natural outgrowth of a mechanical process. When asked about the place of God in his picture of the creation, Laplace, the originator of the theory, said, with all the irreverence of the French Enlightenment, "I have no need for that hypothesis."

In spite of the difficulties in the earliest version of the contraction

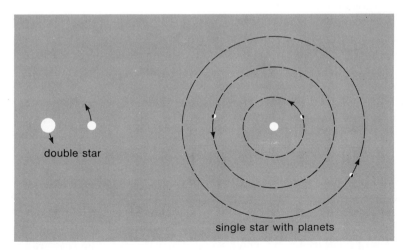

FIGURE 11.1.
Angular momentum can be in the orbital motion of two stars around each other, or else in the motion of smaller but more distant planets around a single star.

picture, it seems very likely today that the planets really were formed during the same process of condensation that produced the Sun. The modern picture of the process still assigns a key role to angular momentum; but now, instead of creating a difficulty, the conservation of angular momentum explains why the planets had to be formed. The difficulty, in fact, is in explaining how the Sun could form at all. In the modern view, a star forms from a contracting cloud of dust and gas that was originally very much larger. The degree of contraction is so great, however, that almost any initial rotation whatever would be transformed into an extremely rapid spin by the time the cloud contracted down to the size of the Sun. The spin-up is so great, in fact, that one might expect the configuration to become rotationally unstable long before it reached a size as small as that of the Sun.

Thus the conservation of angular momentum presents a general difficulty in the theory of star formation. If a star is to form at all, it must interact sufficiently with its surroundings to transfer at least some of its angular momentum to surrounding material that never falls into the star. The whole problem becomes much easier, however, if the angular momentum can somehow be accommodated within the star and its resulting system. There are two ways in which this can easily be accomplished. One way is for the star to split in two and become a double star, in which the angular momentum is in the motion of the two components around each other. The other alternative is a less even breakup: a star plus a system of planets, whose orbital motion contains the angular momentum. Because of the angular-momentum problem, the formation of a single star without planets seems very much less likely than either of these other possibilities.

The modern picture avoids most of the difficulties of the old nebular hypothesis, by postponing the stellar birth of the Sun until a quite late stage in the contraction of the cloud. During most of the contraction, the cloud is the "solar nebula," a cool, ever-denser cloud of material in which the spin of the central condensation continually drags against the outlying material, slowing the center down. Thus the angular momentum ends up in the outer parts of the solar nebula, where the planets will form. Furthermore, there is no problem in forming planetary condensations out of this cool material, which does not have the dissipative tendency of hot gas that has been pulled out of a star.

With this general picture, we can try to understand some of the other characteristics of the Solar System. One of its striking properties is that the planets are few in number, and their orbits are nearly circular and nearly in the same plane. This is easily understood as result of the process of coalescence by which each planet formed. The scattered material that went to form a planet did not condense in a single, instant event. Instead, the process must have begun with small, local condensations. These moved in orbits that were far from circular, and as a result

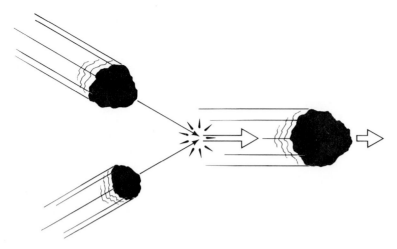

FIGURE 11.2.
When two bodies collide and merge with each other, the resultant, larger body moves with the average motion of the two.

each body ranged inward and outward and thus had a chance to collide and merge with other bodies. The separation of the existing planets is presumably an indication of how large this range was. In the process of coalescence, inward and outward motions merged and canceled each other out, as did upward and downward motions. The one thing that necessarily persisted was the angular momentum, and the result in each case was a single planet, moving in a nearly circular orbit close to the plane of symmetry of the original solar nebula.

In this picture, the size of each planet would depend on a number of factors. One factor is, of course, the amount of material in the region whose contents coalesced to make the planet up. Another important factor is temperature and the way in which it affects the different chemical elements that make up the cosmic mix. As we saw when discussing the Earth, and then the outer planets and their satellites, the heavy elements will condense into solid materials at a relatively high temperature, whereas carbon, nitrogen, and oxygen form compounds that remain gaseous until the temperature drops much lower. Thus it is that the inner bodies of the Solar System are made mainly of rock, whereas a mixture of "CNO ices" prevails in the outer parts. The location of the boundary between rocky bodies and icy bodies also indicates, incidentally, that the Sun had already reached approximately its present luminosity at the time when the planets formed, because the temperature in the Solar System would otherwise have been different at that time, and the rock-ice boundary would have been at a different distance from the Sun.

Jupiter and Saturn still pose a puzzle, however. They are made up largely of hydrogen and helium, gases that will condense into solid material only at temperatures that are improbably low for any stage in the history of the Solar System. It must be that they formed along with the Sun, as small but stable condensations in the original cloud of gas and dust that was to make up the whole Solar System. Such arguments, however, lie on the border between hypothesis and speculation; they will be sharpened only when we have a much better picture of the relative chemical compositions of the Jovian planets.

The smaller bodies of the Solar System pose a different set of problems. The asteroids are one puzzle. It is not at all clear why, in this wide gap between the rocky bodies and the icy bodies, the asteroid zone remained as a swarm of small bodies instead of forming a single planet. The comets, on the other hand, are not hard to understand. As the outermost debris of the Solar System they are, of course, icy; and in their vast, low-density region there was little opportunity for small bodies to collide and coalesce into larger ones.

Another interesting question is the origin of satellites, and why some planets have many, others few or none. It is likely that the formation of the satellite system of each planet was a small-scale rerun of the formation of the Solar System itself. As the planet contracted, conservation of angular momentum made it spin faster. Condensation and fragmentation were slow, however, and the protoplanet was able to transfer its angular momentum to a cloud of surrounding material that eventually condensed and became its satellite system.

In the condensation of planets one additional factor operated, however: tidal friction due to the gravitational force of the Sun. This phenomenon is extremely sensitive to distance from the Sun. Thus it was able to stop the rotations of Mercury and Venus and prevent them from forming satellites, whereas at the distance of Jupiter and beyond, solar tidal friction had almost no inhibiting effect whatever.

Tidal friction also explains why the hierarchy stops at the satellites of the planets—why the satellites do not themselves have subsatellites. Each satellite is simply too close to its parent planet. In every case tidal friction has stopped the axial rotation of the satellite, apparently at such an early stage that no surrounding cloud was ever left behind to form subsatellites.

Although many of the characteristics of the Solar System seem to lend themselves to simple explanations, some others—such as the rings of Saturn, the inclination of the Uranus system, or the size of the Earth's satellite—should probably be regarded as accidents in a complicated process whose workings we may never succeed in tracing in detail. Indeed, even the glimmer of understanding presented here should be looked at as the outline of a plausible picture rather than as any sort of real theory.

LIFE IN THE UNIVERSE

Here, on this one planet circling around one star, there are living organisms. Has this scene repeated itself elsewhere in the universe? The first question to ask is, "What scene?" What is life, and how did it come into existence? Although the real answers to these questions still lie in the fascinating future of biophysics and biochemistry, we can see enough of the outline today at least to examine the astronomical side of the question. Basically, life is a process of organization in which chemical compounds, by their structures, build systems that can protect themselves against their environment and reproduce their kind. Every aspect of the process depends, in one way or another, on the nature of various organic molecules. The arrangement of their molecular structure carries the master plan and the detailed blueprints; their chemical attachments provide the collection and transport of materials; and their chemical reactions produce the physical results of growth, of motion, and of everything that distinguishes the animate from the inanimate.

All these processes depend on molecules that show a high degree of complexity, in both their structure and their behavior. The nature of these molecules immediately sets restrictions both on the chemical processes that are capable of maintaining life and on the physical circumstances in which they can be carried on. It is very likely that the chemistry of life elsewhere in the universe would be very similar to ours, no matter how different the circumstances under which it arose. The reason is that among all the chemical elements only one, carbon, can join its atoms into the infinite variety of chains and rings that make up organic molecules. Silicon, the element whose behavior is most similar to that of carbon, possesses this ability to some small degree, and chemical technology has indeed produced silicone compounds with new and useful properties; but the versatility of compounds of silicon is incomparably less than that of carbon compounds. If we ever encounter living forms elsewhere in the universe, it is almost certain that their chemistry will be that of carbon. The basic organic rules, and many of the common substances themselves, will be those with which we are familiar, although the details of the chemistry, and of biological structures, are likely to be very different.

Along with the element carbon, one simple compound plays an important role in our own organic chemistry: water. All our delicate chemicals processes take place in water solution, in which the different molecules move about and encounter each other readily. This reliance on water is a heritage of our origin. Life on Earth began in the oceans; and when our ancestors crawled out on the land, half a billion years ago, they took a part of the ocean with them, in the cells of their bodies. Thus a large part of our body chemistry is conditioned by the liquid compound that happened to cover most of the surface of our planet.

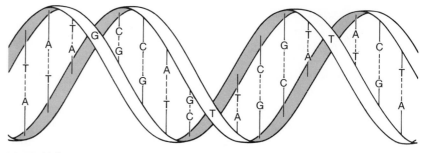

FIGURE 11.3.
The "double helix" of the DNA molecule, which carries the coding that is the basis for the copying of living cells. The symbols A, C, G, and T stand for four particular groups of atoms whose arrangement carries the information that controls how the cell is built. The processes of life depend on the delicate balance of chains of atoms such as these. (From R. J. Britten and D. E. Kohne, "Repeated Segments of DNA," Copyright © 1970 by Scientific American, Inc. All rights reserved.)

For life as we know it, water is essential; and this requirement restricts the conditions under which this particular kind of life can occur. One can speculate about whether a parallel development could have taken place in some other liquid. A liquid seems necessary in any case, in order for the various reagents to move about freely and come into contact with each other; but could this happen in some liquid other than water? The liquid would have to be an abundant substance, however, and this requirement restricts it to the simple compounds of the abundant chemical elements. After water, the best candidates are ammonia (NH_3) and methane (CH_4). At the temperatures to which we are accustomed, however, these are gases; they condense into a liquid phase only at quite low temperatures.

Temperature plays an important role throughout the consideration of conditions that might be suitable for life. The difficulty at low temperature is the slowness of chemical reactions. Their rates are extremely sensitive to temperature, so that life processes in a very cold solution, such as liquid ammonia, would be immeasurably slower than those that we know. One has to ask whether such sluggish life would be able to react fast enough to cope with any threatening changes in its local physical environment.

High temperature poses a different sort of problem. In a warm substance the molecules move faster, and they therefore collide harder. These collisions tend, all too frequently, to break up the molecules, so that complicated chemicals find it hard to survive at high temperature. (This is a fact that we make use of every day, when we boil water to kill the microbes in it, or when we cook food to break down the long molecules that make it tough.)

There is thus only a restricted range of temperature in which life can exist. If the vital reactions take place in water solution, then the permissible range runs only between the freezing point and the boiling point of water—from 273 to 373 degrees, on the absolute temperature scale. Somewhat lower temperatures might be permissible if the solvent were a liquid other than water; but the range cannot be lowered too far, lest the reactions run too slowly. At the higher end the restriction seems even tighter. The fragility of molecules is independent of what liquid they may be dissolved in, and it seems very unlikely that any organisms could live at temperatures that were far above that of boiling water.

Thus, to support life, a body must have a surface temperature not grossly different from that of the Earth, and a gravity sufficient to keep liquids like water from boiling off into space. It must also have, at least on some parts of its surface, a covering of liquid, preferably water. Where else in the universe are these conditions met? Certainly not on the stars, which are too hot, nor in interstellar space, where there is no way of holding liquids. Clearly, if life exists, it is on the surface of some planet or satellite, gently warmed by the star around which it circles.

We can begin by considering the Solar System. The first criterion for a body is the presence of an atmosphere—or at least the ability to hold one. This test rules out Mercury, Pluto, the asteroids, and nearly all the satellites. The temperature criterion eliminates Venus, whose surface is much too hot. Mars is an enigma. In its present state, with only the thinnest of atmospheres and no liquids, it could not possibly give birth to life. Yet there are indications, as we have seen, that the winding canyons and gullies on Mars were carved by water at a time when its atmosphere was quite different from its present state of poverty. The polar caps contain enough frozen material to make a new atmosphere; perhaps during its clement eras Mars is host to life.

In the outer parts of the Solar System, the surfaces of all bodies are very cold. This certainly rules out the possibility of life on a large satellite like Titan, even though it has some atmosphere. For the large outer planets themselves, though, the situation is not nearly so clear. On Jupiter, for instance, the uppermost cloud layer, whose surface we see, is also quite cold; but the heat balance of the planet indicates that deeper layers are warmer. Far down in Jupiter's soupy atmosphere, there is probably a layer in which water is in a liquid state, mixed with all the exotic constituents of that atmosphere. No analysis of the Jovian atmosphere has probed down to that layer in any detail at all, but it would certainly be rash to exclude the possibility of living organisms in such a layer. The same is true of Saturn, and it might apply also to Uranus and Neptune.

In short, the prospects for life elsewhere in the Solar System are by no means hopeless. On Mars there are probably periods when conditions are suitable for life—provided its eggs or spores (or whatever) can survive the long inclement periods in between. On Jupiter, and perhaps on the

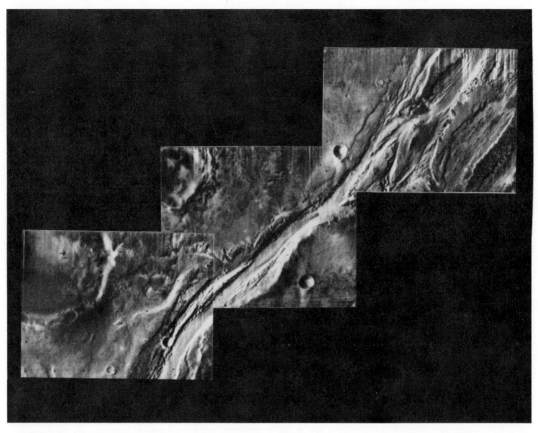

FIGURE 11.4.
Part of a winding pattern on Mars that appears to have been made by an ancient flow of water. Note the more recent crater on top of the stream bank, at the lower left. The section shown here is about 75 kilometers (45 miles) long. (National Aeronautics and Space Administration photograph by Mariner 9.)

other major planets too, there may be a layer in which conditions are suitable. Finding life is another matter, however. On Mars it may be possible for space missions to examine and analyze surface materials to search for some trace of dormant life. But on Jupiter the layer in question is so deep beneath the clouds that it may be a very long time before we can probe it directly.

Thus there are other places in our own Solar System where it is conceivable that life might come into being. Elsewhere in the universe, the problem hinges similarly on the existence of suitable planets. This is a question about which we still have no direct information. The detection of planets going around other stars is a task that is still beyond the reach of Earthbound observing equipment. If, for example, we tried to observe a planet like Jupiter going around a star like the Sun, at the distance of the nearest neighboring star, we would be unable to detect it by any method that is now available. It would appear to be only a few seconds of arc away from its parent star, and only a hundred millionth as bright; so it would be completely lost in the glare of the star. Our best chance would be to observe the tiny orbital motion of the star around the center of mass of itself and its invisible planetary companion. This motion would amount to only a small fraction of a second of arc, however; and it is very unlikely that we would be able to detect it.

Astronomers have, in fact, examined many nearby stars in a search for such motions, but none has yet been reliably detected. In several cases discovery claims have been made, but the initial claim has never been substantiated by further intensive studies. Every such case has been close to the limit of accuracy of modern instrumentation; clearly, what is needed is higher accuracy. For this, we will probably have to wait a decade or more, for observations made by the first large telescopes that are placed in orbit, free of the interference of the Earth's atmosphere.

Thus we can only guess about planets going around other stars. The guess need not be a wild one, however; as we saw in discussing the origin of the Solar System, considerations of angular momentum suggest that most single stars should have planets around them. Since about half of all stars are single, it seems likely that planets are very common in the universe. Even in our own galaxy, among 100 billion stars there should be a billion planets or more that have conditions suitable for the genesis of life.

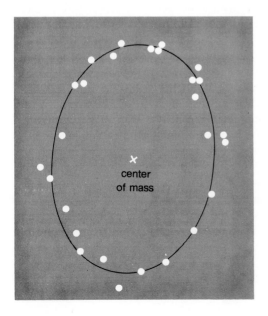

FIGURE 11.5.
The orbit of the star BD +66°34A around the center of mass of itself and its unseen companion. The dots represent deviations of the visible star from its mean position. These observations are close to the highest precision attainable; the major axis of the ellipse is a quarter of a second of arc. (From J. L. Hershey, *Astronomical Journal,* **78** (1973), 938.)

The remaining question is, if physical and chemical conditions are suitable, how likely is it that life will arise? A few decades ago, the origin of life seemed quite mysterious, and most scientists would have considered it a rare accident. Current thinking has swung quite the other way, however. The basic difficulty had been to transform inorganic materials into simple organic chemicals, from which more complicated molecules can readily be made. The difficulty has now disappeared, in the face of evidence from two different directions. One of these is a set of laboratory experiments, in which a mixture of water, methane, and ammonia, to simulate a primitive planetary atmosphere, is subjected to various stimuli. The first experiments used electrical discharges, like those in bolts of lightning; but more recent experiments have used other sources of energy, with very similar results. The outcome is that organic molecules are produced, in a surprising degree of complexity and variety. Apparently the supposed boundary line between inorganic and organic is easy to cross, through simple and natural processes.

The second line of evidence comes from one of the most hostile of environments, interstellar space. There radio astronomy has, by means of its increasing ability to probe new parts of the spectrum, discovered a rich variety of molecules, most of them of a type that older chemistry would have associated exclusively with organic material. This does not suggest for a moment, of course, that there is life in interstellar space; but it does show how readily complex molecules are formed, even under conditions of low temperature and density that an earlier age would have considered hopelessly hostile.

Still more advanced molecules are found in another natural source, the meteorites. Here trace quantities have been found of amino acids, which serve in living material as the building blocks of biochemistry. Given the natural existence of such molecules, it is relatively easy to imagine that under more favorable conditions they could arrange themselves into simple living structures, and thence into more complex ones. It thus seems a reasonable likelihood that life will come into existence wherever the physical conditions and the composition mixtures are suitable. If this is true, then other life in the universe must be common.

This conclusion rests on two basic premises. One is that many stars have planets that are suitable for the genesis and nurture of life. The second premise is that if the conditions are right, naturally occurring substances will naturally arrange themselves into the complex molecules that replicate themselves in living organisms—that life is a process that creates itself in an almost inevitable way. Neither premise is certain; both depend on an exercise of judgment, reaching out from a solid but narrow base within science into the gray mist of questions that are fascinating but elusive.

The question, "Why are we here?" is a deep human problem whose implications go far beyond the confines of science. As such, it illustrates one of the dilemmas of science: the great human questions are never subject to categorical answers and scientific proof, yet for many of them a valid discussion must give an appropriate weight to scientific arguments. How can the scientist, bred as he is to the role of finding certainties, communicate faithfully the solid facts, the attractive probability, and the mere informed speculation? And the humanist, approaching from the other direction, faces his own dilemma: what shall he believe? What part of the scientist's assertions carry an incontrovertible stamp of authority, and at what point do his conclusions begin to be clouded by his mechanistic view? In an intellectually divided society, there is no solution to this dilemma of communication. The only answer is to expand the philosophical precept, "Know thyself"; in our complex world, the motto should be "Know each other." The impact of science on our world has been overwhelming, and our future needs both scientists who are concerned about human values and humanists who understand science.

INTELLIGENCE AND COMMUNICATION

As we contemplate our position in the universe, there are two questions to ask. The first concerns whether life exists elsewhere, but the second focuses on a deeper mystery: do the depths of space conceal other intelligent civilizations? About this we can only speculate, but the role of science is to keep the speculations as realistic as possible. The situation becomes particularly puzzling, however, when we consider the suddenness of our own appearance on the scene. The development of the human species has taken only a thousandth of the history of the Earth, and civilization has existed for less than a hundredth of that brief interval. In other words, if an extraterrestrial visitor had examined our planet in any but the last 1/500,000 part of its existence, he might have found it teeming with life, but there would have been no sign of the artifacts that we call civilization.

Even more perplexing, how will the Earth look in the future? Although civilized society has controlled the surface of the Earth for thousands of years, it is only in the past century that technology has produced phenomena, such as city lights and radio signals, that can be observed from a distance. How much more will the appearance of the Earth change in the next hundred years, or the next thousand, or the next million years? Since our own development has been so sudden and so rapid, how can we know what to look for elsewhere? And, most awesome of all, should we not expect that somewhere in the universe are civilizations that have developed incomparably beyond our own youthful level?

If our previous arguments are correct, about the frequency of planets and the ease with which life can appear, then our conclusion about the high frequency of life must be equally correct. The remaining basic question is, if life exists, will intelligent civilization necessarily develop? The answer is not at all clear. The appearance of our species on the Earth looks in many ways like an accident; nothing in our earlier evolutionary history points toward the rapid developments that began only a few million years in the past.

Our earliest primate ancestors were little mouse-like animals who scurried through the grasslands of fifty million years ago, grubbing for seeds and nuts. They soon felt competition, however, from another order of mammals who had evolved a clever new physical feature. The order that we call rodents developed front teeth that never stop growing, so that they are never worn down. It may have been pressure from these better-equipped chewers that drove our primate ancestors to seek their livelihood in the trees. There they developed dextrous hands and feet, for climbing and swinging through the branches, so that the few species that later descended to the earth had an unusual ability to manipulate objects that they handled. Some two million years ago, one of these species began a unique evolutionary development. Its members began using tools. As this ability became valuable, evolution favored brain development. During the same period, in a process that has unfortunately left no material remains, the same creatures developed speech communication—another valuable survival characteristic. Between the two, dexterity and speech led to an unprecedented evolution of brain power. At some imperceptible stage in this development, Man became a rational creature.

It may well be that the development of intelligence was an accident; certainly nothing indicates why it should have happened in this particular era, on this particular branch of the tree of life. If this is so, then it is very difficult to guess how many or how few planets may have developed intelligent life. The one thing of which we can be sure, however, is that if other civilizations exist, it is very unlikely that ours is the most advanced.

In addition to the biological arguments, one other sober factor must be introduced into a discussion of the development of civilizations: the possibility that nearly all civilizations end by destroying themselves. Already, at an early stage of our own technological development, we control a nuclear power that is capable, if raised in anger, of wiping out mankind and perhaps all life on the Earth. No one can be sure that we will succeed in avoiding the ultimate disaster. Nor can we judge how another civilization would develop or behave, but it may be that there too the development of technology marks the minute before midnight. If this dismal hypothesis were true, the lifetime of each civilization would be short, and any given moment would find very few civilizations in the brief interval between detectability and death.

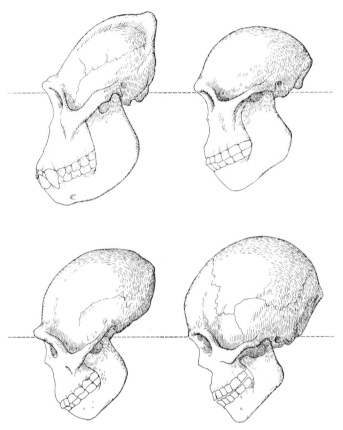

FIGURE 11.6.
In a few million years, the evolution of the brain has made Man
what he is. In these drawings the dashed line distinguishes
the brain part of the skull from the jaws part. (After S. L.
Washburn, "Tools and Human Evolution," Copyright © 1960 by
Scientific American, Inc. All rights reserved.)

Whatever the ultimate answers may be, today we have no idea at all
how many other civilizations may exist on planets circling around remote
stars. We can ask, however, about the possibility of detecting them if
they do exist. Seeing them directly, by visual observation, is of course
out of the question; as we have seen, we cannot even detect their planets,
let alone examine the planetary surfaces.

Nor can we hope, within even the widest visions of modern technology,
to travel to other stars. With any conceivable amount of propulsion
energy, the transit times are hopelessly long. At typical spacecraft velocities, it would take tens of thousands of years to travel even to the nearest
neighboring star. Conversely, if we wished instead to cover those four

FIGURE 11.7.
The all-too-familiar mushroom cloud serves to remind us that civilizations are capable of destroying themselves. (U.S. Air Force.)

light years in twenty years of travel time, this would require an amount of energy that would cost more money than there is on Earth—even if we could find the fuel to generate it from. To make it worse, the cost of putting energy into the motion of a spacecraft goes up much more rapidly than the energy itself. Interstellar travel belongs to the realm of science fiction, not science.

Radio observations might offer a chance of detecting a civilization, however. Their likelihood of success would depend very much on whether the other civilization was broadcasting a recognition signal or whether we were merely trying to detect the presence of stray radio signals. The latter case would, of course, require our receivers to have a much higher sensitivity.

As an example, we can calculate the strength of the radio signals that escape from the Earth—television broadcasts, radar beams, and the like. If a planet going around a nearby star emitted similar signals, would our radio telescopes be able to detect them? The answer is that we would need at least a thousand times as much sensitivity as we now have. This sounds at first like a large gap, but some scientists have argued that billions of dollars ought to be spent to construct radio telescopes for just such a search. The trouble is that the uncertainties are tremendous. First, it is quite arbitrary to take the Earth's present-day radio signals as a standard. A hundred years ago we emitted nothing, and a hundred years from now, who knows? Also, the requirements of the search depend very much on how far we need to look. If radio-emitting civilizations are to be found on stars three parsecs away, the search could be conducted with equipment that might reasonably be built, and there are only a dozen stars to examine. If we had to look ten times as far, however, the radio telescope would have to be ten times as large and would cost a hundred times as much, and there would be thousands of stars to be examined.

Our task would be much easier if the other civilizations were emitting a recognition signal: a strong, recognizable beacon. This is not as easy as it sounds, however. If we ourselves wanted to put out such a beacon, we would not know what other stars to aim it at; hence it would have to be sent in all directions, and the power required would then be tremendous. Nor is there any particular frequency at which a recognition beacon should be tuned; if we were searching for such a beacon, we would have to examine all of the radio spectrum. The one thing that would be unmistakable, however, is the signal itself; instead of the random hiss and crackle of natural radio sources, it would have a regular, and perhaps complicated, pattern. (For this reason the first pulsars caused a flurry of unusual excitement among the small group of radio astronomers who discovered them. While the properties of these flashing radio sources

FIGURE 11.8.
Some scientists have suggested that this tremendous "Cyclops" array of radio telescopes be constructed, to detect radio signals from extraterrestrial life. The array depicted here is several miles across. (From C. Sagan and F. Drake, "The Search for Extraterrestrial Intelligence." Copyright © 1975 by Scientific American, Inc. All rights reserved.)

were being studied more carefully, they were referred to locally as LGM-1 and 2; the LGM stood for "little green men.")

Even if astronomers joke about it among themselves, the search for extraterrestrial life is one of the perplexing and frustrating problems of astronomy. It is perplexing because of the arguments, outlined above, that life ought to exist in abundance. It is likely that many stars have planets, that many planets are suitable for life, and that many do have life on them. It also seems likely that some of these life systems should have developed intelligence and civilization; if so, then some of those civilizations should be greatly superior to ours. But where are they? One can speculate, of course, that a superior civilization is both skillful enough and wise enough to hide from us barbarians; or one can argue pessimistically that all civilizations live for only a moment, destroying themselves as soon as they discover nuclear energy. In any case, we still *live* alone, but we may not *be* alone.

The frustration is to ask such deep questions while knowing so few answers. Indeed, contact with a superior civilization elsewhere in the universe would be the most important event in human history; but there is little that we can do to accomplish this end, or even to show whether it is feasible. This is a hard conclusion to accept, in an age of technology that has sent a man to the Moon as a symbol of its ability to achieve what it wishes. Perhaps this discussion should serve as a reminder that there are problems that do not yield to a mere decision to attack them massively.

REVIEW QUESTIONS

1. How is the age of the Solar System calculated? Why are no rocks this old found on the surface of the Earth?

2. What is angular momentum? Why is it such a problem for theories of the origin of the Solar System?

3. Outline the processes that are currently thought to have been important in the formation of the Solar System.

4. In terms of atoms and molecules, what is life? What physical conditions does it require?

5. Why are most bodies in the Solar System incapable of sustaining life in any form? On which bodies might life of some sort be possible? Discuss the likelihood of life on each of these, and our chances of detecting it.

6. What are the reasons for believing that life is common in the universe?

7. How might we hope to detect life elsewhere in the universe?

8. Contrast the time-scales for the age of the Earth, the development of the human species, the appearance of civilization, and the development of modern technology.

12 | INTRODUCING THE STARS

The sky is filled with stars. The Sun is a star, our nighttime sky has thousands of stars, and telescopes show us numbers of stars that are almost literally countless. The Milky Way, our own galaxy, contains a hundred billion stars; and the other galaxies, which extend as far into space and time as we are able to see, are each made of billions of stars. The light that we see coming from a galaxy is actually the combined light of its individual stars, each too faint to see as an individual but adding up to a beacon that we see across the millions of parsecs.

Stars are the basic objects of the universe. But what is a star? It is a massive ball of hot gas, held together by its own gravitation, shining because it is hot, and living on the burning of nuclear fuel at its center. The key to a star is its energy. From the hot surface of the star, energy streams in all directions into the black depths of space, never to return. This outpouring would soon cool the star off, if new energy were not created all the time, deep in the center of the star. There lies the source of power in the universe: nuclear energy at the centers of stars. Most of this energy comes from a slow conversion of hydrogen to helium. Since the stars are made largely of hydrogen, they have plenty of fuel; it is enough to keep most of them burning for billions of years.

The Sun is a typical star. Some are larger; some are smaller. The Sun's lifetime—4½ billion years in the past and about an equal time in the future—is also an ordinary stellar lifetime. Again, however, some stars have shorter lives and some longer. Change is everywhere. Stars die, and in many places new stars are born out of the interstellar gas and dust.

Motion, change, birth, and death—the study of astronomy is not just the contemplation of the celestial scene; it is the following of the celestial drama. And ultimately, of course, astronomy is the understanding of why the universe is as it is, and how it has become that way.

Before understanding, we must have facts, however. What are the stars? The preceding paragraphs have sketched a picture that is far from obvious. All that we see is a black sky scattered with bright points of light. To have probed their nature and their history is one of the triumphs of the human intellect. It has needed patient observation, deep thought, and occasional brilliant insights. The actual unfolding of stellar astronomy has been halting and sometimes frustrating, but in the end it has created a very satisfying picture, in which the stars are as they are for the simplest of reasons. Remote and strange as they seem, the stars are simply acting out the laws of physics, and it is through this insight that we come to understand them. The process of understanding is itself a fascinating tale, and in the following chapters we shall follow it through, step by step. From a simple first look, we shall put the stars in a perspective in which we can see their physical nature; and each new step will piece previous steps together, as the nature of the stars unfolds itself before us.

THE ENTERING WEDGE

When the astronomer faces the stars, his first problem is to measure their distances. The brightness of a star, however carefully measured, tells us only how bright it *looks*; the star itself may be intrinsically faint but near, or it may be a blazing giant, reduced to apparent mediocrity by its great distance. Only a calculation of the distance will tell, and only then can the astronomer begin to study the real physical nature of the star. It is no wonder, then, that so much of the effort of modern astronomy goes into measuring the distances of stars. The methods are many, but most of them depend on some previous knowledge of the nature of the stars. At the start, however, we need a method that stands on its own. Such a method is trigonometric parallax.

When Copernicus first suggested, four and a half centuries ago, that the Earth circles about the Sun, it was pointed out that the annual back-and-forth motion of the Earth should make the stars appear to oscillate in position, that is, to exhibit annual parallax. Copernicus answered this objection, correctly, by arguing that the stars were so far away that the motion was too tiny to measure. It was 300 years, in fact, before observing techniques improved sufficiently to detect the annual parallaxes of stars; but in the century and a third since then, it has become possible to measure parallaxes of thousands of stars.

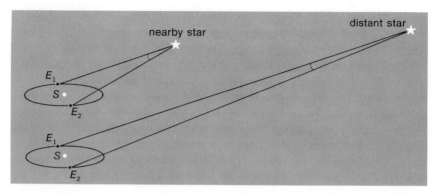

FIGURE 12.1.
A more distant star has a smaller parallax. The nearness and the angles have
been exaggerated for this drawing; actual stellar parallaxes are extremely small
angles.

The measurement is simple in principle: a nearby star should appear
to shift back and forth by a large amount, and a distant star by only a
small amount. The distance of each star is, in fact, inversely proportional
to the size of its parallactic shift. This relationship gives a name to the
unit by which we measure stellar distances: a *parsec* is the distance at
which the *par*allax is one *sec*ond of arc. The size of the angles that must
be measured causes some difficulty, because the largest stellar parallax
is less than a second of arc. (A second is a sixtieth of a sixtieth of a degree,
roughly the apparent size of a dime at a distance of two miles.) In practice, the accurate calculation of stellar parallax requires careful measurement of the position of a star on about fifty photographs taken over a
span of two or three years. Even so, the result is too inaccurate to be of
much value when the parallax is less than about a fortieth of a second,
that is, when the star is more distant than about 40 parsecs.

To apply a knowledge of stellar distances to a measurement of stellar
brightness, we need first a convenient scale on which to specify brightness. For this, astronomers use a scale born in tradition and later sharpened to modern scientific accuracy: the magnitude scale. Magnitudes
go back more than two thousand years, to the star catalog of Hipparchus,
in which the brightest stars were described as of "the first magnitude";
those not so bright were termed "second magnitude," and so on. Quantitative measurements, during the nineteenth century, showed that stars
of each magnitude class were about 2.5 times fainter than those of the
preceding class. Because the fifth power of 2.5 is close to the round number 100, it was decided that 5 modern magnitudes should correspond to
a brightness ratio of exactly 100. Thus, a difference of one magnitude
corresponds to a brightness ratio of $\sqrt[5]{100}$, or 2.512; two magnitudes
correspond to a ratio of $(2.512)^2 = 6.31$; and so on.

This sort of scale, in which the unit corresponds to a certain ratio in the quantity measured, is called a *logarithmic* scale. The magnitudes of two stars, m_1 and m_2, can be related to their brightnesses, b_1 and b_2, by the formula

$$m_2 - m_1 = 2.5 \log(b_1/b_2).$$

Notice that the *m*'s and the *b*'s appear in the opposite order; small-numbered magnitudes are bright, and large-numbered ones are faint. The faintest stars that can be photographed with the largest telescopes are of the twenty-fourth magnitude. At the bright end, magnitude zero precedes first magnitude, and objects that are even brighter have negative magnitudes. The brightest star, Sirius, has a magnitude -1.5, and the magnitude of the daytime Sun is -27.

TABLE 12.1. STELLAR MAGNITUDES AND BRIGHTNESS FACTORS[a]

Difference in magnitudes	Number of times fainter
1	2.5
2	6.3
3	16
4	40
5	100
10	10,000
15	1,000,000

[a]When combining magnitude differences, *add* the number of magnitudes, but *multiply* the number of times fainter. For example, if one star is 6 magnitudes fainter than another, it is 250 times fainter. The magnitude difference of 5 + 1 = 6 corresponds to 100 × 2.5 = 250 times.

Except for its backward direction, the magnitude scale is similar to other scales of sensation. Sound intensity, for example, is measured on the decibel scale, which is also logarithmic. These two examples illustrate a general (if only approximate) law of psychology, that a sensation varies according to the logarithm of the stimulus. Thus, the magnitude scale is related not to conscious choice, but to the way our eyes and brain naturally respond.

Magnitudes as directly measured at the telescope are called *apparent magnitudes.* They describe how bright a star looks as seen from the Earth. Much more interesting physically, however, is the question of how bright the star really is, independent of its distance from the Earth. We can specify this intrinsic luminosity by imagining all stars to be brought to a standard distance from us. The distance chosen as the standard is 10 parsecs, and the apparent magnitude that a star would have at that distance is called its *absolute magnitude.*

In order to convert a star's apparent magnitude to its absolute magnitude, we must know both the star's distance from the Earth and how its brightness would be affected by a change in distance. The relationship between brightness and distance is described by an inverse-square law: the brightness of a given object varies inversely as the square of its distance from the observer. Thus, moving a star 10 times as far away would make it look 100 times fainter. In equivalent terms, *multiplying* its distance by 10 would *add* 5 to its magnitude. Similarly, increasing its distance by a factor of 1.585 would correspond to adding 1 magnitude. This line of reasoning leads to an equation connecting a star's apparent magnitude *m,* its absolute magnitude *M,* and its distance *r:*

$$m - M = 5 \log r - 5.$$

Any specific distance, therefore, corresponds to a particular difference between apparent and absolute magnitudes. Astronomers call this difference the distance modulus, and they sometimes quote it in place of the distance itself. Thus, a distance modulus of 5 magnitudes corresponds to a distance of 100 parsecs, a distance modulus of 10 magnitudes corresponds to a distance of 1,000 parsecs, and so on.

TABLE 12.2.   DISTANCE MODULI AND DISTANCES

Distance modulus	Distance
−5 magnitudes	1 parsec
0	10
5	100
10	1,000 (1 kiloparsec)
15	10,000 (10 kiloparsecs)
20	100,000 (100 kiloparsecs)
25	1,000,000 (1 megaparsec)
30	10,000,000 (10 megaparsecs)

The distance-modulus equation is a two-edged tool. As used above, it gives the absolute magnitude of any star whose distance is known. But when turned in the other direction, this relation becomes the astronomer's commonest method of measuring distance: whenever any star's absolute magnitude is known, comparison with its apparent magnitude immediately yields the star's distance from us. The key to stellar distances is thus to know the absolute magnitudes of recognizable types of stars. First, astronomers must classify the stars, learning to recognize specific types. Then the absolute magnitude of each type must be found,

FIGURE 12.2.
In this picture are two star clusters, whose appearances contrast very much, largely because of their very different distances from us. M35, whose stars are scattered over the central part of the picture, is only 500 parsecs away, whereas NGC 2158, the more concentrated cluster at the right, is more than 2,000 parsecs away. Its stars appear much fainter. (Copyright, National Geographic Society–Palomar Observatory Sky Survey. Reproduced by permission of Hale Observatories.)

either from examples found side by side with other stars of known absolute magnitude, or from members of the class that are close enough for a direct measurement of trigonometric parallax. After a stellar type has been thus calibrated, its representatives serve as stellar yardsticks wherever they are; once they are found, their magnitudes tell their distances. All the large distances of the universe are measured by this photometric method, and much of our attention in the following chapters will be directed to the calibration and exploitation of particular stellar types. Such are the stepping stones of the universe.

THE MOVING STARS

The permanence of the stars is proverbial. The moon and planets wind among them, but year in and year out we see constellations that were old when the Greeks wrote about them. Yet changes in their pattern do take place; several stars have moved conspicuously from the positions marked for them on ancient star charts, and precise modern observations have measured the motions of many thousands of stars. Stellar motions, like the magnitudes of the stars, are more than just a set of data; they offer another way of estimating stellar distances, they help to label types of stars, and they provide a key to understanding much of the structure of the Milky Way.

As a star sails its steady course through space, we see its motion in two ways—or to be more precise, we see separate parts of its motion in two separate ways. Motion across our line of sight shows as a change in the star's apparent position; but motion toward or away from us shows only as a change in the star's spectrum, a shift of wavelength called the *Doppler effect*. Light can be thought of as a train of waves flowing from the source to the observer: if the source is approaching the observer, the

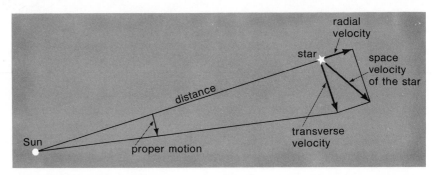

FIGURE 12.3.
The radial and transverse parts of a star's motion are observed in quite different ways.

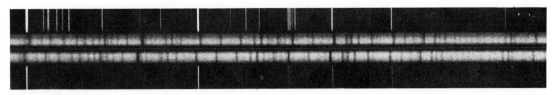

FIGURE 12.4.
These two stellar spectra (middle) illustrate Doppler shifts in wavelength. They are spectra of the same star, taken at different points in its orbit around another star. For each the comparison spectrum (bright lines above and below) serves as a standard of wavelength. (Lick Observatory.)

waves are bunched together, so that the wavelength looks shorter. Similarly, recession of the source stretches out the waves and increases the wavelength. The same phenomenon makes the whine of a jet plane high-pitched as it approaches us and lower when it passes us and recedes. The amount of the shift, it turns out, depends on the velocity of the motion relative to the velocity at which the waves travel. Because the velocity of light is so great, however, even stellar velocities are relatively small and produce only tiny Doppler shifts. Thus radial velocities, as they are called, can be measured only with great care, and the errors in measurement are usually considerable. Typical stellar velocities are in the neighborhood of 20 kilometers per second, but astronomers are often content with an accuracy in their calculations of a few kilometers per second. Because the measurement of each radial velocity requires laboriously photographing the spectrum of a single star, such data have accumulated very slowly; the radial velocities of only about 25,000 stars have been measured so far. On the other hand, the Doppler effect does not depend on a star's distance from us; any star bright enough for spectrography can thus have its radial velocity measured.

But transverse motion is a different matter: here the observed quantity is a change in the star's direction in the sky, which is conspicuous only if the star is nearby. Thus, proper motions, as they are called, have been measured for some three hundred thousand of our neighbors; the more distant stars appear to be practically fixed in position. Again, the limit is set by our measuring accuracy, and accuracy has improved over the decades. Even so, proper motions can be usefully measured only for stars within one or two hundred parsecs of the Sun.

It is interesting to know that the stars move; but the importance of this fact lies, of course, in how we can use it to study the basic nature of the stars and of the system that they outline. Stellar motions are valuable in two ways. First, as we shall see in a later chapter, the knowledge of where stars are going and where they have come from tells us a great deal about our Galaxy, the greater stellar system that they make up. Second, the perspective effect in stellar motions, as stars drift by us, shows in a general way which stars are near us, and how near they are.

230

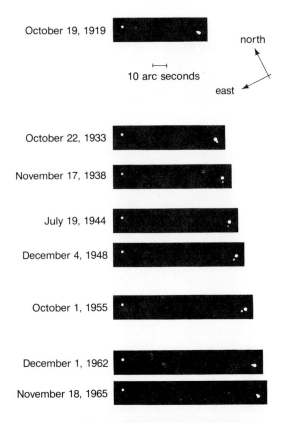

October 19, 1919

north

east

10 arc seconds

October 22, 1933

November 17, 1938

July 19, 1944

December 4, 1948

October 1, 1955

December 1, 1962

November 18, 1965

FIGURE 12.5.
This sequence of photographs shows the motion of the star at the right relative to the lefthand star, which is more distant and has almost no proper motion. The moving star, which is one of our near neighbors in space, is also a binary star (see Chapter 14), and the orbital motion of the two close components is easy to see. (From *Principles of Astrometry,* by Peter Van de Kamp. W. H. Freeman and Company. Copyright © 1967.)

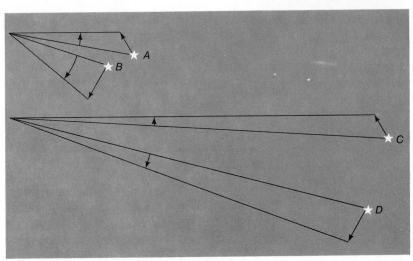

FIGURE 12.6.
When stars are farther away, their transverse velocities correspond to smaller proper motions. In this drawing, stars *A* and *C* have the same velocity, and stars *B* and *D* have the same velocity. In each case, it is clear that the more distant star has a smaller proper motion.

This perspective effect of motion is familiar to everyone. As we drive rapidly down the road, objects that are right at the roadside appear to rush by us. More distant things drift by more slowly, while a mountain on the horizon hardly seems to change its direction at all. So it is with the stars. Nearby stars tend to show large proper motions, but distant stars have motions that are usually too small to measure. We can use this effect in two ways. First, the detection of proper motions allows us to pick out individually our few near neighbors for further study, from among the countless other stars that are equally faint and look superficially the same. The second application is to use motions actually to measure the distances of the stars, relying on the fact that proper motions tend to be large or small, on the average, in a way that depends on the distance of the group of stars that we are considering.

This is simply the perspective effect that was just described, but what we do now is to try to apply it statistically. Some individual stars move through space rapidly and others slowly, but for any set of them we can calculate an average speed. Although any individual may have a speed that is far from the average, if we take a group of 100 stars and find their average speed, it is likely to be very close to the average of any other group of 100. This is true for actual speeds through space, but the apparent proper motions that we will see depend on the distances of the stars from us. Thus if one of two groups of stars is ten times as far as the other, the average size of its proper motions will be only a tenth as large. Average proper motions therefore tell us average distances. If, now, we

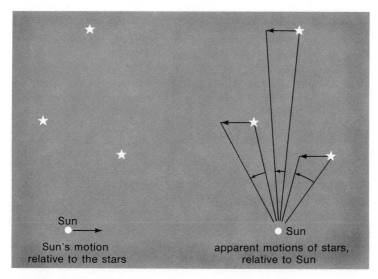

FIGURE 12.7.
As the Sun moves past other stars, they appear to us to be drifting by in the opposite direction. The angular motion, which is what we see, is greater for nearby stars than for distant ones.

carefully choose a set of stars that are all of the same type, then from their average distance and their average apparent magnitudes, we can find the average absolute magnitude of that type of star—an important physical property.

Because this method treats stars statistically rather than as individuals, and because the movements that we observe are a kind of parallactic displacement, we call this the method of statistical parallax. It has the advantage that proper-motion displacements accumulate and increase in size over the years and decades, unlike the annual parallaxes of stars. For this reason, statistical parallaxes can be calculated for stars that are too far away for a good measurement of annual trigonometric parallax. Included among these are some important but rare types of stars that have no representatives close enough for a measurement of trigonometric parallax.

Amid this discussion of the motions of other stars, it is natural that we should ask about the Sun: how does it move through space? Although this seems a simple question, it turns out to be difficult to define the solar motion. If another star shows a radial velocity of approach, should we say that we are moving toward it or that it is moving toward us? The answer is that both are right; it is only the relative motion that counts.

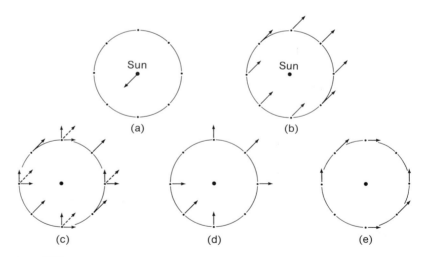

FIGURE 12.8.
Effects of solar motion on the apparent motions of the stars. In (a) is shown the motion of the Sun with respect to stars all around us. Part (b) shows the corresponding apparent motions of the stars with respect to the Sun. These are resolved into radial and transverse parts in (c), while (d) and (e) show the radial and transverse parts of the motions of those stars separately.

There are no marker posts in space; there is nothing to allow us to say that one star is moving and another is stationary. What does make sense, though, is to take as a reference standard the average motion of a large number of stars around us. Thus we consider the average motion of the neighboring stars to be zero, and measure the solar motion with respect to this "local standard of rest."

The solar motion, which is approximately 20 km/sec toward a point near the star Vega, is observable both in the radial velocities of neighboring stars and in their proper motions. What we see is just the perspective effect that was described before, but now we concentrate on the direction of the motions. Near the point in space toward which we are moving—the solar apex—the stars appear, on the average, to be moving toward us; around the antapex, the average motion is away from us. Thus, the average radial velocity of the stars varies around the sky in a systematic way, and from this, the solar motion can be found.

Again, knowing the solar motion not only satisfies our curiosity, but also contributes to our knowledge of the other stars, again by a statistical parallax. Since a star's apparent motion depends on its total velocity relative to us, part of its proper motion is just a reflection of the solar motion. When we average the proper motions of a large number of stars, what we see remaining is just the effect of the solar motion, and the size of this systematic motion is just like a parallax; it tells us the average distance of the stars whose motions we have observed.

REVIEW QUESTIONS

1. Explain why the parallaxes of stars decrease with increasing distance. How large is the parallax of a star that is 20 parsecs away?

2. If a star is 100 parsecs away, why is a trigonometric parallax of little use in determining its distance?

3. Explain the difference between a scale of magnitudes and a scale of brightness. Why is a magnitude scale natural, and why is it useful?

4. Show that a factor of 10 times in brightness corresponds to exactly 2½ magnitudes.

5. Why is it appropriate to call an absolute magnitude "absolute"?

6. (a) A certain recognizable type of star is known to have an absolute magnitude equal to +4. In a certain star cluster these stars have apparent magnitude 19. How far away is the cluster?

   (b) In a galaxy a million parsecs away, how bright (i.e., of what apparent magnitude) are stars of absolute magnitude −6?

7. Distinguish between the radial velocity of a star and its proper motion. Contrast, in as many ways as you can, the way in which the two are observed and the uses to which they can be put.

8. What does "solar motion" mean? Why is the concept hard to define?

9. What is a statistical parallax? What is its relation to the solar motion? Why are statistical parallaxes so valuable?

13 | SORTING OUT THE STARS

An obvious first step in studying the stars is to sort out the various types. In this chapter, we shall see how a simple classification, followed by some straightforward reasoning, leads to a clear picture of how old the stars are, how they are born, and how they die.

If we had complete freedom, we would certainly choose for study some of the fundamental physical properties of a star, such as its mass and its diameter. Unfortunately, these quantities are hard to calculate, and we know them reliably for only a few stars. As a matter of practical necessity, our study of the stars begins with quantities that are easily measured for a large number of stars. One of these became available as soon as we learned to measure stellar distances: the absolute magnitude of the star. A second intrinsic characteristic of a star that is easily observed is its spectral type. The spectrum of any star has in it many dark lines, marking wavelengths that are absorbed by the chemical elements in the star's atmosphere. The same chemical elements are in the atmospheres of all the stars, but the temperatures of the stars differ, and this strongly affects the strengths of their spectral lines. Thus the spectrum reveals a second intrinsic characteristic of a star, its temperature.

The classification of stellar spectra began in earnest just before 1900, when photography began to be used extensively in astronomy. The scheme was very simple: astronomers divided stellar spectra into as many classes as they could easily distinguish, arranged them in a sequence so that neighboring types resembled each other, and assigned letters to the classes. Only later was it found that some of the distinctions were not real and that the classes could be rearranged in a more smoothly

O6

B3

A0

F2

G2

K5

M5

FIGURE 13.1.
Spectra of some stars, illustrating the chief classes. In each reproduction, the central strip is the spectrum of the star. Above and below each stellar spectrum is a comparison spectrum, made by shining the light of a standard substance into the spectrograph. The comparison lines serve as standards of wavelength. (Hale Observatories.)

progressing order. From this erratic development came the present-day set of main spectral classes: O, B, A, F, G, K, and M. A few stars do not fit into this scheme, and additional classes, given such designations as R, N, S, and W, have been created to contain them; but the great majority of stars fall into the seven main classes. With more careful examination, it has turned out that stars can be more finely placed in the spectral sequence; so the classes are now divided into tenths, designated by the numerals 0 through 9. A part of the sequence, therefore, is denoted thus: . . . , A8, A9, F0, F1, . . .

In addition to this awkward sequence of letters, an anachronistic term has clung to spectral classes: at a time when some astronomers mistakenly associated the spectral types with successive stages in the life of a star, the types near the beginning of the sequence were called early and those at the other end were called late. The names have persisted, and we still say that G7 is a "later" type than G6.

When we examine the spectral types and the absolute magnitudes of stars simultaneously, certain regularities appear. These correlations were first found by the astronomers Hertzsprung and Russell, after whom

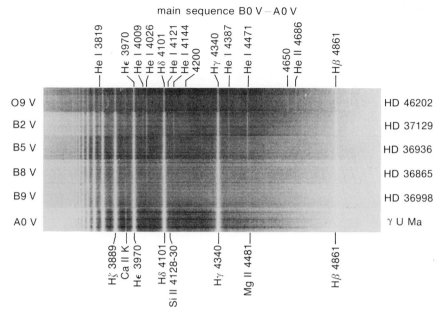

FIGURE 13.2.
This series of spectra of standard stars is taken from an atlas that is intended to assist astronomers in classifying stellar spectra. (They are negative prints because the spectra that the astronomer will classify are the original negatives that were exposed in the spectrograph.) Lines are marked whose strengths are used in the classification process. (Kitt Peak National Observatory.)

the basic graph of stellar characteristics is called the HR diagram. In such a diagram spectral types are plotted from left to right and absolute magnitudes from top to bottom, with the brightest, most negative magnitudes at the top. The characteristics of each star are represented by a point opposite the appropriate spectral type and absolute magnitude.

If points are plotted on an HR diagram for all the stars in the solar neighborhood for which spectral types and absolute magnitudes are known, the points do not scatter all over the diagram. Instead, they lie along several extended regions, or sequences. The majority of stars lie along a diagonal line called the "main sequence." Many others lie along a branch that runs across toward cooler spectral types; by contrast with the "red dwarf" stars of the lower main sequence, these brighter stars are called "red giants." Unfortunately, some later-discovered stellar types have disfigured this terminology. Above the red giants are the "supergiants," and a sequence extending up from the main sequence toward the red giants has been given the name "subgiants." Another important sequence lies below the main sequence and nearly parallel to it; these stars are called the "white dwarfs."

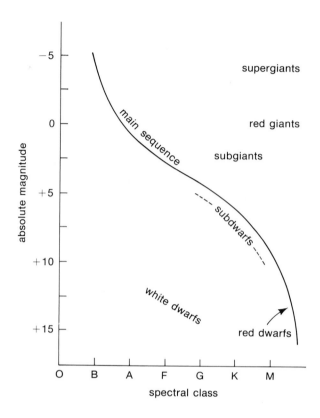

FIGURE 13.3.
The HR diagram.

MAKING SENSE OF THE PICTURE

The HR diagram of all the stars in the Sun's neighborhood shows tantalizing regularities, but still it is confusingly complex. The complexity is due to the fact that the stars that are now passing through our neighborhood in space really originated in different places at different times; they are far from being a homogeneous population. When we do look at homogeneous populations, however, the meaning of the HR diagram reveals itself in a clear way.

Homogeneous populations can be found in star clusters, which are groups of stars that are held together by the gravitational attraction of the stars for each other. Analyses have shown that a star may occasionally escape from a cluster but that the cluster never captures new members. Thus, the stars in a cluster must have been together since they were born, and they must have been born together.

Such a group, born at the same time out of the same material, turns out to have a simple HR diagram. Furthermore, different clusters have HR diagrams that, although different, can be placed in a regular se-

FIGURE 13.4.
The open cluster M11. See also Figure 13.5.
(Hale Observatories photograph by the author.)

quence of types. At one extreme, some clusters have only a main sequence, stretching all the way up to the brightest blue stars. Others lack
the very top of the main sequence but contain some red supergiants. Still
other clusters have a main sequence that terminates lower, with a slight
turnoff at the top, and supergiant or giant stars that are not quite so
luminous. Clusters with lower main-sequence turnoffs have less of a
gap between the main sequence and the giant stars; and at the extreme,
the turnoff is near the Sun's position on the main sequence, and the
turnoff runs smoothly upward into a subgiant and giant branch.

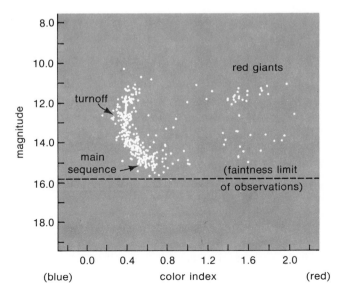

FIGURE 13.5.
Color-magnitude array of the cluster M11.
Individual stars are represented by dots. In
this cluster, as in most, color indices were
measured instead of spectral types, since
they are easier to measure for faint stars. The
scattered stars below the red giants and to
the right of the main sequence are probably
background stars that happen accidentally to
be superimposed on the cluster. (After
G. Hagen, *Publications of the David Dunlap
Observatory,* University of Toronto, Vol. 4,
1970.)

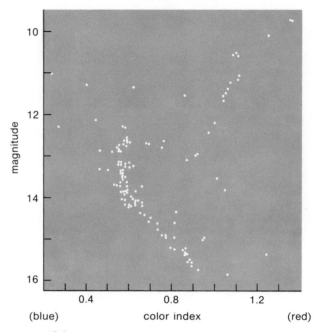

FIGURE 13.6.
Color-magnitude array of the cluster array of the cluster
M67. Very few foreground or background stars are included
here, because proper motions were used to eliminate stars
whose motion is obviously different from that of the cluster.
M67 is a much older cluster than M11 (Figure 13.5); the
subgiant and giant branches wind continuously away from
the main sequence. The few stars in a horizontal line above
the turn off are probably in a late stage of evolution. (After
G Hagen, *Publications of the David Dunlap Observatory,*
University of Toronto, Vol. 4, 1970.)

When astronomers first arranged the HR diagrams in this order, in
the early 1950s, it became clear that the differences between the clusters
were related to their ages. A cluster with a full upper main sequence is
very young. Older clusters have a main sequence whose upper end turns
off, and the older the cluster, the fainter the turnoff point. This is simply
because the brightest stars have the shortest lives; in an old cluster, the
bright stars are already dead.

The lifetimes of stars are easy to understand in terms of the exhaustion
of their fuel. A star cannot shine without cost. A tremendous amount of
energy streams from its surface into space, and the star would soon cool
off and go out, were this energy not continually replaced by a flow of
energy from the interior. The burning that produces this energy, at the
center of the star, is nuclear rather than chemical, but it still needs fuel.

The star's fuel is hydrogen, and the nuclear burning turns it into helium, through reactions that we shall study later (Chapter 18). In nuclear burning, a little fuel goes a long way, and the stars are made mostly of the fuel element hydrogen; but even so, nothing lasts forever.

Calculating the lifetime of a star is also easy. If we know how much fuel it has (its mass) and how fast the fuel is being used (its luminosity), then the quotient of the two gives the length of time that the fuel will last. Actually, modern astronomers calculate a little more accurately than this, taking account of the fact that the nuclear reactions at a star's center can utilize only a part of the star's total fuel, but the principle is the same: the lifetime of a star is determined by its mass divided by its luminosity.

The stars differ tremendously in this ratio of mass to luminosity. The most luminous stars put out about half a million times as much light as the Sun, but have only about fifty times the Sun's mass. Such a star is therefore burning up its irreplaceable fuel ten thousand times as fast as the Sun and will live only one ten-thousandth as long. By contrast, the slowest-burning dwarf stars have less than a ten-thousandth of the Sun's luminosity, but they have almost a tenth of its mass. They are using their fuel so conservatively that their lives are almost unlimited compared with that of the Sun.

Calculations of stellar structure and evolution have shown that the total lifetime of the Sun is about 10 billion years. Because radioactive age-dating of meteorites has shown that the Solar System is roughly 4.6 billion years old, the Sun is now almost halfway through its life. The shortest stellar lifetimes are about a million years, and the longest are perhaps 10^{14} years. Stars of the latter type have not yet aged at all, because the Milky Way itself appears to be only 10 or 20 billion years old; but the shortest lifetimes are strikingly short and reveal an important fact. If a star's total lifetime is 10^6 years, it must have been born less than 10^6 years ago—a twinkling of an eye for the Milky Way. Stars are being born "today"!

Star clusters of different ages show how the HR diagram would look at successive stages in the life of a single cluster. At the beginning, there are stars of all luminosities, and they lie all along the main sequence. Then the most luminous stars reach the end of their lives and can no longer shine like main-sequence stars. They must change. What they change into is clear from the HR diagrams of clusters. A star cluster from which the very top of the main sequence is missing has, at nearly the same luminosity, red supergiant stars. Apparently, these are what the missing main-sequence stars have developed into. In an older cluster, stars from farther down on the main sequence have also reached the end of their main-sequence lives and have turned into red supergiants or giants. The stars that were supergiants at an earlier stage in time have

FIGURE 13.7.
This region, known in catalogs as Messier 16, is a mixture of glowing gas, dark clouds of dust, and newborn stars. The cluster of bright stars at the upper right has an HR diagram that shows it to be a very young group. Some of the small dust clouds silhouetted against the bright gas may be on the verge of forming even newer stars. (Hale Observatories.)

now disappeared from the bright parts of the HR diagram. Apparently, they have turned into some kind of object that is no longer luminous enough to be conspicuous. As time goes on, the disease of aging eats its way down the main sequence, as stars from the top of the main sequence become red giants and then disappear.

More important than the mere dating of clusters is the insight that their HR diagrams give us into the detailed course of evolution for each star. Not only is it clear that red supergiant or red giant stars have evolved from the main-sequence types that have most recently disappeared, but the number of such stars that we see also shows how long they spend in this later evolutionary stage. The reason for this is easy to

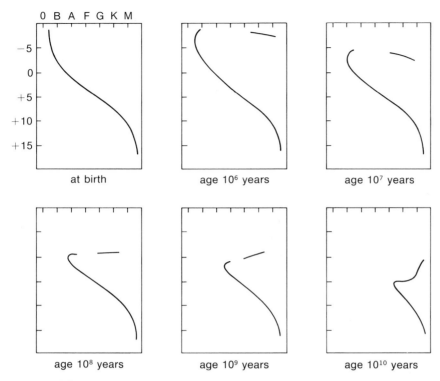

at birth   age 10^6 years   age 10^7 years

age 10^8 years   age 10^9 years   age 10^{10} years

FIGURE 13.8.
Schematic HR diagrams of clusters of different ages. As time passes, the
upper-main-sequence stars die first. The point at which the remaining main
sequence turns off is an index of the cluster's age.

see: if a star passes through a particular stage of evolution rapidly, then
we are likely to find few stars, or none at all, at such a stage, whereas
any snapshot of the universe is likely to reveal many stars at a stage in
which they tend to linger. Thus the preponderance of main-sequence
stars shows that a star spends most of its life on the main sequence. The
gap to the right of the upper main sequence corresponds to an evolutionary stage through which those stars pass rapidly, and the common-
ness of red giants indicates a stage in which stars linger.

Evidently, stars of different sizes do not evolve in the same way. The
most luminous stars, after their central fuel is exhausted, change rapidly
into stars that are much redder. For less luminous stars, the rapid change
is not as great, and for stars at the Sun's part of the main sequence, evolution progresses smoothly away from the main sequence, through the
subgiant region, and up into the red-giant part of the HR diagram.

Further questions need to be answered. What does a star become
after the red-giant or supergiant stage of its life has ended? And how

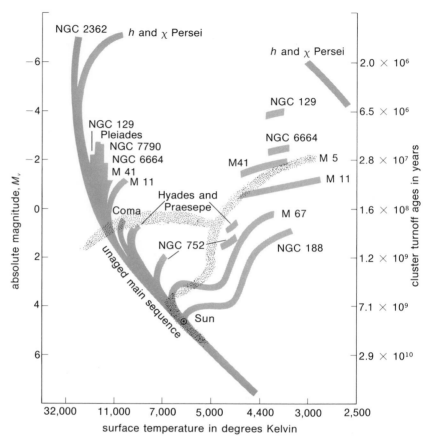

FIGURE 13.9.
In this composite HR diagram, several individual diagrams, such as those in Figures 13.5 and 13.6, have been schematized as lines and combined by fitting together the unevolved parts of their main sequences. The diagram shows vividly the differences in turnoff points. The youngest cluster included is about 10^6 years old, the oldest about 10^{10} years. The dotted line represents a globular cluster; the differences will be discussed below. (Adapted from "Pulsating Stars and Cosmic Distances," by R. P. Kraft. Copyright © 1959 by Scientific American, Inc. All rights reserved.)

are we to account for the remaining prominent sequence in the HR diagram, the white dwarfs? It is an easy step to connect the two questions and to suggest that the red giants will next become white dwarfs. This is indeed the correct answer, even though it does not follow directly from the straightforward kind of observational reasoning that we have been using. The trouble is that white dwarfs are stars of very low luminosity, too faint to find in most of the star clusters that are well-studied. We know them only as members of the random field in the Sun's neighborhood of space. Their identification as an end-stage of stellar evolution comes not from the empirical approach that we are following here, but rather from a physical understanding of stellar evolution, an understanding of what is actually happening inside the stars.

It is amazing how far the empirical approach carries us in charting the course of stellar evolution. For a real understanding, however, we will need to examine the physical workings of a star. We shall return to this problem in Chapter 18.

STELLAR POPULATIONS

Star clusters can be arranged in a clear progression of ages, which seems to have an end. The main sequences of the oldest clusters turn off at an absolute magnitude of about +4, and theoretical studies of stellar evolution say that this will happen at an age of about 10 or 15 billion years. The fact that no older clusters are found suggests that this is the age of the Milky Way itself.

Among the oldest clusters, however, we find a strange difference of types. The richest clusters—those called globular clusters—have a main-sequence turnoff at the same absolute magnitude, but the upper parts of their HR diagrams are very different from those of the "open" clusters that we have been considering up to now. The red-giant branch of a globular cluster is brighter than that of an open cluster, and the HR diagrams of most globular clusters also have a horizontal branch at about absolute magnitude zero.

According to our previous reasoning, which connects turnoff point with age, these differences cannot be explained as differences in age. Something else must differ. Indeed, something else does differ: in the spectra of globular-cluster stars, most of the spectral lines are weaker than the corresponding lines in the spectra of open clusters. Again, a proper quantitative evaluation of this difference requires a theoretical analysis of stellar spectra, but the nature of the answer is clear enough: the chemical elements responsible for those lines must be less abundant in the stars of a globular cluster.

When careful abundance analyses are carried out, the conclusion is that about three-quarters of the material of the universe is hydrogen. The remainder is mostly helium; the other chemical elements, to which astronomers refer loosely as the "heavy" elements, make up only a tiny fraction of the total. It is in the size of this fraction that globular clusters differ. The heavy elements make up only a fraction of 1 per cent of their material, whereas in open clusters, just as in the Sun, the heavy-element abundance is 2 or 3 per cent.

Most of the spectral lines that indicate the abundance of heavy elements are actually due to metallic elements, including such well-known

FIGURE 13.10.
The globular cluster M13. Clusters such as this one belong to the halo of the Milky Way. (Hale Observatories.)

PLATE 1.
The planet Earth, as seen from the Apollo 13 spacecraft.
(National Aeronautics and Space Administration.)

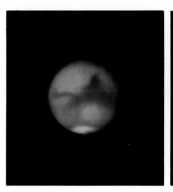

PLATE 2.
Mars and Jupiter, as seen from
the Earth under excellent observing conditions. The sharpness of
the photographs is limited by the
unsteadiness of the Earth's
atmosphere. (Lowell Observatory.)

PLATE 3.
Two views of Jupiter from Pioneer 11. The spacecraft approached Jupiter from below its equator
(picture above), swung round the planet, and departed on a northward path (picture on facing page).
The Great Red Spot can be seen in the picture above. Note the difference between the belted pattern
of Jupiter's equatorial regions and the irregular patchiness in higher latitudes. (National Aeronautics and
Space Administration.)

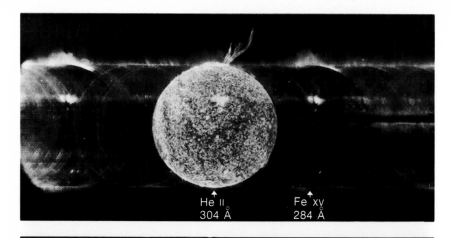

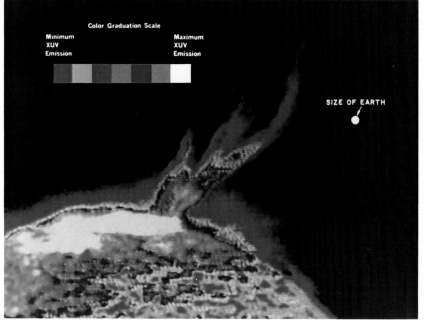

PLATE 4.
(Top) This spectral photograph of the Sun was made from outside the Earth's atmosphere, during the fourth Skylab flight, in January 1974. The extreme ultraviolet part of the sunlight was imaged by a spectroheliograph that formed a separate image of the Sun in the light of each emission line. (The overlapping of images cannot be avoided in such an instrument.) The images at two prominent wavelengths are marked. The ionized helium (He II) emission comes from gas at about 80,000°. Besides the structure of the whole chromosphere, the helium image shows an active region just above the middle of the disc and another at the top, and a spectacular prominence. The emission from 14-times-ionized iron (Fe XV) comes from the 2,500,000° gas above active regions.
(Bottom) The brightness distribution in part of the helium image from the top picture. "XUV" is an abbreviation for "extreme ultraviolet." (Naval Research Laboratory.)

PLATE 5.
The emission nebula M16. The gas glows red, because so much of its energy is in the first line of the hydrogen Balmer series, in the red part of the spectrum. Contrast the blue color of the newly born upper-main-sequence stars. (Hale Observatories. Copyright © by the California Institute of Technology and the Carnegie Institution of Washington.)

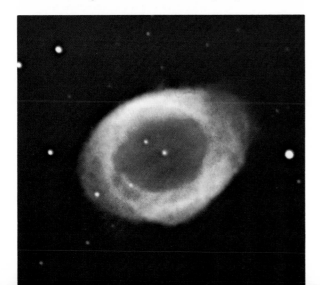

PLATE 6.
The Ring Nebula, M57, is a planetary nebula. The central star has completed the red-giant phase of its evolution and has thrown out a shell of gas, which glows just like an Hɪɪ region. In the red parts, the red line of hydrogen dominates, whereas the green parts have strong radiation from the forbidden "nebular lines" of oxygen. (Kitt Peak National Observatory. Copyright © 1974 by the Association of Universities for Research in Astronomy, Inc.)

PLATE 7.
This color photograph of the Crab Nebula emphasizes what a strange object it is. The white region is a superhot gas, in which light is emitted by electrons gyrating in a magnetic field. The red filaments are cooler regions in which we see emission from hydrogen gas. (Hale Observatories. Copyright © by the California Institute of Technology and the Carnegie Institution of Washington.)

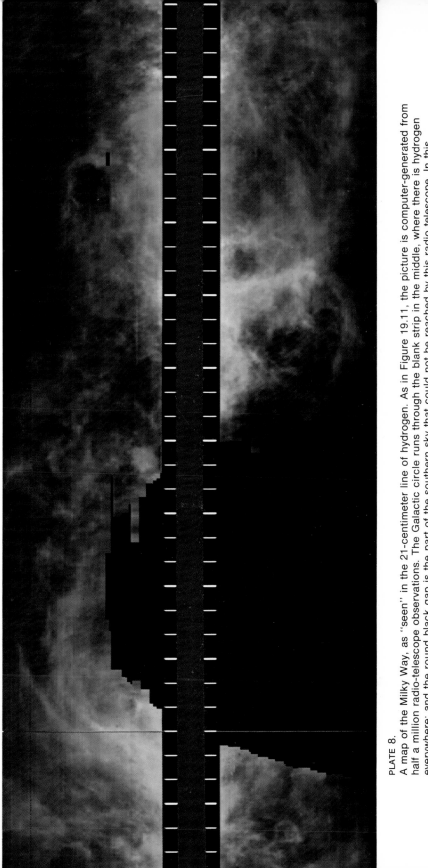

PLATE 8.
A map of the Milky Way, as "seen" in the 21-centimeter line of hydrogen. As in Figure 19.11, the picture is computer-generated from half a million radio-telescope observations. The Galactic circle runs through the blank strip in the middle, where there is hydrogen everywhere; and the round black gap is the part of the southern sky that could not be reached by this radio telescope. In this version of the picture, color is used to indicate velocity: regions colored red are moving rapidly away from us; those colored blue are moving rapidly toward us. The differential rotation of the Milky Way is clearly visible; the red and blue regions correspond to those labeled "+" and "−" in Figure 20.7. (Courtesy of C. E. Heiles, Radio Astronomy Laboratory, University of California, Berkeley.)

PLATE 9.
The central region of the Orion Nebula is shown here in a photograph specially prepared to render the colors as faithfully as possible. The light of the bright, hot inner region is dominated by the green "nebular lines" of oxygen, whereas the red edge at the lower left has more radiation from the red line of hydrogen. (Courtesy of J. S. Miller, Lick Observatory.)

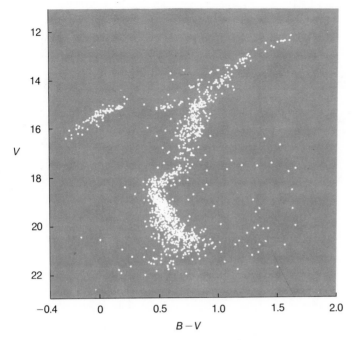

FIGURE 13.11.
A color-magnitude array of the globular cluster M5. (In Figure 13.9 this is compared with the HR diagrams of clusters of the galactic disc.) The scattered points in the lower part of the diagram are probably foreground stars that happen to be superimposed on the cluster. (After H. C. Arp, *Astrophysical J.,* **135** (1962), 311. The University of Chicago Press.)

metals as iron; hence astronomers often refer to the entire class of elements heavier than hydrogen and helium as "metals."

Differences in chemical composition exist between all stars, not just between those in clusters. Groups of stars with particular characteristics of composition are referred to, in general, as populations; but the term is most commonly used to distinguish Population I from Population II. Population I consists of the stars of approximately solar, or "normal," metal abundance, such as occur in open clusters, and Population II consists of the stars of low metal abundance that are found in globular clusters.

Unfortunately, the term "population" has itself gone through an evolution. When population differences were first recognized, the most conspicuous difference was between Population II and the youngest part of Population I. Then the study of cluster ages showed that Population II was old, which suggested that the chief difference between the two populations was age. The full picture became clear only when it was

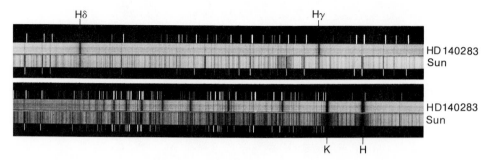

FIGURE 13.12.
A comparison of the spectrum of a Population II star with that of the Sun. Notice how greatly the strengths of the lines differ. The other star has much lower abundances of the heavy elements than does the Sun. (Lick Observatory.)

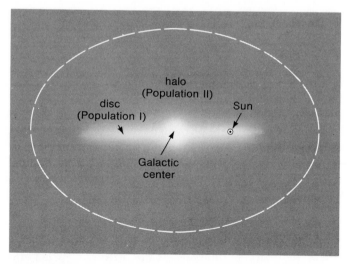

FIGURE 13.13.
A cross section of our Galaxy, showing the arrangement of the two populations. Stars of the disc remain confined to their own region, but halo stars also pass through the disc.

shown that Population I covers the entire range of age, whereas Population II consists of those old stars that also have low metal abundance. The terms "Population I" and "Population II" are still not used in the same way by all astronomers. There is a growing tendency, however, to reserve them to distinguish differences in metal abundance, and these terms will be so employed in this book.

Strangely, Population II seems to consist of old stars only. No young stars of low metal abundance have ever been found. It appears that the birth of such stars ceased in the distant past, a fact that we shall make use of later when discussing the evolution of the Milky Way.

The two populations are also distributed differently in the Milky Way. In the flattened disc that contains most of the stars, including the Sun, nearly all the stars are of Population I. The stars of Population II are spread through a larger, less flattened region called the halo. As a result, the nearby stars that we study in detail are nearly all members of Population I.

THE STELLAR CENSUS

The kinds of stars in a population make up one of its characteristics, but another important characteristic is the number of stars of each type. In a star cluster, in which all the stars are nearly the same distance from us and in a compact region of the sky, it is relatively easy to calculate the number of more luminous stars, the ones that are easily studied, although the number of faint stars is less certain. In particular, we have almost no observational information about white dwarfs in clusters. They must be there, as the final stage of stars that have ended their main-sequence and red-giant lives; in practice, however, our knowledge of white dwarfs comes from those that we have found in the general field of stars in the solar neighborhood.

The trouble with stars of low luminosity is that, unless they are very close to us, their apparent magnitudes are too faint for convenient study. Even worse, such stars are difficult even to find among the thousands and thousands of other equally faint stars that are, in fact, more luminous but farther away. No one could hope to measure parallaxes of all the faint stars in the sky in order to find out which of them are nearby stars of low luminosity. Fortunately, however, such stars can be picked out in another way. Because the proper motion of a star of a given transverse velocity depends on its distance, nearby stars can be expected to have large proper motions. Not every one will, because some stars will just by chance have a very small transverse velocity, but a search for large proper motions should turn up the great majority of our near neighbors in space.

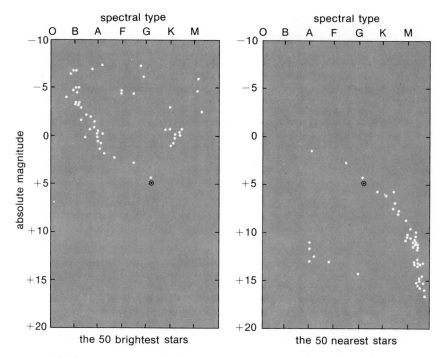

FIGURE 13.14.
HR diagrams of the 50 brightest stars (left) and the 50 nearest stars (right). The Sun is also shown in each diagram by the symbol ⊙. The righthand diagram indicates the luminosity function, the true relative frequency of the various stellar types. The lefthand diagram indicates how we select out the luminous stars when choosing according to apparent magnitude.

Such searches have been carried out. They are done by comparing two photographs of the same region of the sky taken as many years apart as possible. The two photographs are put into a device called a blink comparator, which allows an observer to look at them in rapid alternation. The photographs are carefully lined up so that each star looks almost exactly the same when attention is switched from one photograph to the other—unless the star has moved appreciably, in which case it appears to jump back and forth. In this way, thousands of stars of large proper motion have been discovered. Lists of such stars are then brought to the attention of the observers who measure parallaxes, because only a direct determination of distance can show what the absolute magnitude of the star really is.

Not all stars of large proper motion have had their parallaxes measured yet, but our census of neighbors is probably almost complete out to a distance of 5 parsecs from the Sun. This is an infinitesimal fraction of the Milky Way, but it is large enough to permit us to assess how numerous the stars of low luminosity are in space.

The more luminous stars require different techniques. Not many of them are found very close to us, but a considerable number have been included in the systematic studies that have been made of the spectra and parallaxes of all bright stars. For each type, we can form an estimate of the volume in which the stars have been discovered, and from this, the number per unit volume can be calculated.

The resulting count of the number of stars of each luminosity is called the luminosity function. In a sense, it is the true stellar census because, in the luminosity function, we do not overrate a star just because it happens to be bright.

This unbiased census reveals that the great majority of stars are faint dwarfs. Of the sixty stars within 5 parsecs of the Sun, more than half are lower-main-sequence dwarfs of spectral class M. With the K stars included, the red dwarfs make up three-quarters of the total. Another 10 per cent are white dwarfs. Only three of the sixty stars in this volume of space are more luminous than the Sun.

By contrast, the result is the opposite when we choose stars according to apparent magnitude. Of the twenty stars of magnitude 1.5 or brighter—those that we loosely call the first-magnitude stars—only one is less luminous than the Sun, and several are supergiants. The reason for the difference is easy to see: the more luminous stars are included in this first-magnitude sample even when they are far away, whereas a red dwarf of spectral class M could look this bright only if it were closer to us than any star actually is. Thus, the luminous stars are most readily noticed—indeed, they are often more useful to study—but most stars in the universe are inconspicuous dwarfs.

REVIEW QUESTIONS

1. Draw an HR diagram, labeling the coordinates carefully, Sketch in, and label by name, the principal regions and sequences of stars.

2. Why does a star cluster have a simple HR diagram, whereas the HR diagram of the solar neighborhood is so complicated?

3. Why does the lifetime of a star depend in such a simple way on its mass and its luminosity?

4. Explain how and why the turnoff point of a cluster's main sequence indicates its age.

5. What are Populations I and II? Contrast their stars in age and in composition.

6. What is the luminosity function? Why is it difficult to determine its faint end reliably?

7. How do the 50 nearest stars differ from the 50 brightest stars? Explain the difference.

14 | BINARY STARS:
TWO ARE BETTER THAN ONE

Although most stars look like single points of light, many are in fact multiple. Sometimes the magnification of a telescope shows the star as a pair of points rather than a single point, but more often the stars are too close together to be seen separately, and we can detect the duplicity only by more indirect means. And it sometimes turns out that a star is not just double but multiple: three or more stars together. Astronomers call all of these, whether double or multiple, *binary stars.*

Binaries are very common. In the census of nearby stars, nearly half are binary. This is not just a peculiarity of the dwarf stars that make up this local sample; when we range farther afield and study the more luminous stars, we find that binaries are also common among them. As we saw in discussing the origin of the Solar System and the possibility of life elsewhere in the universe, multiple stars are a very plausible product of the contraction process that makes stars out of clouds of interstellar material.

Binary stars are a topic for astronomical study just because they exist in the universe, but they are particularly important as sources of basic information about stars in general. The reason is that we can often measure quantities for a pair of stars that would be difficult or impossible to measure reliably for a single star. Foremost among these quantities is the mass of the star. Intuitively, mass means to us the amount of material that is in a star, but observationally we can measure it only by its gravitational effect. The orbital motion in a double star gives us an ideal opportunity to measure masses. According to Kepler's third law, all that we need is to measure the size of the orbit of the two stars around each other and the period of time that they take to go around, and we can then calculate the sum of their masses. The separate masses then follow from comparing the motion of each component relative to the center of mass of the pair; the more massive component moves proportionally less.

From observations and calculations such as these, binary stars provide nearly all our knowledge of stellar masses. This means, of course, not just the masses of the particular stars that happen to be members of binaries; from them we deduce the masses of stars in general. If we study a binary, for instance, that includes a main-sequence star of spectral class A5, we can reasonably expect that the mass of the latter will also be the mass of every other main-sequence A5 star.

The simplest kind of binary is a double star, where the two stars go around their common center of mass in elliptical orbits, just like the Earth and the Moon. More complicated binaries always consist of a hierarchy of pairs. Thus a triple star is a double in which one component is itself a much closer double—like the Earth and Moon going around the Sun—and a quadruple star is a double in which each component is a close double. The most complicated binary that is known has six components; it is a double double that has going around it a distant companion that is also double.

There is a simple reason why complicated binaries have this pairing hierarchy: the alternative is unstable. A pair is a stable configuration; and even if the members of the pair are themselves close pairs, this has no appreciable effect on the stability. But if three or more stellar bodies are at comparable distances from each other, the gravitational interactions will almost always throw at least one of the bodies out of the configuration completely.

Here binaries differ from star clusters. In a cluster the stars are not paired off. Each star circulates among all the others, and stars are indeed thrown out. This process depletes a star cluster, but for the richer aggregations it is gradual enough that they can persist for a great length of time. But only a binary configuration is truly stable.

Binary stars are observed in several different ways, and it is convenient for astronomers to classify the binaries themselves according to our method of observation, even though there is no essential difference between them. There are three basic classes. A visual binary is one in which, with sufficient telescopic magnification, we can *see* the component stars as separate points of light. A spectroscopic binary is one whose binary nature is shown by the periodic variation of radial velocity, as the two stars revolve around each other. An eclipsing binary is a pair whose orbit we look at nearly edge-on, so that the total light of the pair is periodically dimmed as one star passes in front of the other. Nearly all

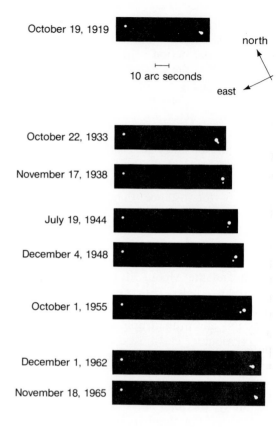

October 19, 1919

10 arc seconds

north

east

October 22, 1933

November 17, 1938

July 19, 1944

December 4, 1948

October 1, 1955

December 1, 1962

November 18, 1965

FIGURE 14.1.
This is the same sequence of photographs as in Figure 12.5, reproduced here to show the orbital motion of the binary star at the right. (From *Principles of Astrometry,* by Peter van de Kamp. W. H. Freeman and Company. Copyright © 1967).

spectroscopic or eclipsing binaries are unresolved single points of light, whose binary nature evidences itself only by the periodic variations in velocity or brightness (or both).

VISUAL BINARIES

Visual binaries tend to be stars that are close to us in space, since they will fall in the visual-binary class only if the two components can be seen to be separated in direction. A pair that is well-resolved at a distance of a few parsecs would blend hopelessly into a single blur if it were a thousand parsecs away. Similarly, visual binaries tend to be pairs whose real separations are large, because this also favors their being resolved.

Thousands of visual binaries are known. Most of them have been catalogued by simply examining one star after another and noting which ones are double. The eventual value, of course, is not in merely discovering a binary; it is in observing the binary for a long enough time—typically, for decades—to determine the period and the size of the orbit.

Finding the orbit is not a trivial calculation, since it is an ellipse of un-known shape and unknown orientation in space, but the problem can be solved.

One serious difficulty does stand in the way of fully analyzing a visual binary: to determine the masses we need to know the size of the orbit in linear measure, but the orbit size that we observe is angular. To con-vert it to a linear scale we need to know the distance of the star from us, and often this is only inaccurately known. Since the formula for stellar masses involves the cube of the orbit size, any uncertainty in distance is magnified in the calculation of the mass, and is often the limiting factor in our knowledge of the mass.

Although most binaries have to be examined carefully to distinguish them from single stars, a few nearby pairs have such wide separations that they look like quite separate stars, mixed into the general field. Their relationship to each other is recognized only by the fact that they are moving in unison through space. Nearby stars tend in general, of course, to have large proper motions that stand out; and when two stars moder-ately near each other in the sky have large proper motions that are iden-tical, both in size and in direction, then the conclusion is unmistakable that they are gravitationally bound to each other and stay together always. The search for common-proper-motion pairs, as these are called, has turned up some of the lowest-luminosity stars known, which are companions to brighter stars in our neighborhood but would never have been noticed if their proper motions had not indicated the relationship. Because these stars are otherwise so inconspicuous, knowledge of them would have been very hard to fill in in any other way.

SPECTROSCOPIC BINARIES

Almost as soon as spectra of the stars began to be photographed, periodic changes in the radial velocities of some stars were noticed. The most striking cases are those in which two equal stars revolve rapidly around each other—say, at 100 kilometers per second in a period of a few weeks. The single point of light that we see is the combined light of the two stars, and the spectrum is likewise the sum of the two spectra. But at the times when one star is moving toward us in its orbit and the other is moving away, we see one set of spectral lines Doppler-shifted toward shorter wavelengths and one set shifted toward longer wavelengths. As a result the spectral lines look double. Then at times when the orbital motion is across the line of sight and produces no Doppler effect, the two spectra exactly superimpose and the lines are single.

Not all spectroscopic binaries show double lines. In some pairs one star is much fainter than the other, and its spectrum contributes too little to be seen separately in the combined total. The orbital motion of the

256

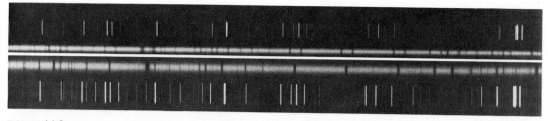

FIGURE 14.2.
These two spectra of the same binary star were taken at different times in the orbital cycle. When the upper spectrum was taken, one star was moving toward us and the other away from us; the lines of the two stars are well-separated by their Doppler shifts. At the time when the lower spectrum was taken, the motion of both the stars was across the line of sight. There was no Doppler shift then, and the lines blend into a single spectrum. (Lick Observatory.)

brighter star still shows up, however, as a periodically varying Doppler shift of its lines.

In both situations we can measure how the velocity changes with time and calculate many of the characteristics of the orbit. Unfortunately, one ambiguity remains and cannot be avoided. Because the Doppler effect is caused only by motion along the line of sight, our knowledge is limited to that part of the motion. As a result we can find only the line-of-sight extent of the orbit; and its total extent—which is what we need to know to calculate the masses—remains unknown. The ambiguity can

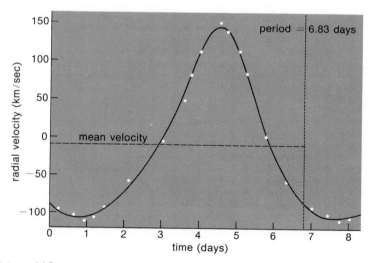

FIGURE 14.3.
The radial-velocity curve of the spectroscopic binary Beta Scorpii. In this spectrum, the lines of the fainter stars are seen only weakly; so only the brighter star has a well-determined velocity curve. The dots represent individual measurements of velocity, whereas the curve shows velocities in the elliptical orbit that fits best.

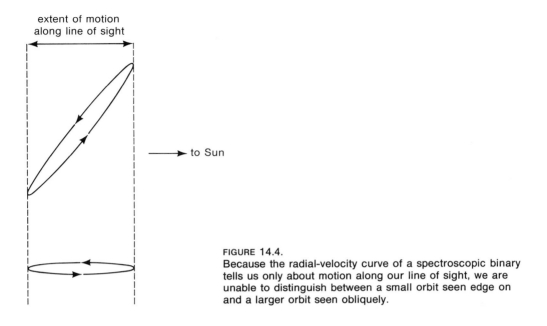

extent of motion
along line of sight

→ to Sun

FIGURE 14.4.
Because the radial-velocity curve of a spectroscopic binary tells us only about motion along our line of sight, we are unable to distinguish between a small orbit seen edge on and a larger orbit seen obliquely.

be removed, and the masses found, only if the spectroscopic binary can also be observed as some other kind of binary. Spectroscopic binaries that are also visual binaries are unfortunately quite rare. The former tend to be the pairs with high velocities, where the Doppler shifts are large enough to measure easily; but in accordance with Kepler's third law, these are the closer pairs, which are the least likely to be visually resolvable. Many spectroscopic binaries eclipse, however, as we shall see next.

ECLIPSING BINARIES

Stars are opaque bodies; so when one star passes behind another its light fails to get through. Thus when two stars have an orbit that we happen to see nearly edge-on, they periodically eclipse each other. Even though we never see anything but the single point that is the combined light of the pair, this total light dims every time that an eclipse takes place. By plotting the brightness of the star against time—in what is called a light curve—we can clearly "see" the eclipses. From analyzing stellar eclipses, we can find out more facts about stars than we can get in any other way.

First of all, with a little thought and by drawing some simple diagrams, one can see that the light curve of the eclipse tells a great deal about the sizes of the stars, relative to each other and relative to the size of the

FIGURE 14.5.
The light curve of the eclipsing binary
Algol (Beta Persei). The deep
minimum occurs when the smaller,
brighter star is eclipsed. Note the
small secondary minimum, when the
bright star hides part of the surface
of the faint star.

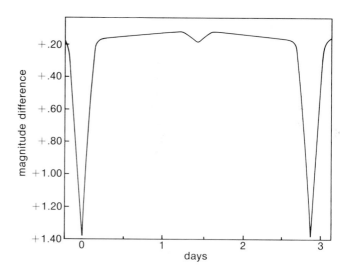

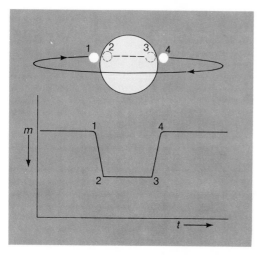

FIGURE 14.6.
A small star going behind a large one has a brief
partial phase and a long total phase.

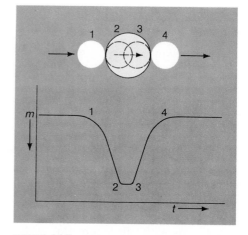

FIGURE 14.7.
Two stars of nearly equal size can have
only a brief total phase.

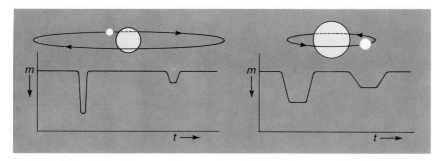

FIGURE 14.8.
If the sizes of the stars are small compared with that of the orbit, they spend only a small fraction of their time in eclipse; if their radii are a larger fraction of the orbit size, they spend a larger fraction of their time in eclipse.

orbit. Complications are caused by the unknown inclination of the orbit to the line of sight, but the inclination affects the shape of the light curve somewhat differently from the other unknowns, and with accurate enough observations of the light curve the whole problem can be solved unambiguously.

It is a clever trick to solve the light curve of an eclipsing binary, but the result is not really satisfying in itself. The radii of the two stars can be found, but only in relation to the size of the orbit, which is itself completely unknown. As with a spectroscopic binary, a full solution of the observations leads only to a partial astronomical answer.

However, these two unsatisfying halves—the spectroscopic and eclipsing observations—can be put together into a very satisfying whole, provided we have both sets of observations for the same star. The analysis of the eclipses gives the orbital inclination. Knowledge of this removes the ambiguity of the spectroscopic result and tells us the size of the orbit. Kepler's third law, with the observed period, then gives the sum of the masses. If both sets of spectral lines are observed, we know how much each star moves and thus can find the individual mass of each star. Finally, applying the known orbit size to the eclipsing results, we can find the actual radius of each star.

Combined spectroscopic and eclipsing observations are, in fact, often available. Not all spectroscopic binaries are oriented in such a way as to show eclipses, but any eclipsing binary is sure to be observable as a spectroscopic binary, since the edge-on aspect guarantees that we will see a large motion to and fro along the line of sight. Eclipsing binaries are our most valuable source of data about fundamental properties of stars.

MASSES AND RADII OF STARS

Good measurements of masses and radii of stars are still relatively few—
dozens, rather than the thousands by which we count stellar motions
and parallaxes. The results show such strong regularities, however, that
these few data can be readily combined with our theoretical insights to
give a broad understanding of the nature of the stars.

It takes only a glance at the known stellar masses to see that they correlate strongly with the luminosities of the stars. Intuitively, it is clear
that in general the more massive stars should be brighter, but what is
impressive is the tightness of the correlation. Astronomers speak of the
mass-luminosity relation, in which one quantity actually determines
the other.

Most stars obey the mass-luminosity relation, but a few do not. Some
red giants fit poorly, and the white dwarfs are a glaring exception. They
have nearly as much mass as the Sun, but are hundreds of times less
luminous.

The way in which we understand the regularities here and resolve
the contradictions is an excellent example of how an interplay of theory
and observation makes science work. First, the existence of a simple
relation between mass and luminosity is a direct conclusion from the
theory of stellar structure. This will be discussed in some detail in Chapter 18; briefly stated, material that is put together to make a star settles
down into a structure that depends only on how much mass is put in.
The same amount of material will always produce the same kind of star,
with all its properties the same, including the luminosity. Thus it is easy
to see that there ought to be a mass-luminosity relation. But the very
argument that we use—that stars should have a unique structure—makes
it clear that the conclusion should apply only to the "normal" part of a
star's life, on the main sequence of the HR diagram. We have seen that
the other sequences and branches in the HR diagram result from changes
of internal structure, due to the exhaustion of the star's original fuel
supply. Thus there is no justification for applying the simple theoretical
argument about mass and luminosity to the complicated structure of
an evolved star. We now sharpen our statement about masses and luminosities: for main-sequence stars there is a simple, close relation between
mass and luminosity, but for stars that are off the main sequence the
simple relation does not apply.

The radii of the stars also show great regularity, and this relationship
has no serious complications. The radii correlate very closely with position in the HR diagram, so closely that we can draw lines through the
HR diagram that mark out the sizes of the stars. At the upper right, stars
have very large radii; an extreme red supergiant is large enough to engulf
the orbits of most of the planets around our Sun. At the lower left of
the HR diagram, the radii are very small; a white dwarf is hardly larger

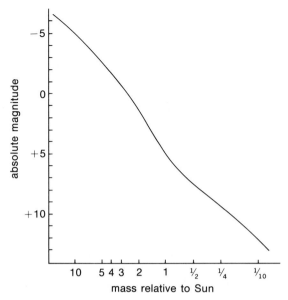

FIGURE 14.9.
The mass-luminosity relation for main-sequence
stars. (The luminosity scale is expressed in absolute
magnitude.) Luminosities of stars go approximately
as the third or fourth power of their masses. Red
giants tend to lie above the main-sequence curve,
and white dwarfs lie much below it.

than the Earth. We shall see in Chapter 16 that this close tie-in of the
radius of a star with its position in the HR diagram is a direct consequence of the way in which the surfaces of bodies radiate; it does not
even involve the internal structure of the star.

MORE ABOUT ECLIPSES

Careful study of stellar eclipses can tell us even more about some stars.
Consider, for instance, the case of a small star that passes in front of a
large one, in what is called an annular eclipse. The small star can obscure no more of the large one than its own area covers, and there would
thus be a stage of constant minimum light—if the disc of the larger star
were uniformly bright all over. It is not, however; the center, where we
see light coming out perpendicular to the star's surface, is brighter than
the edge of the disc (sometimes called the limb), where we see light that
is emitted at a grazing angle to the surface. This "limb-darkening" causes
the annular phase of an eclipsing-star light curve to have a gradual dip
in the middle. There are thus cases in which limb-darkening can be

observed directly, even though the star is only an unresolved point of light, and this knowledge is of great help in understanding the detailed way in which radiation filters up through the atmosphere of a star.

Another problem that leads to unexpected fruits is the way in which stars distort each other tidally when they are close together. The distortion can be seen in the light curve. We see the stars from a changing direction as they go around, and an elongated star does not look equally bright from all directions. Thus there is a small but measurable change in brightness even when no eclipse is going on.

This distortion is not of great interest for its own sake, but it allows us to probe the interiors of stars in a unique way. A spherically symmetric star attracts exactly like a point mass at the center of the star, but a distorted star does not. As a result, two distorted stars that revolve around each other do not follow a simple, permanent elliptical orbit, in the manner of Kepler's first law. What happens is that the orbit gradually turns in its own plane, so that the direction of the major axis (which is called the line of apsides) slowly turns. At one time it will be pointing toward us, and decades later it will have turned across the line of sight. When the orbit is eccentric enough, this orientation shows up as an unequal spacing of the two eclipses in the light curve. The rate of apsidal motion depends, on the other hand, on how distorted the stars are, and this depends in turn on the extent to which the material of a star is evenly spread or is concentrated toward the center. Thus apsidal-motion studies, in the few cases where this effect can be observed, allow us to probe the internal densities of stars by means of observations that we cannot make even for the Sun.

Binary stars are the source of nearly all our information about masses of stars, and most of our observational data on radii also come from them, along with glimpses of such esoteric details as the distribution of brightness over the surface of a star and the distribution of density in its interior. In the metaphor of astronomy as a detective story, eclipsing binaries represent what is perhaps the cleverest chapter. An astronomer studies a speck that his most powerful telescope cannot show as anything more than a single point of light, and from the behavior of this light he shows that this bright point is really two stars, whose masses, sizes, and motions he is then able to figure out. And, as in the detective story, all is clear after he explains how he figured it out.

REVIEW QUESTIONS

1. Why are binaries worth studying? What basic facts do they tell us about stars in general?

2. How common are binaries? How do we know?

3. What important stellar data can be obtained from observations of a visual binary? Where do the uncertainties arise?

4. Why do we get only incomplete information from analyzing observations of a spectroscopic binary? Under what circumstances can the uncertainty be resolved?

5. Why are eclipsing binaries so important? Why is it essential to have spectroscopic observations of them?

6. Why are most visual binaries nearby stars of large separation? Why are most spectroscopic and eclipsing systems close pairs?

7. Explain how limb-darkening can be studied in an annular eclipse.

8. What is apsidal motion? What causes it? Why is it important?

15 | VARIABLE STARS:
INCONSTANCY AS A VIRTUE

Not all stars are simple and single; likewise, not all stars shine with a constant brightness. We have just seen how stellar eclipses can cause brightness changes, but in this chapter we shall look at changes of brightness that take place within the star itself. As with binaries, we pay attention to variable stars not only "because they are there," but especially because of the broader contribution that they make to an understanding of the universe. This understanding has two parts. First, some types of variable stars are useful to us, particularly for measuring astronomical distances. Second, certain variable stars can be identified with a particular stage of stellar evolution, and therefore offer a clue about what the star is doing at that stage of its development.

Thousands of variable stars have been catalogued. Some were noticed accidentally, but most variables have been found in systematic surveys in which an astronomer compares photographs taken of the same region of the sky at different times. Just as in the search for stars of large proper motion, the survey is carried out with a blink comparator, which enables the observer to look alternately at identical parts of two different photographs. The constant stars look the same on the two photographs, but a variable star seems alternately to brighten and to fade, as the observer "blinks" from one photograph to the other.

Many decades ago astronomers adopted a system in which each variable star is named by one or two lettters followed by the name of the constellation that the star is in: U Geminorum, RR Lyrae, etc. In many constellations, however, so many variables have been discovered that the capacity of the lettering system has been exhausted, and new variables are simply assigned V-numbers, such as V1057 Cygni.

Stars vary in brightness for many reasons, but most variable stars belong to two broad classes: pulsating and explosive. A pulsating star is unstable in size; it alternately swells and shrinks, usually in a regular, periodic way. With the changes in size go changes in brightness. The explosive variables are even more unstable; their variation results from an explosive release of energy. Some stars are able to go through this performance repeatedly, but for some others the result is catastrophic.

PULSATING STARS

Most stars are in complete balance throughout their interiors, but in a few types of stars there is a layer just below the surface that oscillates, alternately storing up energy and releasing it. The surface swells and shrinks, and the star gets alternately brighter and fainter. The oscillation is fed by the star's internal store of energy, and continues for as long as the star remains in that stage of its life.

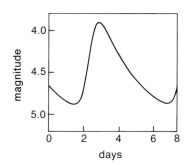

FIGURE 15.1
The light curve of Delta Cephei.

The best-known and best-studied group of pulsating stars is the cepheids, which are named after the first star of this type that was ever recognized as a variable: Delta Cephei. (The brightest stars, already known by a Greek letter and a constellation name, are not given a new name if they are discovered to be variable.) Delta Cephei, which varies by about a magnitude, repeats its changes again and again with a period of five days. Periods of other cepheids range from a little over a day up to more than forty days. Some have amplitudes of variation greater than that of Delta Cephei and some vary rather less, but all have light curves of a generally similar shape.

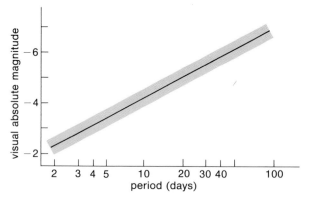

FIGURE 15.2.
The period-luminosity relation for cepheid variables. The relation is not absolutely precise; individual stars scatter throughout the shaded band. Nevertheless, cepheid distances are among the most reliable in stellar astronomy.

The property that has made the cepheid variables famous—and extremely useful—in astronomy is their *period-luminosity relation*; the length of the period tells us how great is the luminosity of the cepheid, that is, what its absolute magnitude is. From the apparent magnitude we can then calculate how far away the star is from us. Because so much of astronomy depends on measuring distances, stars of known absolute magnitude are real treasures. What makes the cepheids particularly valuable is that they are quite luminous stars and can be seen far away, and they occur in other nearby galaxies. Furthermore, their variation in brightness makes them easy to recognize as cepheids—indeed, it even makes them easy to pick out in the first place, among thousands of equally faint stars. As a result, cepheid variables are one of the most important links in the chain of distances by which astronomers have laid out the structure of the universe.

The period-luminosity relation is an excellent example of a tool that has taken a long sequence of efforts to develop. Cepheids are rare stars, scattered here and there in the universe; how did astronomers ever find this astonishing simplicity among them? The question actually has two parts. First, how do we know that there is such a clear-cut connection between period and luminosity? Second, how do we determine what the absolute magnitude is for a star of a given period?

The fact that a period-luminosity relation must exist became clear to astronomers as soon as they looked carefully at the Magellanic Clouds, which are our nearest neighboring galaxies, about 60,000 parsecs away. In each of these galaxies are hundreds of cepheids, and in a graph of

FIGURE 15.3.
The Small Magellanic Cloud. Among its individual stars are hundreds of cepheid variables. (Cerro Tololo Inter-American Observatory.)

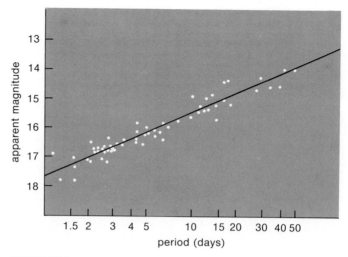

FIGURE 15.4.
The relation between magnitude and period for cepheids in the
Small Magellanic Cloud. (After H. C. Arp, *Astronomical J.*,
65, (1960), 426.)

period against apparent magnitude they all lie close to a neat line. If
the apparent magnitudes are a simple function of period, so then are the
absolute magnitudes, since all these stars are at the same distance from
us, and their apparent magnitudes must all differ from their absolute
magnitudes by a single number that depends on that distance. If we can
find the distance of the Magellanic Clouds from us, we will then know
the relation between period and *absolute* magnitude. Thereafter we can
use this relation to determine the distance of any cepheid variable that
is observed anywhere in the universe.

In actual practice it is not easy to measure the distance of the Magellanic Clouds in an independent way; what we do instead is to determine
the absolute magnitudes of a few cepheids in the Milky Way around us.
We then assume that the same relation between periods and absolute
magnitudes applies in the Magellanic Clouds as applies here. This operation, known to astronomers as "finding the zero point," puts the richly
calibrated period-luminosity relation of the Magellanic Clouds onto a
scale of absolute magnitudes and thus lets it serve as a yardstick to the
universe.

Finding the absolute magnitudes of even a few nearby cepheids is not
an easy job, however. The trouble is that they are not near enough to
us. Even the closest cepheid, Polaris (the "North Star"), is more than
100 parsecs away, and at that distance its trigonometric parallax is too
small to give an accurate distance. Even statistical parallaxes, based on
the proper motions of cepheids, give a poor result because the motions

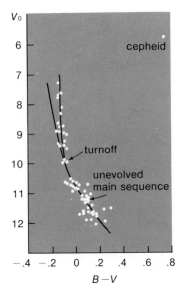

FIGURE 15.5.
Color-magnitude array of the open cluster NGC 6087, from observations by J. D. Fernie. The magnitudes and colors have been corrected for interstellar absorption and reddening (see Chapter 19). The solid line is the zero-age main sequence, the fit to which allows apparent magnitudes to be converted to absolute magnitudes. The absolute magnitude of the cepheid is then known. (After J. D. Fernie, *Astrophysical J.*, **133** (1961), 6. The University of Chicago Press.)

are so small. Our best absolute magnitudes come from a few cepheids that are members of open clusters. In a cluster we can plot the color-magnitude array of the stars, mark out the main-sequence stars, and assign to them the absolute magnitudes that go with the standard main sequence in the HR diagram. The absolute magnitude of the cepheid in the cluster can then be found by comparison with the main-sequence stars. Only a few such cases are known, because cepheid variables are rare, but on them rests the zero point of the period-luminosity relation, and hence the distance scale of the universe.

Thus astronomers have put together the facts about cepheid periods and luminosities. It is quite a different problem, however, to understand why these stars behave as they do. Here we can imagine successive levels of understanding. First, the way in which a star pulsates can be studied without considering the deeper question of why it pulsates. Just as a bell has a natural frequency with which it vibrates when struck, so too a star has a natural oscillation frequency. If it is compressed, it will respond by expanding; the expansion overshoots and a contraction follows; and so it goes, in periodic repetition. The length of time that such a cycle takes depends on the characteristics of the gas and on the strength of the star's gravity. A theoretical analysis of the oscillation shows, in fact, that its period should depend mainly on its mean density, that is, mass divided by volume. Thus the theory says that there should really be a period-mass-radius relation; the reason that we see it as a period-luminosity relation is that the directly concerned physical quantities, mass and radius, are closely correlated with the luminosity of the cepheid, which is the thing that we more easily observe.

This kind of analysis can be carried out for any star, and we can calculate the period with which it would pulsate *if* something were to "excite" it. The more difficult problem is to understand why stars pulsate at all; what is it, inside a cepheid variable, that keeps the pulsation going? One fact is clear: the mechanism must be one that reduces the normal tendency of a compressed gas to dissipate energy and to bounce back less than fully. In fact, to create a pulsation we need a mechanism that will overrespond, so that a compressed gas will extract energy from its surroundings and rebound even more vigorously. Although the details are by no means completely understood, astronomers have a fairly good notion of how this happens inside a pulsating star. The key is in the ionization of hydrogen and helium, the atoms that make up most of the material of the star. Ionization is the separation of electrons from an atom; they are held to the atom by considerable force, and it takes energy to remove them. Conversely, an ionized atom that catches an electron gives up energy as the electron gets reattached. Deep inside a star, the atoms of hydrogen and helium are all ionized, whereas at the surface they are nearly all neutral, that is, they possess their normal number of electrons. Somewhere below the surface is a layer in which they may be either neutral or ionized, depending on the immediate conditions of temperature and pressure. Such a gas has an unusual ability to store and release energy, as it ionizes and recombines, and the degree of ionization also has a potent control over the flow of the large amount of radiant energy coming up from the interior. If the ionization zone is neither too deep in the star nor too close to the surface, all these factors combine to store and release energy at just the right stages to maintain a pulsation.

Since a pulsating star must have this particular kind of structure, it is not surprising that pulsation is confined to a narrow region of the HR diagram. The cepheids inhabit an "instability strip" that extends upward through the yellow giant and supergiant region. The evolutionary tracks of massive stars pass through this region after they have left the main sequence; and any star that happens to be within the instability strip in the HR diagram pulsates, and is thus a cepheid, until it has evolved out of that region. A star's evolutionary track may carry it back and forth several times across the instability strip; during each crossing it is a cepheid. The chief difference between cepheids is that the more massive ones, originating higher on the main sequence, cross the instability strip at a higher luminosity. Their radius at that evolutionary stage, and their mass, are such that the pulsation period is correspondingly longer: hence the period-luminosity relation that is so precious to the observers.

Other populations have their pulsating stars too; all that is required is that the star pass through a stage in which its structure contains this

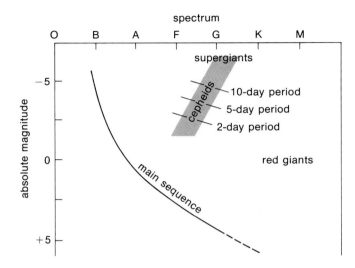

FIGURE 15.6.
The cepheid instability strip
in the HR diagram.

FIGURE 15.7.
In this photograph of the globular cluster M55, three of the RR Lyrae stars are
marked. They were discovered by comparing photographs taken at different
times. (Harvard Observatory.)

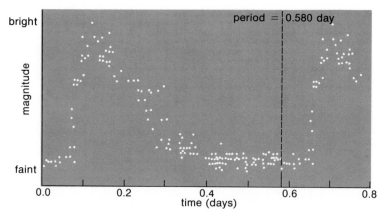

FIGURE 15.8.
The light curve of the star marked "1" in Figure 15.7. Each dot
represents a magnitude estimate made on a different photograph. After
the period was found out, from analysis of the observations, each
observation was plotted according to its position in the repetitive cycle.
The photographs were taken over a span of 54 years.

instability. Thus the cepheids belong to a young population, but the
old Population II of the galactic halo has stars that behave in a similar
way. They are called RR Lyrae stars, after the first such variable star
to be studied. RR Lyraes can be found all over the halo; but their properties are best studied in globular clusters, where we see them in immediate association with stars of many other types and can thus study
their relationship more easily. Most globular clusters have RR Lyrae
stars, and they can be picked out with relative ease by blinking several
photographs of the same cluster and looking for variability.

   When RR Lyrae stars are plotted in the HR diagram of a globular
cluster, their nature instantly becomes clear. They are on the horizontal
branch; in fact they *are* a region of the horizontal branch. Within a well-
defined segment of that branch, every star is an RR Lyrae, and no RR
Lyrae star lies outside that region. (This part of the horizontal branch
is, in fact, often misnamed the "RR Lyrae gap," by observers of HR
diagrams who, to avoid the effort of measuring the variable stars, simply
leave them out.) What clearer evidence could there be that the pulsation
is just a consequence of the structure that a star has at that particular
stage of its evolution? To make the relationship even clearer, the RR
Lyraes lie just where the cepheid instability strip would cross the horizontal branch if the instability strip were extended downward. Even
though the RR Lyraes are of a different population, a different age, and
a different mass from the cepheids, they fall within that general range
of structure where pulsation sets in, and pulsate they do. Compared with

cepheids, however, RR Lyrae stars have a much higher density, and this makes them pulsate faster. A typical period is half a day.

The RR Lyrae stars have been introduced here in their population setting in globular clusters, where it is clear from the HR diagrams that their absolute magnitude is between 0 and $+1$. Historically and logically, however, the relationship goes the other way. There is no other easy, direct way to measure the distances of globular clusters and, therefore, the absolute magnitudes of the stars in them. In actual practice it is the RR Lyraes themselves that are the key. We find out the absolute magnitude of RR Lyrae stars as a class, by studying the ones in the general field in the solar neighborhood, and then we use the RR Lyraes to measure the distances of the globular clusters. Like the cepheids, no RR Lyrae star is close enough for a reliable trigonometric parallax (the closest, RR Lyrae itself, is over 100 parsecs away), but their motions are large and a fairly good statistical parallax can be found. There is no question of a period-luminosity relation here; the location of the RR Lyraes on a *horizontal* branch in the HR diagram shows that they all have nearly the same absolute magnitude.

Just as the cepheids are the best distance indicators in a young Type I population, so the RR Lyrae stars are the best distance indicators in Population II. Unfortunately, being much less luminous than the cepheids, they can not be seen out to as great a distance, but they do allow the positions of globular clusters to be plotted in the galactic halo, and the unattached RR Lyraes of the field are our best probe of the extent of the halo itself.

The RR Lyrae stars serve one other purpose: they are a sensitive population indicator. The Type II population is by no means homogeneous. Different globular clusters show notable differences in the shapes of their HR diagrams, and their RR Lyrae stars also differ, in period-lengths, shape of light curve, etc. The cause of the underlying differences has not yet been tracked down, although it is probably in the chemical abundances of the material out of which the stars are made. In any case, the RR Lyrae stars contribute one of the sensitive details in the factual picture.

In addition to the RR Lyrae stars, some globular clusters have a few pulsating stars that mimic the "classical" cepheids of the young Type I population. They are called Type II cepheids, to distinguish them. They also have a period-luminosity relation, but for a given period a Type II cepheid is about 1.5 magnitudes less luminous than the corresponding classical cepheid. The difference should not be surprising. The Type II cepheids are subject to the same ionization-zone driving mechanism, and they therefore pulsate, but they are much less massive stars than the classical cepheids; so they have different natural periods. In fact, we do not know exactly what stage of evolution the Type II cepheids are at, but their evolution has brought them into the instability strip and thus they pulsate.

Another important class of pulsating stars falls among the M stars, in the giant-to-supergiant range. These stars pulsate slowly, with periods of a year or so; they are called, simply enough, long-period variables. The visible light of a long-period variable changes by a large amount, 5 magnitudes or more. They are cool stars, however, with the bulk of their radiation at infrared wavelengths, and that part of the radiation does not vary nearly as much.

The best-known long-period variable is Omicron Ceti, which is more commonly known by its old Latin name Mira, meaning "wondrous." Hence the long-period variables are also called "Miras."

Long-period variables make themselves conspicuous, and we see them out to great distances because of their high luminosities, but they are actually rare stars in space. As a result we know relatively little about them, beyond the fact that they are stars at a late stage of evolution. One fact is clear: they occur in both stellar populations. Type I Miras lie in the galactic disc and have low velocities; they are probably a late stage in the evolution of a star of several solar masses. Unfortunately none is known in a star cluster, presumably because stars of this type are rare. If we knew some Miras in clusters, we would know much more than we do about their pedigree. Type II Miras occur as high-velocity stars around us, and some are also found in globular clusters, where they are at the bright tip of the red-giant branch. The similarity between Type I and Type II Miras may again be a case of mimicry, however; they are certainly stars of very different ages and masses. Unfortunately, we do not know what physical mechanism is responsible for their pulsation. Ultimately Miras may be another key to stellar populations and evolution, but today they are still a puzzle.

The long-period variables do not repeat their light variations quite as regularly as the cepheids, but they are still reasonably regular. Not so the stars higher up in the HR diagram. Practically all the cooler supergiants of spectral class M are somewhat variable, some of them in a quite

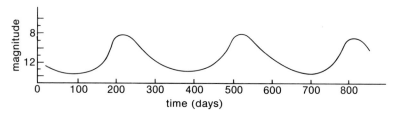

FIGURE 15.9.
The light curve of a typical long-period variable. The variation repeats approximately, but not exactly, from cycle to cycle.

erratic way. We know even less about the nature and cause of their variation than we do about the Miras. These stars are intriguing, nevertheless, because some of them show evidence of a significant rate of mass loss, which may be an important general phenomenon in stellar evolution.

EXPLOSIVE STARS: NOVAE AND THEIR KIN

In another broad class of variable stars, the cause of the variation is an explosive event. Typical is the *nova* (Latin, "new"; plural, *novae*), named for the fact that a new star suddenly appears where none was seen before. The star really was there before, but it was faint; what has happened is a tremendous and rapid brightening. A typical nova flares up by about 12 magnitudes, which means 60,000 times, in a few days. It remains bright for a few weeks and then gradually fades over the months. The final return to its original brightness takes decades.

It is certainly proper to refer to a nova outburst as an explosion. Doppler-shifted lines in the spectrum show expansion velocities of 500 or 1,000 kilometers per second; what happens is that the star suddenly blows its surface layers right off. The material is driven off into space in all directions, with the underlying star quieting down only slowly afterward.

As violent as it may sound, however, a nova explosion is only a superficial thing for the star. Novae have been observed before the explosion (by checking old photographs of the region) and followed carefully after-

FIGURE 15.10.
Two photographs of Nova Herculis 1934. One was taken in 1934 when the nova was near its brightest, the other 14 months later, when it had subsided almost to normal brightness again. (Lick Observatory.)

276

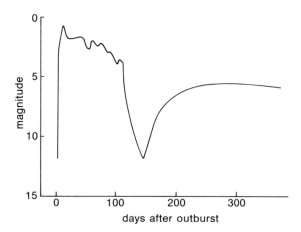

FIGURE 15.11.
The light curve of Nova Herculis 1934.

ward, and it is clear that the star eventually settles down to its original magnitude again. Furthermore, the amount of material thrown out, which can be estimated from the spectral lines that show the expansion velocities, is only an inconsequential fraction of the mass of the star.

Even so, the explosion of a nova involves a large release of energy, which must have a cause. As is usual in astronomy, the immediate question is the specific process, but our deeper objective is to understand what sort of star a nova is and what its development can tell us about the behavior of stars in general. First of all, what kind of star becomes a nova? The first part of the answer comes from measuring the color of a nova long before or long after the explosion; it is a blue star. The next step is to find its absolute magnitude. Again the distance measurement

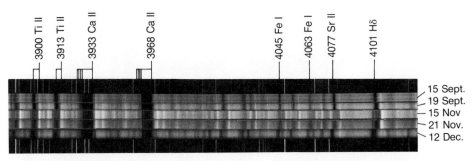

FIGURE 15.12.
Spectra of Nova Delphini 1967. Note the changes in the spectra, and the complex structure of the lines, some of which are marked by wavelength and element. (Ca = calcium; Fe = iron; H = hydrogen; Sr = strontium; Ti = titanium.) In the multiple dark lines that are marked, each component comes from a shell moving outward with different velocity. (Lick Observatory spectra, courtesy of L. V. Kuhi.)

is difficult; in this case the best measurement probably comes from observing nova outbursts in other galaxies whose distances we know. Distance measurements for novae in the Milky Way have been made in less direct ways and give a similar answer, though. From knowing the maximum luminosity of a nova and the amount by which it has brightened, we deduce that the faint, quiescent stage has an absolute magnitude of around +5. Thus novae inhabit the sparse region of the HR diagram that lies between the main sequence and the white dwarfs.

But the real key to our understanding of novae—such as it is—was the discovery that novae are close binaries. That is, the nova is really a pair of stars, which revolve around each other nearly in contact. Novae have now been observed as spectroscopic binaries, and some have even been observed to eclipse. In the picture that emerges, the star on which the explosion occurs is a dwarf of high density, whose lifetime as a normal main-sequence star has already ended, whereas the companion is a main-sequence star of more normal size. The key to the outburst probably lies in the nature of this configuration. On the one hand, the dwarf is made up of compressed, burned-out material of the type that (as we shall see in Chapter 18) is unstable in the presence of new fuel. The companion, on the other hand, is made of material of normal composition, and it is so close to the dwarf that tidal forces draw out material from the companion and cause it to spill into the dwarf. The details are still vague, but the mixture is potentially explosive enough to explain a nova outburst.

There is an intriguing possibility that novae are related to a whole class of binary stars in which interaction between the pair of stars diverts the normal course of their stellar evolution. Within the class of explosive variables are others that resemble novae but have a less violent outburst that is repeated with some frequency. The "recurrent novae" are so called because they repeat their outburst every decade, or every few decades. The "dwarf novae," or U Geminorum stars, have a much less violent flare-up (typically, 4 magnitudes) several times a year. All are close binaries.

These systems all contain a highly evolved star; if we are to trace their past development—how they got that way—we should look for possible

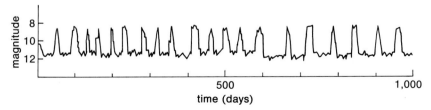

FIGURE 15.13.
A typical light curve of a U Geminorum star.

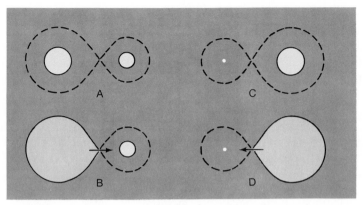

FIGURE 15.14.
Stages in the evolution of a close binary. A. Neither star has yet
begun to evolve. The dashed line shows the "sphere of influence"
within which each star is capable of holding material gravitationally.
B. The larger star has evolved to the point where it has filled its
limiting surface and begun spilling material onto its companion.
C. Stage B has ended; the star at the left has ended its evolution as
a small, high-density dwarf, while the other star has become larger
because of the material it has acquired. D. The second star has
evolved and filled its limiting surface; material now begins to spill
onto the high-density dwarf, where it can presumably lead to nova
explosions.

progenitors from which these explosive stars could have evolved. What
were they in their earlier stages? Excellent candidates exist. The general
lists of variable stars include a large number of eclipsing binaries that
consist of two main-sequence stars going around almost in contact with
each other. They are called the W Ursae Majoris stars, named, as usual,
after the first one to be found and studied. The question that interests
us here is, how will the member stars of a W Ursae Majoris binary
evolve?

The thing that is different about a close binary is that its components
exert strong tidal forces on each other. Around each star is a limiting
surface beyond which it cannot hold onto material at all. Somewhere
on the line between the two stars, the limiting surfaces meet at a balance
point. On one side of this point, material will fall into one of the stars;
on the other side it will fall into the other star.

Initially each of the stars is smaller than its limiting surface; and
except for a small tidal stretching, each star has an equilibrium and a
structure just like that of a single star. As stars always do, the more massive member of the pair will use up its fuel faster, converting its central
hydrogen to helium, and eventually it will reach the end of its main-
sequence life. The next stage of stellar evolution calls for its surface to
become larger, and here is where it begins to differ from a single star.

As soon as the actual surface of the evolving member of the pair becomes as large as its limiting surface, its evolution can no longer proceed like that of a single star. Instead of growing larger—which it cannot, because of the tidal forces—it instead spills material over the balance point into the other star. Evolution continues at its center, however; the star continues trying to grow and succeeds only in transferring more material to its companion. Meanwhile the central burned-out region, in which the hydrogen has all been turned to helium, increases; the equilibrium of the star adjusts to make this core denser. One can presume that the final result is a dense burned-out core, and that the star loses the remaining material of its envelope in some way, just as astronomers believe that a single star does, at the last stage of evolution.

We now have a binary with one dense component and one normal star. The normal star will eventually evolve too, and in fact its evolution will be speeded up by the increased mass that it has acquired from its companion. When it too begins to expand, it will soon reach its limiting surface and begin spilling material back into the dense star. This may well be the stage at which nova outbursts take place.

This picture of the evolution of close binaries is highly speculative, but this is the kind of responsible speculation on which scientific hypotheses are built. It begins with an observed fact, the existence of close binaries, and asks, how *must* they behave? Two stars close together must have tidal limits, and an evolving star must grow. The uncertainties begin to come in when we try to predict just how a star will behave when it begins to spill over its limits and how its companion will behave when matter is dumped into it. This is the point at which more detailed calculations need to be made, and beyond which the picture is extremely uncertain without the results of those calculations.

SUPERNOVAE

In another class of stars, a much more violent explosion takes place. Because of a superficial resemblance to novae, these stars have been named supernovae, but they actually represent a quite different kind of event. A supernova really blows itself up. The energy released in the explosion is comparable to the energy radiated by an ordinary star in its whole lifetime, and only a small remnant of the supernova is left behind after the explosion.

We know rather little about supernovae, however, because they are so rare. There is probably only about one supernova explosion per century in the whole Milky Way, and most of these occur in distant places, obscured by dust clouds, and are not seen. No supernova has been observed in the Milky Way for more than three centuries. Our knowledge of supernovae comes from observing them in other galaxies, along with

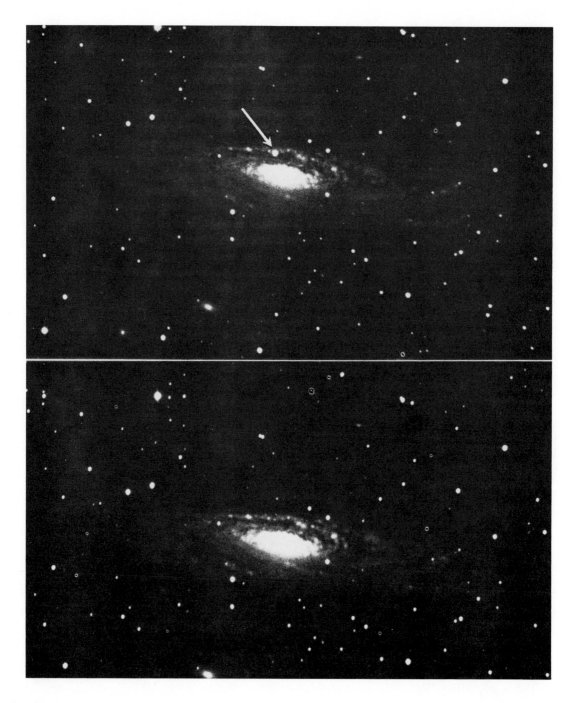

FIGURE 15.15.
In the upper one of these two photographs is a supernova, which appeared in this nearby galaxy in 1959. (Lick Observatory.)

what we learn from studying the remnants left by past supernovae in our own Milky Way.

At the peak of its outburst a supernova becomes extremely luminous; the brightest ones reach an absolute magnitude of -20. This is some ten magnitudes brighter than any normal star. It is a luminosity that is more appropriate for a whole galaxy than for a star; a supernova may, in fact, temporarily outshine its whole parent galaxy.

In attempting to understand supernovae, the first step is to marshal the observed facts. First, the light curve. As with ordinary novae, the period of maximum brightness is measured in days or in weeks, and the decline in months. A more careful examination of the light curves shows that there are at least two distinct types of supernovae. Those that are called Type I settle down, after their initial peak of brightness, into a steady decline of about half a magnitude or so per month. Type II supernovae, on the other hand, do not become as bright as those of Type I, and they decline faster.*

The difference between supernovae of Types I and II is also apparent in their spectra. The Type II supernovae have spectra like those of ordinary novae, but the lines quite understandably indicate higher velocities of expansion. The spectra of Type I supernovae are even more extreme. The lines are so broadened, by Doppler shifts in material moving toward and away from us, that they merge into each other and make the spectrum hard to interpret in detail; but the picture of extreme violence is unmistakable.

In their efforts to gather information on supernovae, astronomers have surveyed many galaxies, blinking photographs taken at different times. More than 300 supernovae have been found over the years. From their statistics it can be estimated that the largest galaxies, those a few times more massive than the Milky Way, each produce several supernovae per century. Smaller galaxies have fewer supernovae, simply because they have fewer stars.

But the basic questions still remain: what kinds of stars become supernovae, and why? The first real clue to this comes from seeing where they occur. Supernovae of Type I are found in all types of galaxies, including those that have only an old stellar population. Thus they appear to belong to an old population, probably old Population I, since it dominates most galaxies and seems to be present to some extent in all types. The stars that are now reaching the end of their lives in that population, however, are middle-main-sequence stars like the Sun. They are very numerous, so that if every one of them went through a supernova stage we would see many more supernovae than we do. Thus we are left with

---

*The designation of Type I and Type II supernovae should not be confused with Populations I and II. It is simply an arbitrary numbering. As we shall see later, the true connection with populations is almost opposite to these numbers.

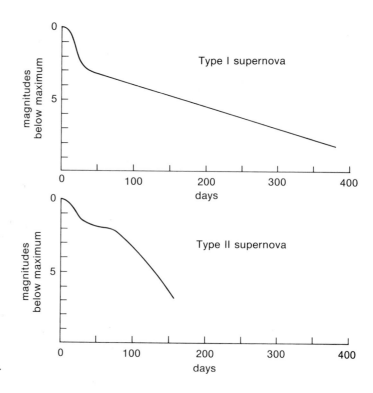

FIGURE 15.16.
Typical light curves of supernovae.

an unsatisfying picture in which supernovae of Type I come either from unspecified rare stars, or else from unspecified rare events in ordinary stars.

Type II supernovae also show clearly what stellar population they come from; but beyond that, little is clear. They occur only in galaxies that contain some stars of a young population, and in spiral galaxies they tend to occur in the spiral arms. Thus it would appear that Type II supernovae come from young stars. Presumably they are a final event in the evolution of a massive star whose total lifetime is short.

After a supernova explosion subsides in a distant galaxy, we have no opportunity to study the remnant that is left, because at such a great distance it is far too faint to see. Only within the Milky Way (and occasionally in the Magellanic Clouds) can we study supernova remnants. Dozens are known; they are recognized by visible or radio radiation, or both, coming from a region of gas that still looks like the aftermath of an explosion in the distant past. In most cases, unfortunately, we have little or no information about the time or the appearance of the actual supernova explosion; all that we see is the remnant. Only four supernovae in the Milky Way are well-known to have been seen. There are a few observations of a supernova in the year 1006, some of a bright one in 1054, extensive observations of the supernova of 1572, and many

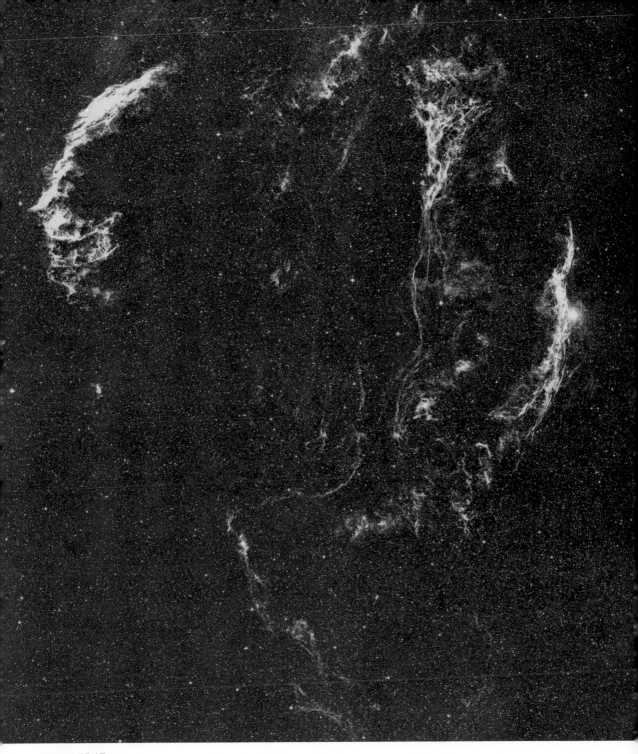

FIGURE 15.17.
This large glowing shell is the result of a supernova explosion tens of thousands of years ago. The gas is still expanding, colliding with its surroundings as it goes. The two most prominent parts are known as the Veil and Network Nebulae. (Hale Observatories.)

284

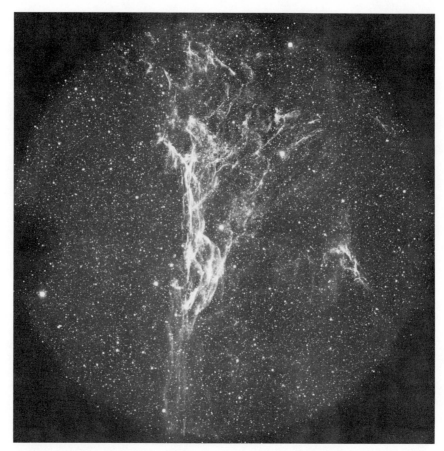

FIGURE 15.18
A part of the Network Nebula (Figure 15.17), photographed with a much larger telescope. (Kitt Peak National Observatory.)

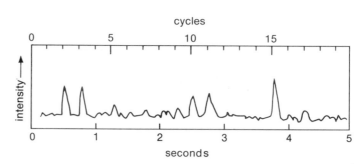

FIGURE 15.19.
The radio intensity of this pulsar brightens sharply, with a period of about ¼ second. The timing is very regular, but it does not pulse noticeably on every cycle.

observations of the one in 1604. The extent and quality of the observations reflect history and culture more than they do the prominence of the star. The supernova of 1054, for example, was noted in China and Japan as having been visible in the daytime, but there is not a single record of it in Europe or the Arabic world. There may also be some luck involved. The strongest radio source in the sky, Cassiopeia A, is a supernova remnant whose expansion rate indicates that the explosion occurred around the year 1700. From its distance, and with allowance for the amount of intervening interstellar dust, we can estimate that the apparent magnitude of the supernova was 0 or brighter—yet it was not seen. Perhaps it reached this brightness only at a time of year when the Sun was close to its region of the sky, so that it would not be high in the sky during the hours of darkness—or perhaps the problem of Cas A indicates that something fundamental is missing from our knowledge of supernovae and their remnants.

Further information about supernovae comes from variable objects of one of the strangest types of all: the pulsars. These are radio sources that turn on and off rapidly and with extreme regularity. A typical period is about half a second, and the pulse of radiation typically lasts for only about a tenth of each cycle. The strength of the pulse varies, but its timing is extremely stable. The nature of pulsars is not at all well-understood, but the most plausible picture interprets a pulsar as a rapidly rotating object with an extremely strong magnetic field that allows its radiation to come out only in a single direction (or a pair of opposite directions). The pulsar is thus like a rotating antenna, or like the beam of a lighthouse, which we see each time it flashes by us.

The very short rotation periods of pulsars show that they must be extremely small bodies; a larger body would break up under the forces of such a rotation. They must be so small, in fact, that astronomers immediately identified them with the smallest of possible stars, the neutron stars, whose nature and place in the scheme of things will be discussed in Chapter 18. But there was already a strong presumption that such a condensed configuration could arise only as a result of a supernova explosion; hence it seemed likely that pulsars were also supernova remnants. This connection was established in a spectacular way when a pulsar was discovered in the middle of the most prominent of all supernova remnants, the Crab Nebula.

By any standard, the Crab Nebula is the strangest object in the sky. It is a glowing gas cloud of unusual shape (which led early observers to bestow its fanciful name) and unusual spectrum. Careful measurements over the decades show its expansion, which can be traced backward in time to give its age. The result, along with its position in the sky, leaves no doubt that the Crab Nebula is the remnant of the supernova of A.D. 1054. Even more amazing was the discovery that in the very center

FIGURE 15.20.
The Crab Nebula. (Hale Observatories; see also Plate 7.)

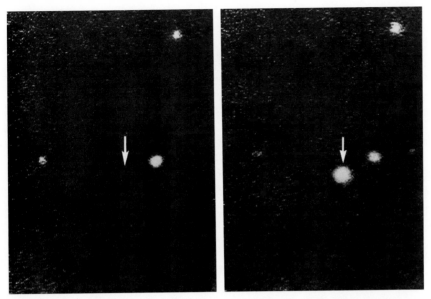

FIGURE 15.21.
These photographs of the optical pulsar in the middle of the Crab Nebula were made with a television camera that could be synchronized with the period of pulsation. At the left the star is invisible, between pulses. The righthand picture was timed to catch it during pulses. (Courtesy of J. S. Miller and E. J. Wampler, Lick Observatory.)

of the nebula is a sixteenth-magnitude star whose visible light pulses with the same period as the radio pulsar. The period of the Crab is a thirtieth of a second, the fastest pulsar known. This is too rapid a flash-rate to see, but it is easy to record electronically.

The star at the center of the Crab Nebula is not only an optical pulsar, but also a pulsed source of X-rays. With its radio, optical, and X-ray pulses, surrounded by a high-energy nebula, the Crab is a totally unique object.

X-ray observations of the sky are relatively new. Radiation at these very short wavelengths does not penetrate the Earth's atmosphere; hence X-ray observations can be made only from rockets and satellites (and to some small extent from high-altitude balloons). Surveys of the X-ray sky show a number of bright objects. Along with the Crab, a few other supernova remnants have been identified as X-ray sources. Some peculiar galaxies are also included in the X-ray list; these will be discussed in a later chapter. But the X-ray catalogs also include some sources that have been identified optically with peculiar-looking stars. Some of these are condensed objects in binary systems, but attempts to analyze their orbits have so far led to more confusion than understanding. X-ray stars tend to have a variable brightness in the X-ray part of the spectrum, and some even pulse. But except for the Crab, it is not at all clear what their connection—if any—is with supernovae or novae.

Indeed, as fascinating as the problems of the X-ray and pulsar bestiary may be, they do not lead us very close to answering the basic question, what makes a supernova? Not even the source of supernova energy is clear. The entry of the pulsars into the question makes gravitational energy a likely possibility. In falling together under its own gravitational attraction, material will release energy. To put it more strongly, in fact, material cannot condense under its own gravitation *unless* it gets rid of a large amount of energy. For the extreme degree of condensation in a pulsar, the amount of gravitational energy is comparable to the energy released in a supernova explosion. Thus it is hard to see how a star could ever become a pulsar without at some time passing through some trauma comparable to a supernova explosion.

As strong as this presumptive argument may be, many astronomers are unwilling to accept it as a full answer, for two reasons. First, it is by no means clear that all supernovae end as pulsars. Second, nuclear energy sources can also provide an adequate amount of energy; and, as we shall see in Chapter 18, there is a strong presumption that high-energy nuclear events are needed to build up, somewhere in the universe, the nuclei of the heavy elements. And here too, there is an argument of necessity: it is hard to imagine how an explosion can have the violence of a supernova *without* producing temperatures and pressures that will set off runaway nuclear reactions.

To sum up, the nature of supernovae is unclear in many ways. There are at least two types, which come from stars of different ages. They leave recognizable remnants behind, some of which contain pulsars. However, most supernova remnants are not observed to contain a pulsar, and most pulsars are not surrounded by recognizable supernova remnants. Similarly, a few supernova remnants are strong X-ray sources but most are not, and most X-ray sources are probably not supernova remnants.

This confused picture of supernovae contrasts blatantly with the neat framework into which we can arrange much of the other material of this book. But this is quite characteristic of scientific problems; along with the neat explanations in some areas, we must face other problems where the clues far outnumber the solutions and enlightenment seems far from our grasp. Yet the chaos of frontier problems should not obscure our basic tenet: the universe is orderly, and by understanding that order we can understand the universe. If there are stars that we now fail to understand, the fault lies not in our stars, but in ourselves.

YOUNG VARIABLE STARS

The pulsating stars and the explosive stars both show phenomena that arise during late stages of a star's evolution. There are also phenomena in a star's *early* life that give rise to variability in brightness. Like the other causes of variability, this is of interest less for its own sake than for its labeling of interesting stellar types. By looking for certain types of variability, we can pick out, for further study, stars that are at the very beginning of their lives, before they even become main-sequence stars. The most prominent type of such variables are those known as the T Tauri stars.

T Tauri stars are G and K stars that vary irregularly in brightness and have unusual bright lines in their spectra. As usual, their true nature becomes clear when we observe them as members of a star cluster, where we can compare them on an equal basis with main-sequence stars that were born at the same time. The HR diagram of such a cluster shows that it is extremely young, and the T Tauris lie above the main sequence. The main sequence of such a star cluster extends very far upward at the

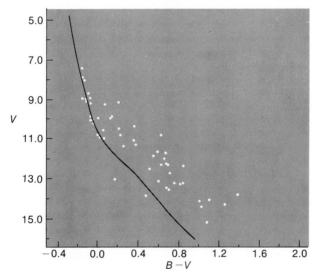

FIGURE 15.22.
Color-magnitude array of the very young star cluster NGC 2264. The zero-age main sequence is drawn in. Below magnitude 11, which corresponds to absolute magnitude +1 in this cluster, the stars have not yet reached the main sequence. (After G. Hagen, *Publications of the David Dunlap Observatory*. University of Toronto, Vol. 4, 1970.)

luminous blue end; since it includes stars that live only a short time, the cluster must be very young. Furthermore, we cannot follow the main sequence very far down; somewhere beyond spectral type A, it disintegrates into a scattering of stars that tend toward the region of the HR diagram that the T Tauri stars occupy. The implication is clear: in this very young group, the most massive stars have already settled down to the main-sequence structure that they will have for most of their lives, but the less massive stars have not yet reached that stage. They are the T Tauri stars, still contracting, still getting hotter, and not yet stable.

Another property of many T Tauri stars also suggests their pre-main-sequence nature. Observations at wavelengths in the far-infrared part of the spectrum show that a large part of their radiation comes out there rather than in the visible spectrum. Apparently these stars are still buried in—or at least veiled by—a remainder of the dust cloud out of which they formed. Much of the star's radiation is absorbed by the dust shell, which then reradiates this energy at the long wavelengths that are appropriate to its cool temperature.

T Tauris are infant stars, and our knowledge of them is similarly in its infancy. Already, however, two different T Tauri stars have been observed to brighten by many magnitudes, each of them making the transition in a few months' time to the next stage in its development. This later stage looks like the beginning of the star's final movement toward a stable main-sequence structure. Its target point on the main sequence indicates, because of the mass-luminosity relation, how great the masses of these stars are. Those masses, in turn, can be assigned to the T Tauri predecessors; and calculations of stellar structure indicate again that the T Tauri stars must be protostars that have not yet reached an equilibrium condition.

THE VIRTUE OF VARIABILITY

We have seen in this chapter how valuable the variable stars can be in helping us to understand the nature and evolution of stars in general. From a practical point of view, this value arises from a combination of two reasons. First, from the details of what varies, and of how it varies, we can often probe secrets that would never be revealed by a taciturn, constant star. Second, the changes of brightness make it easy for us to pick out these interesting objects quickly, among the countless stars that are equally faint. Once found, a variable star costs a lot of effort, in following its changes and deciphering its nature, but the results are well worth the game.

REVIEW QUESTIONS

1. Why are variable stars worth studying? Give as many reasons as you can, and illustrate each of them.

2. Justify the statement that the cepheids are the most important variable stars of all.

3. How do we know that cepheids have a period-luminosity relation, and how do we determine its zero point? (Distinguish the two problems clearly.)

4. In what ways are RR Lyrae stars similar to classical cepheids, and how do they differ?

5. What is a nova? What do we see, and what causes it?

6. Why do we believe that supernova explosions arise in stars of more than one type? What possible causes have been suggested for the explosion?

7. How are pulsars detected? What do we know about their physical nature?

8. Where do T Tauri stars fit into our understanding of stellar evolution?

16 | THE SECRETS OF STARLIGHT: UNRAVELING THE SPECTRUM

The study of magnitudes, colors, and spectral classes gives a broad insight into the nature of the stars; but if we wish to understand them in detail, we must analyze them in detail. The real treasury of information lies in the spectra of the stars. To unlock it, however, we must first understand the general principles of how matter emits and absorbs radiation. As is so typical in astronomy, we must stop and learn some physics, but the physics will pay off directly and immediately in an understanding of the stars.

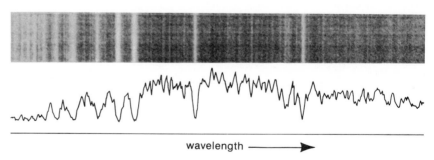

wavelength ⟶

FIGURE 16.1.
A photograph of the spectrum of a main-sequence F star, along with a graph of the same spectrum. The photograph (courtesy of Kitt Peak National Observatory) is reproduced from an atlas used by astronomers in classifying spectra. It is a negative print, to match the original negatives with which astronomers carry out such studies. The graph was made on a machine that measures photographic negatives and plots the measurements.

The spectrum of a star is the array of all its radiation, laid out according to wavelength. For scientific analysis, the most useful form of a spectrum is a numerical graph that tells just how much radiation there is at each wavelength. For illustration, however, and for most general purposes, it is easiest just to look at spectrograms. These are the pictures, made by spreading the light out according to wavelength, that we have already used to classify the stars into spectral types.

In spectral classification, we made use of the strengths of the dark lines in the spectrum. What are these lines, in our new quantitative point of view? Clearly, each line is a small region of wavelengths at which the radiation is much weaker than it is at surrounding wavelengths. In simpler language, each line is a dip in the continuous spectrum. The continuous spectrum also has its significant characteristics; even if we ignore the lines, the amount of radiation varies in a broad general way from one end of the spectrum to the other. This continuous spectrum also differs from star to star, and it is quite natural to look at it separately from the line spectrum. Each is connected in a different fashion with the way in which matter radiates, and each gives us a different sort of insight into the nature of the stars. For the details, we will later return to the line spectrum, but first we shall see how the continuous spectrum sketches a general picture of a star.

THE CONTINUOUS SPECTRUM

All matter is capable of emitting and absorbing radiation at all wavelengths. How much it emits or absorbs, however, depends on what it is made of and on its physical condition. Although the full details of radiation processes are exceedingly complicated, there are some idealized situations that are very simple. One of these, fortunately, is the emission of radiation by an opaque body, like a star, for example. The continuous spectrum of such a body depends almost completely on its temperature and depends very little on the material that it is made of. We can follow this idea to find out how hot the stars are.

Under one idealized circumstance the radiation of a body depends *only* on the temperature. The requirement is simply that all visible parts of an opaque, nonreflecting body be at the same temperature. In the language of physics, such a body is called a black body, and the radiation that it emits is called black-body radiation.

One must avoid being confused here by the word "black." It derives from the body's being opaque and nonreflecting, so that it looks black if illuminated from outside. Blackness does not mean, however, that the body cannot *emit* radiation. To repeat, all bodies radiate. Even if a body (a lump of charcoal, for example) looks black when it is cool, it is perfectly capable of shining when it is hot.

The law that describes the intensity of black-body radiation, at any wavelength and for any temperature, was first discovered in its complete form by the physicist Planck. Planck's law is usually stated as a mathematical formula, but it can most meaningfully be displayed here in graphical form.

Two aspects of Planck's law are immediately evident from the graph: as temperature increases, (1) there is more radiation at all wavelengths, but (2) the predominant radiation is concentrated toward shorter wavelengths. Each of these characteristics can be stated quantitatively in a

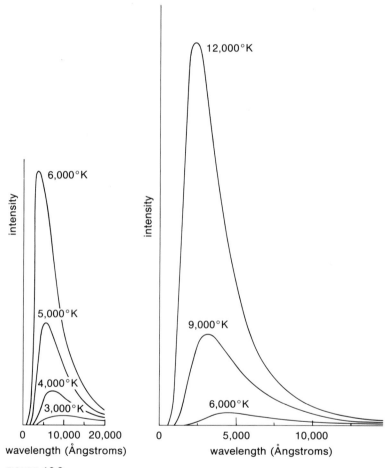

FIGURE 16.2.
Planck curves, showing the intensity of black-body radiation at each wavelength, for a variety of temperatures. Note how with increasing temperature the peak shifts to shorter wavelength (Wien's law) and the total amount of radiation increases (Stefan's law). The increase with temperature is so rapid that not even this small range of temperature can be conveniently plotted on a single scale.

simple way, each was known separately prior to Planck's discovery of the general law, and each allows us to deduce a different property of the stars.

The predominant wavelength of the radiation is given by Wien's law, which says, the wavelength at which the radiation is most intense is inversely proportional to the absolute temperature. Thus at low temperature the strongest radiation is at long wavelengths; at high temperature it shifts to short wavelengths. At room temperature the thermal radiation filling the room, which comes from the walls and from all the objects in the room, is predominantly in the far infrared. Only at much higher temperatures is appreciable radiation emitted at wavelengths short enough that some of it is visible. We call such an object "red-hot"; in actual fact it is only infrared-hot, but there is enough shorter-wavelength radiation to make the object visible. What we see looks red because the small visible part of the radiation is so heavily concentrated toward the long-wavelength, or red, end of the spectrum.

The stars are much hotter than this. Their radiation is centered in the visible part of the spectrum, or even beyond in the ultraviolet. The Sun, for instance, emits its most intense radiation in the middle of the visible spectrum, in the green. (Remember again that this correspondence between sunlight and visible light is not really a statement about the Sun, but rather about our eyes, which were "designed" to use sunlight.)

Radiation like that of the Sun, peaked in the green part of the spectrum, does not look green to our eyes; it looks a yellow-white. This is also an effect of the way our perception is designed. Our senses are set to perceive total sunlight—a mixture of direct sunlight and the blue scattered light of the sky—as a neutral white. Because direct sunlight has somewhat more long-wavelength light than this average, it looks yellowish.

A star that is much cooler than the Sun gives much more radiation as the long-wavelength, or red, end of the spectrum and much less at the blue end; hence its net appearance is red. A less cool star will look orange, and a star that is just a little cooler than the Sun will merely look a deeper yellow. A hotter star looks white, and the hottest stars have the balance of their radiation shifted so predominantly to the short-wavelength end of the spectrum that they look blue.

The colors of the brighter stars are easy to notice, even with the naked eye. For fainter objects the human eye has no color sensitivity, however; so the colors of faint stars are not noticeable. With the light amplification that a telescope gives, the colors of many stars are striking.

Thus the apparent colors of the stars correspond to their temperatures. In astronomical terms likewise, the color index of a star corresponds to the temperature of its surface. (As we shall see later, for distant stars there is usually an additional reddening due to interstellar dust, but this can be allowed for.) This gives us a way to calculate the temperatures of stars directly from simple observations of their colors.

The other aspect of the black-body spectrum, the over-all brightening with increasing temperature, is described quantitatively by Stefan's law: The total rate of radiation from a body, added up over all wavelengths, is proportional to the area of its surface times the fourth power of its absolute temperature. The fourth-power part of the law is what makes stars so prominent in spite of their very modest size; they are so bright because they are hot.

But our big dividend in studying the stars in this way comes from the other dependence that is included in Stefan's law: that the amount of radiation is proportional to the surface area—or, equivalently, to the square of the star's diameter. This means that if we know how much a star radiates, as measured by its luminosity, and know its temperature, which can be deduced from its color via Wien's law, we can calculate its diameter.

This relationship makes a star's diameter depend in a very simple way on its position in the HR diagram, since the coordinates of the HR diagram correspond to the other two quantities in Stefan's law. The absolute magnitude corresponds to the total radiation, and the spectral type corresponds to the surface temperature. As a result, a set of simple diagonal lines across the diagram mark out the diameters of the stars.

This discussion has sketched the physical highlights in a simple way, but analyzing the continuous spectra of real stars accurately is not nearly so easy. The trouble is that the stars do not radiate exactly like black bodies. A stellar surface is not abruptly opaque. Instead, the gaseous atmosphere is a murky medium, and we see a mixture of radiation from different depths, each depth with a different temperature. A correct treatment of the problem has to consider the absorption and emission at each level in the atmosphere, taking into account the variation of temperature, which itself depends on the flow of radiation.

Although the correct treatment of the problem is so much more complicated, it is by no means wrong to have gone through the simplified discussion of stellar radiation in terms of black bodies. This is what the astronomer does, in fact. The black-body treatment, crude as it is, shows him the general behavior of the problem and is easy to work with. Furthermore, many complicated problems can be solved by successive approximations, which start with a rough answer, like the black-body approximation, or sometimes even with a shrewd guess, and produce a result that has improved accuracy.

Another value of approximations like black-body radiation—perhaps even more important for understanding how science progresses—is that an approximate model allows the scientist to think rapidly. In terms of the simple model, he can see the nature of the answers to problems without actually solving them in careful detail. This enables him to employ that most productive of mental processes, intuition. He can scan a large

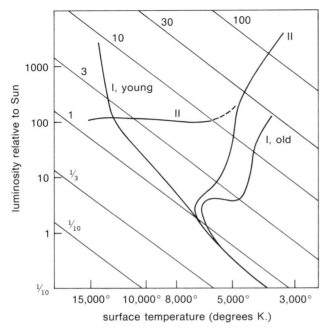

FIGURE 16.3.
If an HR diagram is plotted in terms of luminosity and
surface temperature, then Stefan's law insures that the radii
of the stars are given in a simple way by their positions in
the diagram. The diagonal lines show stellar radii, relative to
the radius of the Sun. For reference, sequences are shown
for a young Population I, for an old Population I, and for
Population II. (After M. Schwarzschild, *Structure and
Evolution of the Stars.* Copyright © 1958 by Princeton
University Press.)

range of possible answer-schemes and choose one that promises to work.
Ultimately this light-hearted romp through the approximate problem
must be followed by the travail of working through the real problem, but
knowing what the answer should look like is often a prerequisite to ever
being able to find it at all. This is the value of scientific intuition.

SPECTRAL LINES AND ATOMS

The spectral lines of a star are rich in information. They not only reveal
the physical conditions in the star's atmosphere, but also tell about the
atoms themselves. But to understand their story we must first understand
how atoms absorb and emit radiation.

An atom is the smallest unit of any given chemical element, and all
atoms of that element have the same behavior, which depends on the in-

ternal structure of the atom. The key to the behavior is that this structure can be temporarily modified, with an accompanying change in the amount of energy that the atom holds. It is this ability to store and release energy that enables an atom to absorb and emit radiation.

The atom itself is made of smaller particles. Nearly all its mass is concentrated in a tiny nucleus, whose nature is characteristic of atoms of that element. The nucleus has a certain amount of positive electric charge; it is this amount which is the essential characteristic of atoms of that element. The nucleus is surrounded by electrons, each having a negative charge, and the number of electrons is just sufficient for their total negative charge to balance the positive charge on the nucleus. The atom as a whole is thus electrically neutral, in its normal state. All electrons are identical, no matter what kind of atom they may be in. An atom can temporarily hold an extra electron or, more commonly in the astronomical context, it can temporarily lose an electron. In either case it is then called an ion.

The nucleus contains an atom's mass, but the behavior of the electrons gives the atom all its other properties: how it reacts or combines with other atoms, and how it interacts with radiation. For radiation the important property of an atom is its ability to hold energy. It is the electrons that do this, since they are held to the atom by the electrical attraction between their negative charges and the positive charge of the nucleus. The attraction holding each electron is like a stretchable spring. Energy can be put in, to stretch it, and the energy can be released again when the electron returns to its normal level. Imagine pulling back an elastic band. It takes work to stretch it, but the energy is given out again when you release it.

(Now that atoms and energy have been mentioned in the same paragraph, we should pause and clarify the various scales of energy. The energies by which the outermost electrons are held to an atom—those that will be involved in our discussion here of atoms and radiation—are quite modest in size. Such energies are readily available in ordinary light or in collisions between atoms at moderate temperatures. The nucleus itself, however, consists of particles that are held together by forces of an altogether different order of magnitude. The pulling apart or recombination of these nuclear particles involves energies that are tremendously larger than those binding the electrons to the nucleus. That scale of energy is nuclear energy. It is often referred to loosely as atomic energy, but we must be careful to distinguish the small electronic energies involved in stellar spectra, as discussed in this chapter, from the very much larger nuclear energies that will be discussed in a later chapter in connection with stellar energy generation.)

An electron in an atom differs from a stretched spring in basic ways, however. Most important, the electron cannot take up amounts of energy of all sizes, in a gradual way; instead, it can assume only a certain number of discrete states. Second, the electrons interact with each other, as well as with the nucleus, so that the energy contained in an atom depends in a complex way on the states of all its electrons. This total energy, however, can assume only discrete values. To change the metaphor, the internal energy content of an atom is like a bank balance that is allowed to take on only certain specific values. Given one of these allowable balances, a withdrawal can be made only if its amount is just right to leave the account at another allowable level, and even a deposit must be refused unless it brings the balance to another allowable level. In an atom, the absorption and emission of energy must occur in jumps, from one permitted level of energy to another.

Here is the intimate level of connection between matter and radiation; just as matter consists of a collection of tiny particles, so does radiation. What seems, at the gross level of everyday sensation, to be a smooth beam of light is in detailed fact a machine-gun stream of tiny, discrete bundles of radiant energy. Each individual energy packet in this staccato flow of radiation is called a photon. Photons are the atoms of radiation, and here too each photon carries within itself the nature of the radiation that it typifies. The identity of radiation is simple, however; as we have seen, it can be characterized by its wavelength. This in turn is related in a simple way to the nature of the photon: the *energy* of each photon depends only on the wavelength of the beam of light of which it is a part.

The relation between the wavelength of a photon and its energy is also a simple one: the energy goes inversely as the wavelength. Thus long-wavelength radiation, such as radio waves, consists of photons each of which has a very small energy; whereas short-wavelength radiation, such as X-rays, consists of photons of high individual energy. This is the reason that radiation of short wavelengths is physiologically damaging; each photon, when absorbed, releases a large enough burst of energy to do local damage. Absorbing gamma rays rather than sunlight is like being hit with hailstones rather than a gentle sprinkle of rain.

Photons differ in energy, and they also differ in number. It is the two together that make up the intensity of the light. In a beam of light of a single color, every photon has the same energy; but a more intense beam delivers more photons per second. Each of the chunks of light still has the same energy; "brighter" just means more of them.

This dual behavior of light is one of the cornerstones of modern physics and also one of its most perplexing paradoxes. Light acts like a wave, particularly in its ability to interfere and reinforce exactly like waves, but it also acts like a stream of particles. The resolution of the paradox is not easy. Before clarification came, material particles themselves had to be shown to have the same dual wave-particle nature, by experiments in which electrons were diffracted like waves; and the resolution—the theory of quantum mechanics—involves recondite physical principles and difficult mathematics.

Quantum mechanics, with the picture that it gives of the structure of atoms, is one of the triumphs of modern physics. It achieves with spectacular success the basic aim of science: to understand a large range of detailed phenomena as direct consequences of a few underlying principles. But in an equally vivid way it illustrates how far the layman's understanding is from that of the scientist. To the theoretical physicist, the principles of quantum mechanics are clear and simple, the language in which it explains the phenomena is familiar, and the resulting structure is a thing of beauty. To the nonphysicist, the principles are like the White Queen, in *Alice Through the Looking-Glass,* "believing six impossible things every day before breakfast"; the language is unintelligible; and the result is one that he can at best take on faith or may at worst reject with hostility. How then are the two to meet? A reasonable compromise is to accept the correctness of the picture and to look at those of its aspects that can be easily presented in an intelligible way.

Let us start with atoms and radiation. The key to the physical picture is that a single atom absorbs or emits a single photon. Any large-scale process of absorption and emission of radiation is the aggregate of a countless number of such individual and repeated actions, but the underlying nature of the process depends on the behavior, and hence on the nature and structure, of individual atoms and photons.

Specifically, the key is energy. When an atom absorbs a photon, the photon ceases to exist and its energy is totally added to the atom's internal store of energy. In emitting light, conversely, an atom creates a photon and gives up exactly the energy of that photon. But the atom has only certain allowed energy levels and can therefore change its energy only by amounts that correspond to differences between those energy levels; hence it can absorb or emit only photons of those particular energies. Finally, since the energies of photons correspond to the wavelengths of the light that they make up, the energy levels of an atom—specifically,

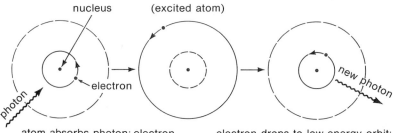

atom absorbs photon; electron   electron drops to low-energy orbit;
jumps to higher-energy orbit.   atom emits a photon.

FIGURE 16.4.
Radiation is absorbed and emitted, one photon at a time, when an atom jumps between higher and lower energy levels.

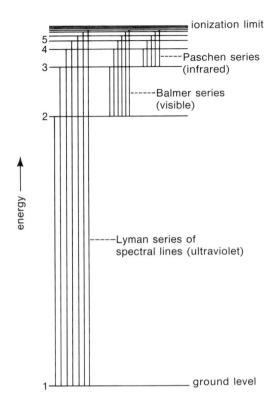

ionization limit

5
4

-----Paschen series
(infrared)

3

------Balmer series
(visible)

2

-----Lyman series of
spectral lines (ultraviolet)

FIGURE 16.5.
The energy levels of the hydrogen atom. The horizontal
lines represent the energy levels and the vertical lines
the transitions between them. The length of the vertical
line indicates the energy of the corresponding photon,
and this in turn determines its wavelength. Long jumps
correspond to short wavelengths, and short jumps to
long wavelengths.

energy ——→

1   ground level

their spacings—determine the set of wavelengths that it can emit or absorb. These wavelengths are the line spectrum of that atom.

This set of wavelengths is the same whether the atom is absorbing
or emitting radiation, since it represents the same set of differences between energy levels. When energy is fed into the atoms of a particular
chemical element, they emit its characteristic set of spectral lines. Correspondingly, when a continuous spectrum of bright light is passed
through them, they absorb—and thus remove—these same characteristic
lines. Both situations occur in astronomy, depending on the circumstances. A gaseous nebula, for instance, has energy fed into it from nearby hot stars, and it emits a bright-line spectrum. The atmosphere of a
star, by contrast, is a gas through which radiation flows up from the
hotter layers underneath; thus it absorbs out a spectrum of dark lines.
But the lines of a given element fall at the same set of wavelengths in
either case.

Each kind of atom has a different set of energy levels. These levels
depend on the number of electrons that the atom contains and the way in
which the electrons interact with the nucleus and with each other.
Hydrogen, the simplest atom, has only a single electron, and its energy

levels have a simple pattern (Figure 16.5). Each set of transitions up from or down to the same lower level forms a distinctive series of spectral lines. These are called the Lyman, Balmer, Paschen, etc., series, after the spectroscopists who first called attention to them. Unfortunately, only hydrogen has such a simple spectrum; other elements have a more complicated arrangement of energy levels, and therefore have more complicated spectra.

SPECTRAL LINES: FINGERPRINTING THE STARS

The spectral lines of an atom are its fingerprints. They indicate the presence of atoms of that element; and even more, they tell what state the atoms are in. Just as a person does not leave a full set of fingerprints on every object that he handles, so a chemical element does not show its full set of lines in the spectrum of every star in which it is present. The lines that we see depend on the conditions in the star's atmosphere, most particularly on the temperature.

This is the reason for the spectral sequence, and more specifically, it is the reason why spectral classes are a sequence of stellar temperatures. The explanation is basically simple. Every spectral line that is absorbed is a transition from a particular lower level of energy to a particular upper level, so that the atom can absorb that line only when it is in the right energy level to begin with. Thus the spectral lines that we see tell not only what atoms are present; they tell what energy levels they are in. In a cool gas, not much energy is available, and most of the atoms are in low energy levels; so in the spectrum of a cool star we see lines that originate from low energy levels. In a hot star, energy is plentiful; many of the atoms are in high energy levels, and we see the lines that come

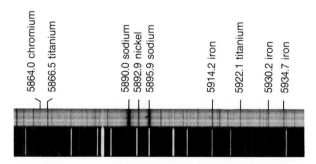

FIGURE 16.6.
A portion of the yellow part of the spectrum of a K1 giant star, along with a comparison spectrum of neon, from which the scale of wavelengths can be established. A number of absorption lines are marked with the wavelength and the elements that are responsible for them. (Lick Observatory spectrum, courtesy of H. Spinrad.)

from those high levels. This effect of excitation, as astronomers call it, makes the spectra of a hot and a cool star strikingly different, even though the two stars are made of the same material.

In understanding what we see in stellar spectra, the other important factor is the size of the energy jump. As we have seen, this determines the wavelength of the corresponding photons. Since our classification of spectra is done in only one part of the visible spectrum, we are really watching only for energy jumps of a narrow range of sizes. Larger jumps give lines at shorter wavelengths, most of them in the unobserved ultraviolet. Jumps between closer-spaced energy levels give long-wavelength lines, which also fall outside our prime region for spectral classification.

It is this narrowness of our view that explains why we predominantly see the nonmetallic elements in hot stars and the metals in cool stars. The key is in the fact that the nonmetals hold their electrons tightly and the metals hold theirs loosely. For a tightly held electron there is a big difference of energy between the lowest energy level and the higher levels. This means that transitions from the ground level to the excited levels, as they are called, have large energies and thus correspond to ultraviolet lines. As a result, a nonmetal at low temperature, with nearly all its atoms in the ground level, absorbs only ultraviolet lines that do not figure in spectral classification. The only lines that a nonmetal can absorb at visible wavelengths are those that come from transitions between various of the more highly excited energy levels; we see these only in stars hot enough for many of the atoms to be in those levels.

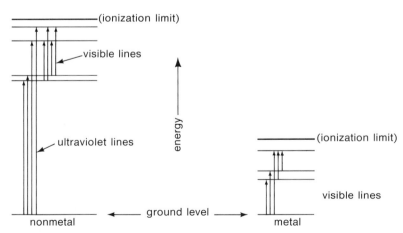

FIGURE 16.7.
These schematic diagrams of energy levels contrast a nonmetallic atom with a metallic atom. Because of the wider energy spacing, the ground-level lines of the nonmetal have ultraviolet wavelengths.

In a metallic atom, by contrast, the loose hold on the electrons makes for a smaller spacing between energy levels. Many transitions from the ground level thus have energies small enough to correspond to visible wavelengths; and these lines show up, of course, in the spectra of cool stars.

It is interesting to note that this same difference between metals and nonmetals—how tightly they hold onto their electrons—also accounts for their more familiar physical properties. It is the ease with which electrons are loosened that makes a collection of metallic atoms behave like a metal. Freely moving electrons give it its shininess and its ability to conduct electricity and heat, and the way in which the electrons help the atoms bind together is responsible for the hardnesss.

Thus the visible spectra of cool stars are full of absorption lines that come from metallic elements. The only nonmetal that is easy to see is hydrogen; on account of its great overabundance, it absorbs lines that come from an excited energy level that fewer than a millionth of the atoms occupy. Also visible in the coolest stars are bands—closely spaced groups of lines—due to molecules, in which atoms are bound together in pairs; at higher temperature the molecules are broken apart.

In the spectra of stars of higher temperature the hydrogen lines are stronger, as more atoms populate the excited energy levels. The lines of the metals begin to weaken, however, because many of the atoms acquire too much energy and can no longer hold it. The spring breaks, and the outermost electron is removed from the atom. The remaining ion holds its remaining electrons more tightly; and few, if any, of its spectral lines have small enough energy jumps to be in the visible part of the spectrum. Thus at higher temperatures, as we progress up through the G and F stars, the metallic lines weaken and disappear, and a few lines of metallic ions are seen. At spectral type A, the hydrogen lines reach their greatest strength; in hotter stars the hydrogen is largely ionized. Helium, which is even harder to excite and ionize than hydrogen, shows up strongly in the spectra of the B stars, and in the O stars even lines of ionized helium are seen.

Thus temperature differences explain nearly all the differences between stellar spectra. How then can the astronomer determine the abundances of the chemical elements in the stars? He must first allow painstakingly for all the effects of temperature on the excitation and ionization of each element; then, by comparing calculated line strengths with the observed ones, he can estimate how many atoms of each element are present. This, then, is a situation where the answers of real fundamental interest never become easy or obvious, no matter how hard we try. The fundamental question is, "What are the stars made of?"; but to answer it the astronomer must first work his way through all the complexities of stellar atmospheres.

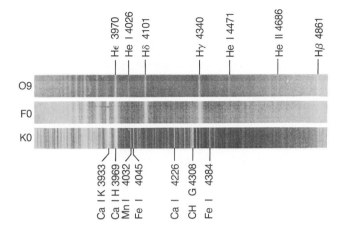

FIGURE 16.8.
The stars whose spectra are illustrated here have very different surface temperatures, and as a result the spectra look very different, even though all three stars have the same chemical composition. Lines marked at the top are characteristic of high-temperature spectra; at the bottom, low temperature. (Spectra from Kitt Peak National Observatory.)

A final point about atoms and radiation: how are we to explain continuous spectra? More specifically, if each photon must be absorbed or emitted by an individual atom, and if the atom has only a *discrete* set of energy levels, how can atoms emit the *continuous* range of photon energies in the black-body spectrum? The answer lies in interactions between atoms and *free* electrons, which do allow a continuous range of energy jumps. In one such process, called photoionization, an atom absorbs a photon whose energy is more than enough to raise it to the highest level at which it can hold onto all its electrons. The result is that some of the energy of the photon goes into detaching an electron, and the remainder of its energy then goes into motion of the electron away from the atom. Thus the atom is able to absorb photons of all energies greater than its ionization limit, corresponding to wavelengths shorter than this limit. (The opposite process can also happen: a passing electron is captured by an ionized atom, and a photon is emitted. This is just the reverse of photoionization, and the result is an emission continuum.)

In addition to these processes of photoionization and recombination, an atom can interact with an electron that is merely passing by. A photon can then be emitted or absorbed, and the energy of motion of the electron decreases or increases accordingly. Because the electron is free both before and after the transition in energy, it is called a free-free transition, in contradistinction to photoionizations and recombinations, which are called bound-free. Free-free transitions can involve any energy change whatever, so they give rise to a continuous spectrum.

Bound-free and free-free transitions allow an atom to absorb or emit a continuous range of wavelengths. What then is the importance of spectral lines after all? The answer lies in the relative strengths of the line and continuum processes. An atom absorbs its spectral lines avidly, but it is much less likely to absorb in its continuous spectrum. Thus a few atoms in the uppermost layers of a star's atmosphere are enough to produce a strong absorption-line spectrum, whereas at other wavelengths the continuous spectrum that is emitted in the denser layer below passes almost unimpeded through the thin outer layers of the atmosphere.

LOOKING SHARPLY: LINE PROFILES

Like the detective, the astronomer cannot afford to neglect any of the clues. A particularly valuable set of clues lies in the detailed shapes of spectra lines. When a spectral line is examined in close detail, it is found not to lie at a single precise wavelength but rather to spread over a range of wavelengths. The nature of this spread—the line profile—tells the astronomer further facts about the conditions under which the line was formed in the star's atmosphere. Again we will have to understand the atoms first, and then we will understand the stars.

The first question of all is, how can we reconcile a fuzziness of wavelength with the basic idea that an atom has only a specific set of energy levels? The answer has two parts: first, the energy levels themselves are not absolutely sharp; and second, Doppler effects in the stellar atmosphere shift the apparent wavelengths at which we see the absorption lines of atoms. Each of these answers will tell us something about the stars.

Part of the fuzziness of energy levels is inherent in the atom itself. It is due to one of the fundamental principles of modern physics: the uncertainty principle. This says that on an atomic scale we can never pin down both the position and the motion of a particle. (Strictly, the uncertainty principle applies to particles of all sizes, but the amount of the uncertainty is such that it matters seriously only for atomic-sized particles.) The more certain we are about where a particle is, the less sure we are of how fast it is moving, and conversely. Basically the uncertainty principle arises from a simple paradox: in order to know a quantity we have to observe it, but the very act of observation disturbs the thing that we are trying to measure. When physicists reason out carefully the problems of measuring an atomic-sized particle by means of another such particle or a photon, they arrive at a quantitative statement of the uncertainties that the result will necessarily have.

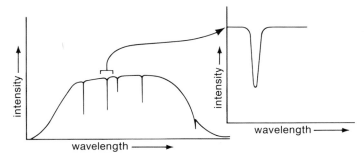

FIGURE 16.9.
When we look at a spectrum on a very fine-wavelength scale,
the absorption lines show characteristic "profiles."

Since its introduction in 1927, the uncertainty principle has had a tremendous impact on man's philosophic view of the universe. Gone forever is the clockwork universe of the nineteenth century, in which the positions and velocities of all the particles at any one moment will determine the future of the universe forever. According to the uncertainty principle, we can never hope to follow the detailed motions of individual particles.

Like many other important ideas, however, the uncertainty principle has been grossly misused in philosophical arguments. It does not say in any crude way, "Everything is uncertain." True enough, individual atomic events are uncertain, but for larger numbers of atoms we can make very firm statistical predictions. In the same way, an individual spin of a roulette wheel is extremely uncertain, but we can make a good prediction about the average results of a million spins. Objects of everyday size involve millions of millions of millions of atoms; so their mechanical behavior has only a negligible uncertainty. But on the atomic level, the uncertainty principle dominates the story, and the whole quantum theory of atomic structure and behavior is one of its consequences.

An alternative form of the uncertainty principle relates the uncertainty in any energy with the length of time for which it is observed, and here we can make a connection with the width of an atomic energy level. An atom does not remain in an excited energy level very long at all; in a tiny fraction of a second it will emit a photon and drop to a lower energy level. The shortness of this time makes the energy level uncertain by some small amount, and this gives a "natural width" to lines that involve that energy level. Thus no spectral line can be infinitely narrow.

In addition to this natural width, other mechanisms broaden spectral lines. An important process in many stellar atmospheres is the occurrence of collisions between the atoms. Under conditions of high density, collisions will further shorten the time that an atom remains in a specific energy level, and this broadens the line to even more than the natural width. The amount of collisional broadening is a measure of the frequency of collisions and hence of the density in the star's atmosphere.

Besides these abrupt collisions, there are slower passages by charged particles—especially ionized atoms—whose electrical force temporarily perturbs the energy levels of the atom, a phenomenon called the Stark effect. While it is in this perturbed state the atom can absorb wavelengths slightly different from the usual ones, and again the line is broadened.

The effectiveness of these last two broadening mechanisms depends on how dense is the gas in the atmosphere of the star, since a denser gas will have more frequent collisions. Density differs greatly between a giant and a dwarf star of the same spectral type. To see this, consider the force of gravity. A dwarf star has a somewhat lower mass, but it has a much smaller radius than a giant star, and the force of gravity at its surface (which goes as mass divided by radius squared) is much greater than at the surface of a giant. The gas of the dwarf star's atmosphere therefore weighs more and is compressed to a higher density by its own weight. Many of the differences between the spectra of giant and dwarf stars, which make possible the spectroscopic determination of absolute magnitudes, result from these density differences and the consequent line-broadening.

The second general type of broadening of spectral lines comes from Doppler shifts. First among these is the shift due to thermal motion of the atoms, which dash about with individual energies of motion that are proportional to the temperature. (This is, in fact, the basic property of temperature.) At any given moment some atoms will be moving toward us and thus have their absorption lines shifted to shorter wavelength; others will be moving away from us, and their lines will be shifted to longer wavelength. Since a whole range of speeds is present, the absorption line is smoothly broadened. Its width is proportional to the average speed and thus goes up with the temperature.

In addition to these motions on a microscopic scale, many stellar atmospheres have a larger-scale turbulent motion of the gas. Again some gas elements are moving toward us and some away from us, with various speeds, and the result is again a Doppler broadening. In such a situation, the width of the line is greater than can be accounted for by temperature alone, and astronomers are thus able to detect turbulent motions in stellar atmospheres.

There is indeed a wealth of information in line profiles, so much that we must take care not to confuse the various causes. When a line is broadened, which of the many posssible mechanisms is responsible? We can answer this question by examining the shape of the line profile. Both natural width and collisional broadening (which astronomers often lump together under the name "damping") give a characteristic profile that has a narrow core and extended "wings." Doppler broadening, by

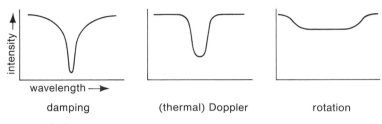

FIGURE 16.10.
Different mechanisms of line broadening produce line profiles of different characteristic shapes, which allow us to distinguish them, and often to separate them when two or more mechanisms operate at the same time.

contrast, gives a broad core and practically no wings at all. The result of putting the two together is a profile that has a broad core, whose width is due to the amount of Doppler broadening, and extended wings, whose extent is due to the amount of damping.

Within each type of broadening it is easy to separate the two contributors. The part of the damping that comes from the uncertainty principle is a known characteristic of the atomic energy levels; so the rest of the damping width is due to collisions and is a measure of the atmospheric density. In the Doppler core, the thermal width is easily predicted, since the star's temperature is known from its spectral class; the remainder of the Doppler width is due to turbulence.

Stark-effect broadening, produced by passage of electrically charged ions, is a special case. It also has its own characteristic profile, but it is distinguished mainly by the fact that it affects the lines of hydrogen much more than those of any heavier atom. Stark broadening is responsible for the tremendous width of the hydrogen lines in the spectra of main-sequence A stars and of white dwarfs; but it is not important for stars of other types.

One other broadening mechanism is of special interest because it allows us to measure a property of the stars that would otherwise be inaccessible: rotation. This is also a kind of Doppler broadening, caused by the fact that one side of the star is moving toward us and the other away from us; but the large-scale systematic motion of a rotating body gives a line profile very different from that of randomly moving atoms or turbulent eddies. The width of this profile gives the speed of rotation of the star.

The study of line profiles is an excellent example of how science works. Here is a detail of stellar spectra, whose observation requires special effort; but it provides a source of information that is new and additional. To interpret the data requires theoretical study, predicting each of the various effects that have been described and calculating how they will combine with each other. The result pays in three ways. First, fitting the details together in a coherent way increases our confidence that we really understand the workings of a stellar atmosphere. Second, the detailed understanding allows us to make more reliable calculations of other problems, such as the abundance of atoms of each type. And third, we are able in this way to observe phenomena such as rotation and turbulence, which would otherwise remain hidden mysteries.

REVIEW QUESTIONS

1. What does it mean to say that a star has both a continuous spectrum and a line spectrum? Distinguish clearly between the two.

2. Explain why the color of a star depends on the temperature of its surface.

3. Why does the diameter of a star depend only on its position in the HR diagram, without regard to its mass or internal structure?

4. Draw a diagram of the energy levels of the hydrogen atom. Indicate the transitions that give rise to lines of the Lyman series, and those that give lines of the Balmer series. Explain why the Lyman lines are in the ultraviolet part of the spectrum and the Balmer lines in the visible.

5. Why are the Balmer lines of hydrogen stronger in the spectrum of an A star than in that of a K star? Why are they stronger in an A star than in a B star? Why do we see stronger metal lines in a cool star?

6. What is the uncertainty principle? What sort of predictions about atoms does it prevent us from making? What sort of predictions does it *not* prevent us from making?

7. What is meant by a spectral line profile? List as many ways as you can in which line profiles give us information about the stars.

17 | THE SUN: OUR OWN STAR

The Sun is the only star whose surface we can observe in detail. As a result, study of the Sun introduces a profusion of new facts and new phenomena. These give rise to questions about scientific problems of two sorts. One problem area is the solar phenomena themselves: Why are the outer layers of the Sun structured as they are? Why do they go through the entertaining antics that solar astronomers follow with such interest? The second type of question takes a broader view: From this opportunity to study one star in detail, how can we gain a better understanding of the nature of stars in general?

It is clearly questions of the latter type that concern the student of the universe, and we shall approach the study of the Sun mainly with this view in mind. Although the solar phenomena themselves are almost infinite in their detail, we can extract from them an over-all, star-oriented picture, while leaving the details in the hands of the solar specialists.

THE SOLAR SURFACE

Even the first glance at the Sun raises a paradoxical question: Why does this sphere of gas appear to have a sharp edge, just as if it were a solid body? The answer is in the density of the gas in the Sun's atmosphere, which increases sharply with depth, simply because the deeper layers of gas are compressed by the weight of those above them. What we perceive as the "surface" of the Sun is just the depth at which the haziness of its atmosphere finally adds up to almost complete opaqueness. Because the Sun's strong gravity makes such a steep density increase, the transition from transparent to opaque takes only a few hundred kilometers of depth. At the distance of the Sun from us, this looks like a sharp edge.

This apparent surface is the level from which the Sun's radiation comes; it is called the *photosphere*. The temperature of the gas at that depth is around 6,000 degrees, and as a result the color of sunlight is characteristic of that temperature. Higher up in the Sun's atmosphere the gas is 1,000 to 2,000 degrees cooler, and in those layers the dark absorption lines are impressed on the spectrum.

Photographs show quite clearly that the center of the Sun's disc is noticeably brighter than the edges. In accordance with a long-standing terminology in which the edge of the Sun is called the limb, this phenomenon is referred to as *limb-darkening*. It does not of course indicate that the Sun is aiming its radiation particularly at us; the situation is rather that every point on the Sun's surface sends out more radiation perpendicular to the surface than at oblique or grazing angles. This is easy to understand; at an oblique angle we do not see to as deep a level, and the cooler gas at this higher level emits less. The magnitude of this effect, and its dependence on wavelength, are of great interest to astronomers who study the atmospheres of stars, because limb-darkening tells much about the different layers of the solar atmosphere and about the way in which radiation filters up through them. Limb-darkening has, of course, been observed in some eclipsing binaries, but the values measured for the Sun are very much more accurate and reliable.

The solar surface also shows variations of brightness on a much smaller scale—so small, in fact, that they can be seen well only when the "seeing" in the Earth's atmosphere is particularly sharp. This is the *granulation*, an irregular mosaic pattern of bright areas separated by darker lanes. The "dark" regions are actually only a little less bright than the rest; they are about 100 degrees cooler and only look dark because of the brightness contrast. The cells are about 1,000 kilometers in size, and successive photographs show that they change in a few minutes. In fact, speeded-up motion pictures show the Sun's surface looking like a simmering pot of soup.

This is an apt analogy, because the surface layers are in a constant churning, as hotter material rises—making the bright cells—and cooler material sinks downward in between. The cause is in the layer just below the surface, where the temperature becomes high enough for atomic collisions to ionize hydrogen. At the surface, nearly all the hydrogen atoms are neutral, and in the hot interior they are all ionized; but in this medium-temperature layer in between, the hydrogen atoms ionize and recombine readily. The energy involved in the ionization process gives

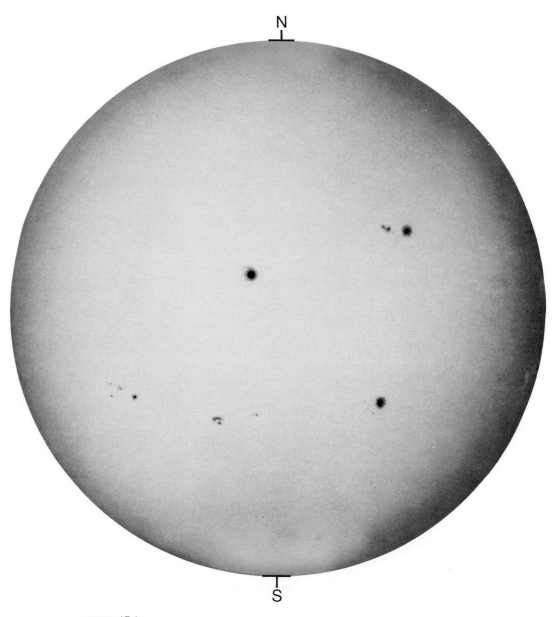

N

S

FIGURE 17.1.
The "disc" of the Sun. A few sunspots can be seen, and the limb-darkening is distinctly visible. (Courtesy of N. Sheeley, Kitt Peak National Observatory.)

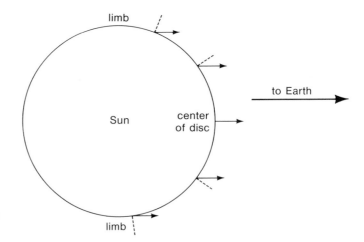

FIGURE 17.2.
The cause of limb-darkening in the Sun. Near the limb we see light that has come out at a grazing angle to the surface.

FIGURE 17.3.
The granulation of the Sun's surface, as photographed by a telescope carried by a balloon. Above the Earth's turbulent troposphere, the telescope was able to take a sharper picture than can be made from the Earth's surface. (Courtesy Project Stratoscope of Princeton University, sponsored by the Office of Naval Research, the National Science Foundation, and the National Aeronautics and Space Administration.)

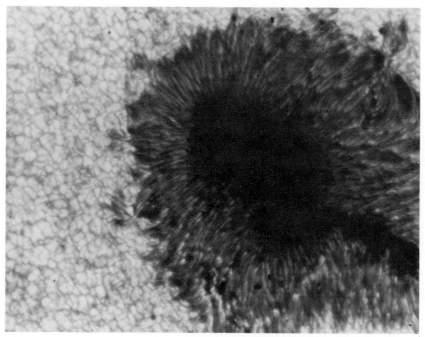

FIGURE 17.4.
A small sunspot, seen in the same detail as the previous figure. (Courtesy Project Stratoscope of Princeton University, sponsored by the Office of Naval Research, the National Science Foundation, and the National Aeronautics and Space Administration.)

the hydrogen gas at this depth an unusual ability to transport energy, and that it does, by means of the vertical churning motions that physicists call convection of heat. Thus the granulation of the solar surface is a direct product of the *hydrogen convection zone* immediately beneath.

Granulation is the normal condition of the Sun's surface. At times, however, larger abnormal dark regions appear; they are called *sunspots.* Like the dark lanes in the granulation, they are not really dark; they are just cooler—but by about a thousand degrees—and they therefore shine less brightly. Lifetimes of sunspots are measured in days rather than minutes, and some manage to persist for months. Their sizes range from pores in the granulation up to groups that look like large strings of islands spread across the Sun's surface. The cause of sunspots is not known, although it is undoubtedly connected with the strong magnetic fields that they are observed to have. (A magnetic field affects the energy levels of atoms, so that its presence can be detected in the spectrum.) We shall see magnetic fields again and again in examining solar phenomena. Although their role is poorly understood, they are certainly important in determining what happens on the Sun's surface.

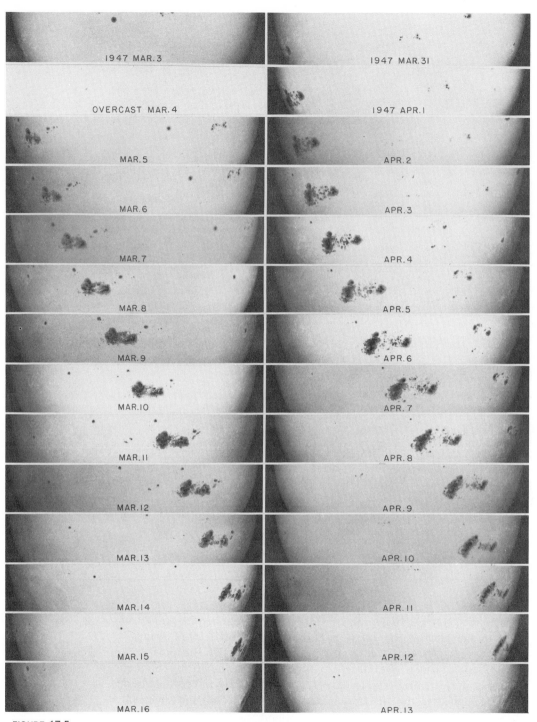

FIGURE 17.5.
These sequences show the solar rotation, as the same large sunspot group is carried around in two successive rotation periods. Note the changes in the form of the spot group. (Hale Observatories.)

Day-after-day observations of sunspots show that they drift across the Sun's disc. They all move in the same direction, at nearly the same speed, so that it is obvious that what we are observing is the rotation of the Sun. A complete rotation takes about a month, so that some of the longest-lived sunspots are seen again the next time round. Since the Sun is not a solid body, there is no reason why it should rotate rigidly, and indeed it does not. The equator goes around fastest, with the higher latitudes lagging behind.

CHROMOSPHERE AND CORONA

Above the photosphere, the Sun's atmosphere thins rapidly at greater heights. This thinner gas is transparent to most wavelengths, but we detect it at the wavelengths that the atoms absorb and emit most readily. Thus the atoms in the cooler layers immediately above the photosphere are responsible for the dark lines of the solar spectrum. About 500 kilometers above the photosphere, however, the temperature begins to rise again in a surprising way. Within a few thousand kilometers it has reached a million degrees! This thin, high-temperature gas extends outward from the Sun in all directions. Close to the Sun it is nearly stationary, but within a few million kilometers it turns into an outward-blowing *solar wind,* which extends outward through the Solar System. As it passes the Earth, 150 million kilometers away from the Sun, the solar wind has a speed of 500 kilometers per second.

The region of hot gas that surrounds the Sun is called the *corona.* Its density is so low that it has very little effect on the Sun's surroundings and is, in fact, difficult to observe. The corona has been known since time immemorial, however, from the beautiful glow with which it surrounds the Sun during the brief moments of a total eclipse. At other times the light of the daytime sky completely drowns out the delicate coronal glow, which is only a millionth as bright as the solar disc.

The light of the corona is mostly ordinary sunlight scattered by the electrons of the ionized coronal gas. (Free electrons reflect light much more effectively than ordinary atoms do.) The coronal spectrum also has emission lines, however. For many decades their source could not be identified, partly because astronomers did not realize that the corona is so hot, and partly because the coronal lines are very unusual spectral lines. At such temperatures the atoms are highly ionized. Hydrogen and helium atoms are able to hold no electrons at all and are therefore unable to emit a line spectrum. Each of the heavier atoms has also lost several electrons. Since a multiply-ionized atom holds its remaining electrons very tenaciously, the jumps between energy levels are quite large and the corresponding spectral lines are in the far ultraviolet. Only a few of these ions have energy levels that produce lines in the visible spectrum.

FIGURE 17.6.
The solar corona, as observed during a total eclipse. (Lick Observatory.)

The two strongest lines come from atoms of iron that have lost 13 and 9 electrons respectively. (A neutral iron atom has 26 electrons.) These are far from being the most abundant atoms; they just happen to be the ones that have strong lines in the visible part of the spectrum. Until space observations made the rich ultraviolet spectrum accessible, coronal spectroscopy was rather like studying an elephant by examining only the tip of his tail.

Although hydrogen and helium are the most common elements in the corona, they are not seen there, because of the high temperature. Where they do show up prominently is in the thin transition region at the base of the corona, where the temperature is high enough to excite these atoms but not high enough to ionize them. This layer, only a few thousand kilometers thick, can be seen at the edge of the Sun during an eclipse. Because of its reddish color, contributed by the strongest hydrogen line, it is called the *chromosphere*. It was in eclipse spectra of the chromosphere that the element helium was first discovered (its name comes from the Greek word for "Sun"), long before it was found on Earth.

Eclipses give the astronomer only a glimpse of the Sun's outer layers, during the few minutes that each eclipse lasts. If he went on expeditions to every total solar eclipse that occurs, he would accumulate only a few hours of observation in his lifetime. Thus it is indeed valuable that instruments have been developed that allow observation of the chromosphere and the brighter parts of the corona, without waiting for an eclipse. The most important of these is the coronagraph, a specially designed telescope that includes a black disc to block out the dazzling light of the photosphere. Used in the dark-blue sky of a high-altitude site, and with color filters that isolate particular spectral lines, coronagraphs make observations every clear day.

Another valuable solar instrument is the spectroheliograph, which selects out a very narrow range of wavelength and takes a photograph of the Sun using that wavelength only. This can isolate the light of a particular element, such as hydrogen, and by use of Doppler shifts the photograph can even be restricted to material within a particular range of velocities.

Observations of this type have shown that the chromosphere has a very complex structure—a network of small spikes, called spicules, spread in clusters over the surface. The corona is similarly complex, but on a scale of much larger radial streamers. In both cases a magnetic field must play an important role in determining the structures.

The most spectacular of solar structures are the *prominences,* which extend from the chromosphere into the corona. Typical prominences stretch along great magnetic arcs coming up from the Sun's surface. Paradoxically, they shine visibly because they are cooler than the surrounding corona; their hydrogen and helium are *not* hot enough to be ionized and hence invisible. Some prominences remain static for long periods of time, many rain down on the Sun's surface, and a few are shot out of the surface.

As we noted at the beginning of the chapter, many of these phenomena are at such a level of fine detail that an analysis of them would contribute little to a general understanding of stars. One major fact stands out, however: the Sun is surrounded by a region of very hot gas that is continually blown out into space. Why should this be? The answer is not at all clear in detail, but the general cause is probably to be found in the waves generated in the photosphere by the motions of the granulation. Some of these propagate upward; as the wave passes into gas of lower density, it can maintain itself only by growing stronger and stronger, until it turns into a shock wave, like a sonic boom, and converts its energy into heating of the surroundings. This happens at the base of the corona, and the heat is carried up from there to the rest of the corona.

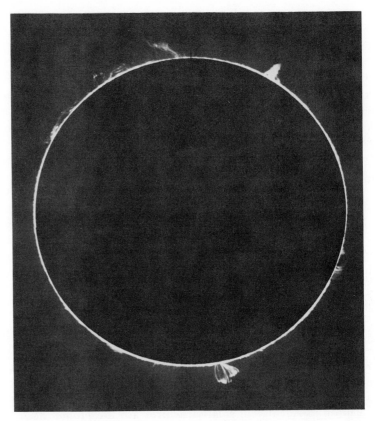

FIGURE 17.7.
A spectroheliogram, with the bright solar disc blotted out. The chromosphere shows as a ring, and several prominences are visible. (Hale Observatories.)

The story would stop there if the heating were only modest. It is so strong, however, that if the corona were simply to absorb all this heat, it would reach an even higher temperature than it is known to have. What happens instead is that the thermal velocities of the atoms become so large that many of them are able to escape from the Sun's gravitation completely. Like a low-mass planet, the Sun is losing atmosphere. The proportions are completely different, however. Instead of a slow leakage, the loss from the corona turns into an outright stampede, and the Sun's whole outer atmosphere rushes outward. The result, however, is less drastic. When a planetary atmosphere escapes, the planet remains bare. The Sun, however, has a gaseous surface, so that as soon as any atmosphere escapes, it is replaced from below. The corona is such a tiny fraction of the Sun's mass that it can be replaced continually, for billions of years, without any serious effect.

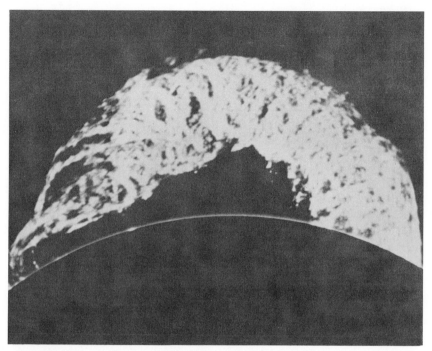

FIGURE 17.8.
One of the largest prominences ever seen. The black occulting disc of the
coronagraph, at the bottom, shows the size of the Sun. This arch prominence
eventually spread completely outside the region that is shown here. (High
Altitude Observatory.)

SOLAR ACTIVITY

The surface granulation, the chromosphere, the corona, and the solar
wind are all phenomena of the "quiet Sun." The Sun is not always quiet,
however. Sunspots and prominences appear and disappear. In addition,
the Sun produces smaller but much more violent events called *flares.*
These are sudden releases of energy in very small regions. A flare causes
a brightening of a small part of the solar surface, especially in the emission lines; and it produces strong X-ray, ultraviolet, and radio radiation,
along with an outflow of high-speed particles. Although these events
last for about half an hour, the original release of energy in a flare seems
to be quite sudden.

All these phenomena tend to concentrate in "active regions," which
are most clearly marked by groups of sunspots. At times the Sun may
have no active regions at all, but at other times there are several active
regions at once. The over-all level of solar activity waxes and wanes in

322

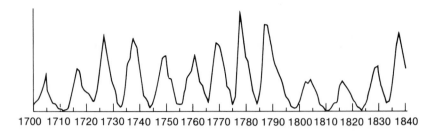

1700 1710 1720 1730 1740 1750 1760 1770 1780 1790 1800 1810 1820 1830 1840

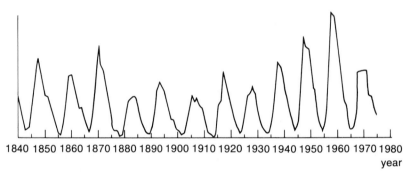

1840 1850 1860 1870 1880 1890 1900 1910 1920 1930 1940 1950 1960 1970 1980

year

FIGURE 17.9.
Sunspot records have been kept for hundreds of years. Note the general
recurrence with an 11-year interval, but note also the degree of irregularity.

a cycle of about 11 years; since the number of sunspots is the easiest-
observed measure of activity, this 11-year periodicity is usually called
the sunspot cycle. Solar activity is about 10 times as great as sunspot
maximum as at minimum, but the amount of activity varies greatly
from one cycle to the next. The last maximum, in 1969, was of ordinary
intensity; but the previous sunspot maximum, peaking in 1958, was the
most intense observed in more than two centuries of solar records.

We do not understand solar activity very well at all. What is clear
is that all the phenomena involve magnetic fields in an ionized gas. In
detail, however, theoretical problems of this type are terribly hard to
solve, and we have little idea of how the material is able to store up
energy and release it in an unstable way.

Similarly, we have no more than a general notion of what might produce the 11-year cycle. This notion is that the differential rotation of the
Sun's surface, in which the material at the equator goes round faster
than that at the poles, drags the subsurface magnetic field along with it
and stretches the field out. After a few years the field has been wound
several times around the Sun. The crowded-up field develops instabil-
ities, which pop up to the surface as sunspots. After these decay, their

own remaining fields drift toward the poles, where they eventually neutralize the original magnetic field and replace it with a weak field of the opposite polarity. The process then starts all over again. In a way this picture is contrived, in order to explain the observed facts, including the fact that sunspots in successive solar cycles have opposite magnetic polarities. It does have the appeal, however, that differential rotation *ought* to wind up the field and this *ought* to lead to some sort of unstable result. At present it is a working hypothesis rather than a real explanation.

Whatever the explanation of solar activity may be, the actual events are followed carefully because of their effects on the Earth, most of which arise from solar flares. The main practical effect is on the ionosphere, which normally makes long-distance communication possible by reflecting radio waves that could not otherwise bend around the curved surface of the Earth. As soon as a strong flare occurs, however, its short-wavelength radiation—especially X-rays—increases the ionization of part of the ionosphere so that it absorbs radio waves instead of reflecting them. This disturbance lasts only a few minutes; but often a much more prolonged radio blackout occurs about two days later, when a burst of solarwind particles reaches the Earth. These charged particles also swing toward the Earth's magnetic poles and produce brilliant auroral displays. Auroras are pretty, but interruptions of radio communications are not. It is concern with communications that is responsible for a large part of the support that government agencies give to solar astronomy.

SOLAR RADIO ASTRONOMY

Quite aside from its effect on terrestrial radio communications, the Sun emits its own radio waves and is easy to observe with radio telescopes. The radio Sun look very different from the optical Sun; it is bigger, and its brightness indicates a temperature of a million degrees. These facts are not at all surprising, however, when we consider that as in the Earth's ionosphere, the electrons of the solar corona interact strongly with radio waves; as a result the level where the gas becomes opaque is well up in the corona. This is the Sun's radio "surface," and *its* size and temperature are what a radio telescope sees. Different radio wavelengths "see" to different depths, but nearly all are above the optical surface.

Solar activity affects the thin material of the corona much more than the denser material of the solar surface, and radio observations are therefore a rich source of information about the active Sun. Near the time of sunspot maximum, bursts of radio emission are frequent, and giant outbursts can be as much as a million times as strong as the steady level of radio radiation that astronomers call "the quiet Sun." Since different radio wavelengths can monitor different levels in the corona, radio observations add a new dimension to studies of rapid disturbances in the corona.

THE SUN AND OTHER STARS

The Sun is the only star whose mass and radius we know with high accuracy. We also know its age, which is simply the age that we calculate for the Solar System by studying radioactive decay in meteorites. We also know its chemical composition better than that of any other star, just because it is so bright that we can analyze its light in greater detail. As a result of all this knowledge, the Sun is the best check-point for theories of the structure of stars, and as such it has been very helpful.

Most of the value of the Sun in stellar astronomy, however, involves those details that we see just because we view the Sun as a disc rather than a point. If the Sun has these characteristics, then presumably other stars do too. There is no doubt that other main-sequence G stars must behave in a very similar way. What about stars of other types, however? Here our experience with the Sun helps in two ways. First, our detailed knowledge of the Sun tells us what to look for in other stars. Second, when we observe something unexpected in another star, we can often try to interpret it as an exaggerated version of some solar phenomenon.

Chromospheres and coronas are a good example. Many stars show emission lines in their spectra, in addition to the absorption lines with which we are familiar. These must come from hotter regions that lie above the star's photosphere, because cooler material would give only dark spectral lines. It appears that these stars have more prominent chromospheres than the Sun. Presumably they also have stronger coronas, but a corona is such a tenuous thing that even a rich corona can be drowned out by the other light of a star. Only in the most extreme cases do we see lines in stars that indicate temperatures as high as that of the Sun's corona. In the late stages of some nova outbursts, however, some of the very same lines are seen that are in the spectrum of the solar corona.

We would understand emission-line stars better if we understood the Sun better. Presumably in every case some process is depositing energy in the thin upper atmosphere of the star, where a small amount of energy can make a large difference to the temperature. What the process is, however, is not at all clear. The Sun is a sobering example, since the gigantic corona seems to have its origin in the inconspicuous phenomenon of surface granulation.

Nevertheless it is a good bet that stellar coronas should be accompanied by stellar winds. Some emission-line stars do, in fact, have expanding atmospheres, as shown by the large Doppler widths of the emission lines, velocities so great that the material must be escaping from the star. For other stars the outflow is not directly observable, but the existence of an emitting chromosphere suggests that, as in the Sun, a flow must begin farther out. It is intriguing to speculate that an exaggerated stellar wind may sometimes play an important role in the evolution of the star. As we shall see in the following chapter, studies of stellar evolution suggest that processes of mass loss must play an important role; if it is strong enough, a stellar wind may be one of these processes.

Another intriguing problem is the relation between rotation and a stellar magnetic field. The Sun, like other cool stars, is a slow rotator. Stars of the upper main sequence, on the other hand, tend to rotate rapidly, some of them a hundred times as fast as the Sun. Some B stars rotate so rapidly, in fact, that emission lines in their spectra are probably due to shells of gas thrown off by the rapid rotation.

With a faster rotation than the Sun, do these stars have correspondingly stronger magnetic fields? The answer is, yes and no. There is a class of hot stars with strong magnetic fields, but many others lack any appreciable field. The magnetic stars are not at all the fastest rotators; in fact, they seem to rotate more slowly than the average for their spectral class. They are nearly all A stars, with peculiar spectra in which the lines of some of the less common chemical elements are unusually strong. The strengthening of these lines is apparently connected with the existence of the magnetic field, in a way that we do not understand. Nor do we know why these stars differ from other A stars. It seems very unlikely, however, that we are dealing here with the same mechanism that is responsible for the sunspot cycle.

Unlike some special types of stars, which contribute to a general understanding of stellar behavior, the peculiar stars introduced here raise far more problems than they solve. Thus they play little role in today's systematic account of the workings of the universe, but they do remind us that there are more things in the heavens than are understood on Earth.

REVIEW QUESTIONS

1. Why does the Sun, a sphere of gas without any real boundary, appear to have a sharp edge?

2. Distinguish between the photosphere, the chromosphere, and the corona.

3. How do we know that the solar corona is very hot? What makes it so hot?

4. What is the solar wind, and what makes it blow?

5. What are sunspots? What causes them?

6. What is the sunspot cycle, and what effects does it have on the Earth?

7. In what ways is solar astronomy a field unto itself, and in what ways does it help us to understand other stars?

18 | THE STRUCTURE OF A STAR

Beneath the surface of a star is an interior that is completely hidden from our view. The hidden region contains nearly all the mass of the star, and hidden there is also the source of energy that keeps the star shining for billions of years. We can study the interior of a star only by purely theoretical reasoning. The basic principles of the theory, however, are amazingly simple, since all that they require is that the star keep itself in balance in two ways: mechanically and thermally. By analyzing these two balances, we can figure out the internal structure of the star.

The mechanical balance simply expresses the requirement that the star hold up its own weight. The star's own gravitation pulls it together and compresses it; but the more the gas is compressed, the harder it pushes back. A balance is reached when the gas pressure is just enough to support the weight. Thus the pressure at each depth within the star must be just enough to support the weight of all the overlying layers.

The thermal balance requires that the temperature everywhere inside the star be just high enough to keep the energy flowing outward. Conceptually we can start by considering the surface of the star, where a large amount of energy continually flows out in all directions into space. This loss of energy would quickly cool the star's surface, if the lost energy were not replaced by energy that flows up from below. The same is true at deeper levels; in each layer the upward-flowing energy must be replaced by an equal amount of energy from below. Ultimately, at the center of the star we must pay the reckoning; somehow new energy must be created to keep the flow going, by the burning of nuclear fuel near the center of the star, a process that we shall take up in some detail later on.

The key to understanding the thermal balance is to follow the flow of energy quantitatively. First of all, energy flows from hotter regions to cooler. Second, there is a resistance to the flow. For radiation, which is usually the mechanism that transports energy through a star, the resistance is opacity. The murky gas inside a star impedes the flow of radiation by constantly absorbing the energy and requiring it to be radiated over again. The greater the opacity, the greater is the temperature difference that is needed to drive it. Thus for a constant energy flow the temperature must rise with depth in the star, at just the right rate to keep the energy flowing upward. Using this principle, the astronomer can calculate the temperature at each level inside the star, just as he uses the mechanical balance to calculate the pressure at each level.

The principles are simple, but the practice is complicated. The precise equation for the temperature depends on the pressures, and the pressure equation depends on the temperature, and finally the pressure and temperature that result at the center must be just such that atomic nuclei under those conditions will generate the right amount of energy. The whole thing becomes a complicated computer process, but the answer eventually does come out, and the astronomer is able to calculate from first principles the entire interior structure of a star.

For a basic understanding of the universe, however, the important question is not how the astronomer behaves but rather how the star behaves. When material is put together to form a star, it adjusts its structure until these balances are satisfied: the pressure at every point balances the weight, the temperature gradient at every point is just sufficient to drive the energy flow, and the central temperature and pressure cause the atomic nuclei there to generate just the needed amount of energy. For a given chemical composition and mass, there is only one configuration that will satisfy all these conditions, and this determines what radius and luminosity the star will have. To put this differently, for a given chemical composition, a star of a given mass must have a specific radius and luminosity; there should thus be a unique sequence of radius and luminosity, depending only on the mass of the star. Translated into the HR diagram, this sequence of radius and luminosity is just the familiar main sequence. Thus, by reasoning about how a star is built, we have reached the conclusion that the main sequence is just the natural result of making stars out of different-sized chunks of the same kind of material. And it is equally clear why there should be a mass-luminosity relation.

At the same time it is easy to see why there should be stars that are not on the main sequence. These are the evolved stars, whose composition has changed at the center because they have burned all their nuclear fuel. With the composition changed in part of the star, the same balance requirements now give a different answer, and the star does not have the same radius and luminosity as a main-sequence star of uniform composition.

STELLAR ENERGY AND NUCLEAR STRUCTURE

To understand the stars fully, we must know where their energy comes from. Chemical fuel would be hopelessly inadequate for a star; a mass of coal equal to the Sun would provide the Sun's radiant energy for only 3,000 years. A more efficient process is gravitational contraction: as the material falls together to form a star, it releases energy. But contraction of the Sun from an infinite size down to its present density would provide energy at the Sun's present rate for less than 100 million years, much less than the length of time that we know life has existed on the Earth. Gravitational-contraction energy is not negligible—in fact, we shall see that it is responsible for the initial heating up of a star—but it is far from adequate for the length of an ordinary star's lifetime.

The only energy source that produces enough energy from a small mass is nuclear energy, and this is what keeps the stars shining. As we have already seen, the process that fuels nearly all the stars is the conversion of hydrogen into helium.

There are many ways of looking at nuclear energy, but certainly the easiest is to think of the nucleus as storing up energy in the forces that hold its component particles together. We can apply the same stretched-spring analogy that we used for the force holding an electron to the atom: stretching the spring stores energy, which can be released when the spring contracts. The forces between particles inside the nucleus are very much stronger than those holding the electrons. Just in the same way, however, pulling the particles into a more loosely bound state requires putting energy into a nucleus, whereas a nucleus whose particles become more tightly bound will release energy.

Physicists can measure the energy that binds a nucleus by an amazing method: they weigh it. The theory of relativity asserts that mass and energy are equivalent, according to the famous equation $E = mc^2$. Thus a nucleus that has given up energy in binding its particles tightly together should have less mass than the sum of the particles that went to make it up. It does indeed, and the difference is big enough to measure easily. Thus physicists are able, by means of mass measurements, to calculate the binding energy of each type of atomic nucleus. When one nucleus is transformed into another, the energy absorbed or released is simply the difference in the binding energies.

When all the quantities are expressed in everyday units, a little bit of mass m corresponds to a very large energy E, just because c, the velocity of light, is such a large number in everyday units. This means, conversely, that even nuclear-sized energies correspond only to small mass differences. Electronic energies also produce differences in mass, but they are a million times smaller and are therefore not practical to measure.

$E = mc^2$ also holds terrors, because it seems to imply that mass actually *can* be converted into energy. If this were possible, the amount of energy released would be tremendous. An ounce of matter would turn into enough energy to lift all the buildings of a large city high in the air. But this is science fiction, not science. Total conversion of mass into energy—annihilation, as it is called—occurs only for the rare "antiparticles" that are produced by some nuclear reactions, never for large-scale amounts of ordinary matter. Actual nuclear reactions convert only a tiny fraction of the mass into energy, so that we should regard $E = mc^2$ as just a convenient way of using the measured masses of nuclei to calculate how much energy a reaction will produce.

Even so, the realities of nuclear energy are impressive enough. Nuclear fuels—controlled or explosive—can produce energy yields as high as $0.002\ mc^2$, so that even a moderate amount of fuel can hold a devastating amount of energy—as our nervous world knows.

To return to the structure of the nuclei themselves, a nucleus is made up of particles of two kinds: *protons* and *neutrons.* Each of these particles has almost the same mass, but they differ in charge. A proton has one unit of positive charge (just as an electron has one unit of negative charge), and a neutron has no charge at all. Thus when protons and neutrons are put together into a nucleus, the nucleus has a positive charge equal to the number of protons, and has a mass, in atomic units, that is equal to the number of protons plus the number of neutrons. (The mass of the electrons that surround the nucleus is almost negligible.)

It is the charge—that is, the number of protons—that determines which chemical element a nucleus belongs to. The reason is simple. Consider, for instance, a nucleus that contains six protons and therefore has six units of positive charge. It will strongly attract any electrons in its neighborhood, because of their negative charge, and it has a good chance of capturing an electron into a bound orbit around the nucleus. This capture process will continue repeatedly, until there are six bound electrons. Thereafter a passing electron will feel the attractive force of the six protons plus the repulsive force of the six electrons, or a net of zero. More simply stated, the atom is electrically neutral over-all and has no further tendency to capture electrons. Nearly everything about the behavior of the atom—how it associates with others of its kind, how it combines chemically with other atoms, and how it absorbs and emits radiation—is governed by the fact that the number of electrons is six, rather than some other number; and the number of electrons goes back to the first cause, the number of protons in the nucleus.

Atoms that have six electrons around a nucleus with six positive charges exhibit the behavior of the element carbon. The usual way of saying this is that the atomic number of carbon is 6. Thus all carbon nuclei contain six protons, but they may in fact contain different numbers of neutrons. The neutrons affect the mass of the atom, although they have almost no effect on its chemical behavior.

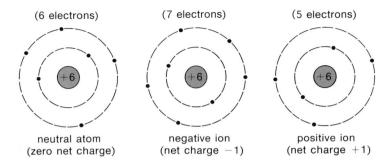

(6 electrons)   (7 electrons)   (5 electrons)

+6   +6   +6

neutral atom   negative ion   positive ion
(zero net charge)   (net charge −1)   (net charge +1)

FIGURE 18.1.
The "right" number of electrons for an atom is determined by the
charge on its nucleus; the natural state of the atom is neutral. If it has
too many electrons (second figure), it will tend to lose one; if it has
too few (third figure), it will tend to catch another electron.

Atoms of the same chemical element that have different masses (that is, with nuclei that contain the same number of protons but different numbers of neutrons) are called *isotopes*. The kinds of isotopes of an element that can exist stably are determined by the way in which protons and neutrons fit themselves together. Some combinations are more stable than others. Thus for carbon, six protons and six neutrons form a very stable nucleus, which is called carbon-12 (C^{12}). Six protons and seven neutrons also form a stable configuration, C^{13}. But six protons and five neutrons will not hold together; in a short time the nucleus ejects a particle and rearranges the remainder in a way that is then stable.

A combination of six protons and eight neutrons gives an interesting result. This nucleus, C^{14}, is almost stable—but not quite. It is likely to hold together for thousands of years; but eventually, at some unexpected moment, the instability asserts itself and a particle is ejected. This slow random break-up is called radioactivity. (See Chapter 11 for the use of radioactive nuclei in calculating the age of the Solar System.) Because of the action of cosmic rays, a small amount of radiocarbon, as C^{14} is called, always exists in the Earth's atmosphere, and all living material has its share of radiocarbon. The C^{14} nuclei gradually decay in dead material, which has no further exchange with the atmosphere; and the amount of radiocarbon in a sample of organic material is an index of the length of time since it was alive. Radiocarbon dating has been of immense value to archaeology.

So it is for carbon. Each other chemical element also has its own characteristic nuclear charge—that is, number of protons in the nucleus. The first and simplest element is hydrogen. Most hydrogen nuclei are H^1— just a proton. Hydrogen has one other stable nucleus, H^2, which has one proton and one neutron. This nucleus is often called by the special name *deuteron,* and a sample of hydrogen whose nuclei are deuterons is called *deuterium,* as if it were a separate chemical element. It is not; it is just "heavy hydrogen." Nevertheless, deuterium is important enough in nuclear affairs to deserve a special name.

A third hydrogen isotope, H^3, is radioactive and lasts for only a few years after having been made by nuclear bombardment.

Similarly, most of the other chemical elements have more than one stable isotope, and a few have long-lasting radioactive isotopes. The heaviest naturally occurring nucleus is uranium-238, which has 92 protons and 146 neutrons. The general structure of nuclei is still not fully understood; but even our incomplete understanding, along with a wealth of experimental data, allows a large number of successful calculations and predictions to be made.

NUCLEAR ENERGY: FISSION AND FUSION

The details of nuclear-energy production depend on the details of nuclear structure, but the over-all picture depends only on the general way in which binding energy depends on the size of the nucleus. A graph of this quantity makes the picture clear. The most tightly bound nuclei are in the middle of the range, and those at either end are less tightly bound. Hence nuclear energy can be produced by moving from either end toward the middle, by building the lightest nuclei into larger ones or by splitting the heaviest nuclei into smaller pieces. The build-up process is called fusion, and the splitting is called fission.

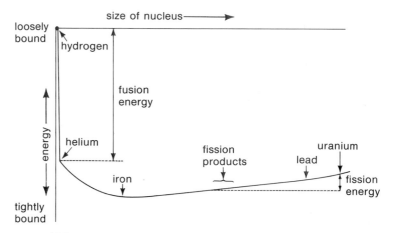

FIGURE 18.2.
This diagram illustrates the binding energy of nuclei of various types. Rearranging the particles into a more tightly bound nucleus results in the release of energy. Note how much more energy is available from fusion than from fission.

Nuclear fission involves principally one of the naturally occurring isotopes of uranium, U^{235}, and some isotopes of plutonium that can be made artificially in nuclear reactors. Whereas other radioactive nuclei eject small particles and make only small changes in their structure, these nuclei will spontaneously split almost in half, releasing in this one step almost all the energy that they are capable of giving up. What makes them usable as a practical source of energy is the principle of the chain reaction. A uranium nucleus that undergoes fission throws out several neutrons, in addition to the two remaining large pieces, and these neutrons are capable of making other uranium nuclei undergo fission at once. Thus in a pure sample of fissionable material an initial fission event will cause several others, each of which in turn causes several others; and in a very short time every nucleus undergoes fission and all the available nuclear energy is released. This will not happen, however, in a small mass of material, because then most of the neutrons will leak out at the edges. Only in a large mass will enough of the neutrons be held in for a runaway chain reaction to occur. The minimum mass for which this will happen is called the critical mass.

The process just described is that of a fission bomb. The bomb contains several separate masses that add up to more than a critical mass. When these are put together, a runaway chain reaction occurs, and all the energy is released.

Fission energy need not be explosive; it can also be controlled. If an inert, non-fissionable substance is inserted between the pieces of fissionable material, it will slow down or absorb some of the neutrons, and the rate of fission can be kept at a controlled level. The heat that is produced can then boil water to drive steam turbines and generate electric power. This is how a nuclear power plant works. The controls must be managed with extreme care, however. If the flux of neutrons were ever allowed to go too high, the nuclear fuel could "go critical," and a nuclear explosion would result, with all its disastrous consequences.

Fissionable materials are rare, however, both on Earth and in the stars, where they make up less than a billionth of the material. The process for which almost unlimited fuel is available is fusion. The primary fuel here is hydrogen, the most abundant of all the elements.

HYDROGEN BURNING IN THE SUN AND ON EARTH

Stars generate energy by a variety of nuclear reactions, but the great majority of them get their energy by converting hydrogen into helium. The basic result of fusing four nuclei of hydrogen to make one of helium can be accomplished in either of two ways: the proton-proton chain and the CNO cycle. In the first, hydrogen nuclei fuse directly with each other;

in the second, nuclei of carbon, nitrogen, and oxygen act as intermediaries. Which process dominates depends mainly on the temperature at the center of the star; the result of either is the same.

The high-temperature gas at the center of a star is a mixture of nuclei and free electrons. Collisions are frequent, and they occur at high speeds. Even so, the electrical repulsion between two positively charged nuclei prevents their coming close together, and they almost always rebound harmlessly. Only the very highest-speed collisions press the nuclei close enough together to allow nuclear reactions to occur, and these highest-speed reactions are very rare. Thus the fuel burns slowly.

The proton-proton chain begins with one of these high-impact collisions, when two hydrogen nuclei, each a single proton, remain bound together by nuclear forces. A nucleus that consists only of two protons would not be stable; but the fusion process ejects one of the charges, in the form of a positron, and converts one proton into a neutron. A positron is a particle that is just like an electron but has a positive charge. As soon as the positron meets an electron, the two annihilate each other and release a photon of radiant energy, which simply makes up part of the energy production.

The main result of the reaction, however, is that two H^1 nuclei have fused to form a nucleus containing one proton and one neutron, or H^2. This nucleus very soon collides with another H^1 and simply fuses; the resulting nucleus, with two protons and one neutron, is He^3. The further reactions can go in a number of different ways, but in every case the net result is fusion with one more H^1 and ejection of another positive charge, to leave a stable nucleus containing two protons and two neutrons, or He^4. Over-all, four H^1 nuclei have fused to make one He^4 nucleus, and there has been an energy release equivalent to the binding energy of the He^4.

Table 18.1 shows the three routes that the proton-proton chain can take. At each branch point, the percentage of the reactions that takes each route depends on the composition of the material, the temperature, and the pressure. Each route, however, produces some neutrinos, which are a bizarre sort of elementary particle. They react very little with other particles, so little that they are able to escape directly from the center of a star, passing easily through all the overlying material. Thus neutrinos from the Sun should be passing through the Earth all the time. Physicists have made attempts to detect them, but with no success. This failure to detect solar neutrinos is one of the sore points of modern astrophysics. Something is wrong somewhere. Either the experiment is being performed wrong, or astronomers are seriously wrong in their picture of stellar structure, or physicists badly misunderstand the properties of neutrinos. At present, the experts in each area believe that their own house is in order, and no one knows where the error will turn out to lie. The problem of the missing solar neutrinos is a sobering reminder that even the securest pictures may eventually need revision. For the time

TABLE 18.1. THE PROTON-PROTON CHAIN[a]

---

(1) $H^1 + H^1 \rightarrow H^2 + e^+ + \nu$

(2) $H^2 + H^1 \rightarrow He^3 + \gamma$

Branch point: next step can be either (3) or (4).

(3) $He^3 + He^3 \rightarrow He^4 + H^1 + H^1$

(End of chain I.)

(4) $He^3 + He^4 \rightarrow Be^7 + \gamma$

Branch point: next step can be either (5) or (8).

(5) $Be^7 + H^1 \rightarrow B^8 + \gamma$

(6) $B^8 \rightarrow Be^8 + e^+ + \nu$

(7) $Be^8 \rightarrow He^4 + He^4$

(End of chain II.)

(8) $Be^7 + e^- \rightarrow Li^7 + \nu$

(9) $Li^7 + H^1 \rightarrow He^4 + He^4$

(End of chain III.)

---

[a]These steps are given in the notation of nuclear reactions; e^- is an electron, e_+ a positron, γ a gamma-ray photon, and ν a neutrino.

being, however, astronomers go on the assumption that our picture of stellar structure is correct.

If so, this chain is how energy is generated in the center of the Sun. Although it produces a large amount of energy, the reaction is really a very slow one, as should be obvious from the fact that the fuel at the Sun's center, subjected continuously to these conditions, still takes billions of years to be consumed. In this respect the Sun is quite unlike the customary sort of engine, in which fuel burns rapidly but only small amounts are injected at one time. In the Sun the fuel is there all the time, but it burns only fast enough to keep the engine from cooling off.

The controlling reaction in the proton-proton chain is the first one, in which two protons fuse to form deuterium. The average proton at the Sun's center lasts for billions of years before it undergoes this transformation. The next reaction, by contrast, is very fast. A deuteron will, at that temperature and pressure, react with a proton in a few seconds. It is this reaction, which yields about 30 percent of the energy of the proton-proton chain, that is the basis of fusion energy on Earth. A small part of naturally occurring hydrogen (about 1 part in 6,000) is deuterium. This can be separated out and a mixture made that is half deuterium and half ordinary hydrogen; the mixture will then serve as fuel for the deuterium fusion reaction. The reaction requires, however, that the nuclei collide with each other hard, as they do in the high-temperature gas inside the Sun. Once the fuel begins to react, the energy that is produced can keep the temperature high, but the gas must first be raised to a temperature of about a million degrees to get the reaction going at all.

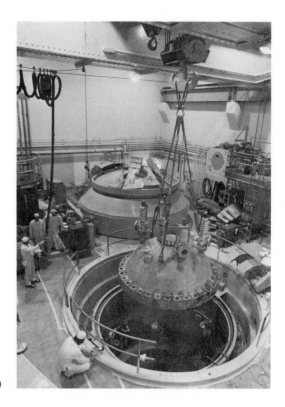

FIGURE 18.3.
The interior of a nuclear reactor.
(Courtesy of Pacific Gas and Electric Company.)

As with fission, the easiest kind of fusion reaction to create is an explosion. In a hydrogen bomb a small fission bomb is exploded to create the million-degree temperature, and this ignites a larger mass of deuterium fuel. Deuterium produces somewhat more energy than uranium, and it is of course much more readily available. The unit of nuclear terror, the megaton, is the amount of energy that is contained in about 50 pounds of a deuterium-hydrogen mixture.

The constructive use of fusion energy, by means of controlled fusion reactions, is man's hope for a continuing energy source in the future. Fossil fuels, such as coal and oil, will be exhausted in the next century or less, and the supply of uranium is comparably limited. But the hydrogen of the oceans is for all practical purposes inexhaustible. The practical problem that has not yet been solved, however, is how to use it. To burn deuterium slowly at a million degrees requires containing it in a controlled way at that temperature. It is the old problem of the "universal solvent" that will dissolve any substance—what do you keep it in? A million-degree gas will very quickly vaporize any substance with which it comes in contact. One possible solution would be to suspend the gas in some configuration of electric and magnetic fields that will keep it compressed and keep it away from the walls of the tank. Another possibility is to burn fuel in short bursts, by rapidly heating a small amount of fuel with the concentrated energy of a laser beam. Much effort has been spent on the objective of controlled fusion, but the problem has not yet been solved.

THE CNO CYCLE

In the Sun and in the other stars of the lower main sequence, energy is generated by direct fusion through the proton-proton chain. There exists another sequence of nuclear processes, however, that is more efficient at higher temperatures and is actually the energy source for the stars of the upper main sequence. The result is the same, the fusion of four H^1 nuclei to make a nucleus of He^4, but in this cycle other nuclei participate in the process. These nuclei become successively carbon, nitrogen, oxygen, and then the original carbon again. Each time around the cycle, a carbon nucleus fuses successively with four H^1's and eventually ejects a He^4. The carbon enables the net result, hydrogen fusing to helium, to happen, but it is not itself consumed; it acts as what a chemist would call a catalyst.

The actual sequence of reactions is shown diagrammatically in Figure 18.4. One can begin reading it at any point, since it is a continuous cycle; but a logical point to start is at C^{12}, where the build-up starts. One by one, H^1 nuclei—that is, protons—are absorbed. Twice, when the resulting nucleus is too proton-rich to be stable, it ejects a positive charge. Otherwise the build-up is continuous until the final stage, at which the product nucleus of helium is ejected.

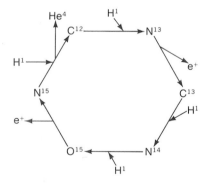

FIGURE 18.4.
Schematic outline of the CNO cycle. The arrows that enter the cycle represent the four hydrogen atoms that are consumed. The emerging arrows represent two positrons (e^+), which are quickly annihilated, and the helium atom that is produced. The C, N, and O symbols represent the successive stages through which the same host nucleus passes again and again.

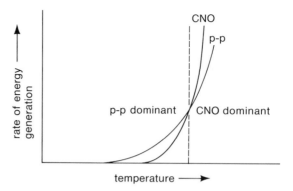

FIGURE 18.5.
Because the energy generation of the CNO cycle
depends on temperature much more strongly than
that of the p-p chain, the CNO cycle becomes the
dominant source of energy at higher temperatures.

The positron-ejection reactions are rapid ones, but the proton-fusion steps are slow at the temperatures and pressures that exist inside most stars, and thus the hydrogen fuel burns slowly. Here again, it is the fact that positive charges repel each other that makes it so hard for a proton to come close enough to fuse. Collisions of high-enough speed are rare. Both here and in the proton-proton chain, the number of collisions at high-enough speed depends very sensitively on temperature. The temperature dependence of the CNO cycle is much stronger than that of the proton-proton chain, however; hence at low temperature the CNO cycle produces only a negligible amount of energy compared with the proton-proton chain, whereas at high temperature the energy production of the CNO cycle completely dominates. The crossover point in Population I stars is at about 18 million degrees, which is the central temperature of a main-sequence star of spectral type F. The lower abundance of the CNO elements in Population II stars has hardly any effect on this question, since the temperature effect dominates so strongly.

One may wonder why these temperature-sensitive reactions do not run away at the center of a star and convert it into a nuclear bomb. Energy generation produces heat, which would tend to raise the temperature, which would increase the rate of energy generation, and so on until complete disaster. What prevents this runaway and keeps a star stable? The answer is in the behavior of a hot gas; the combined effects of temperature and pressure act like a thermostat. Heating a gas causes it to expand, and the expansion then cools it with an effect that is greater than that of the original heating. Thus the center of the star keeps in very good temperature balance; any tendency to overheat would cause an expansion that lowers the temperature again, and any slackening of the heat would cause a warming contraction.

THE LIFE HISTORY OF A STAR

Stars are born out of clouds of interstellar material. The details of the process are not well-understood; what we do know about it will be discussed in the following chapter, along with the general behavior of interstellar clouds. Once the contraction toward stellar size gets underway, however, some things are easy to predict. One is that the protostar, as we may now call it, will heat up. This is just the conversion of gravitational energy into heat; it is perfectly analogous to the work that falling water does in driving a mill-wheel. As in the case of the mill-wheel, what counts is not the speed but only the distance through which the water drops. The difference here is that all the energy released by the falling-together material appears as heat.

Like every other body in the universe, the protostar radiates according to its temperature. When its surface reaches a high-enough temperature, the protostar looks very much like a star. It has a luminosity, continuous spectrum, and line spectrum that are determined by the temperature and density of its surface, according to the simple and inevitable laws of physics. Astronomers plot it on the HR diagram; so we should now properly call it a pre-main-sequence star. Its interior has not yet settled down to a stellar state, but on the outside it looks very much like an ordinary star. Objects of this type are hard to recognize for what they are, but it is almost certain that the T Tauri stars, discussed in Chapter 15, are in this pre-main-sequence stage of evolution.

The contractive stage of a star's life is relatively short, since the contraction process provides it with only a modest total amount of energy to radiate. What happens finally is that the contraction raises the internal temperature higher and higher, until finally the nuclear hydrogen-burning reactions begin at the center. The star has now become a normal member of the main sequence.

The main-sequence life is the major part of a star's lifetime. It is stable and steady; it will remain the same as long as it can go on burning hydrogen to replace the energy that flows out through its surface. Thus the majority of stars that we see are main-sequence stars.

But no fuel-burning process can go on forever; there is only a limited amount of hydrogen in the star's center. In the hotter stars convective currents carry in fresh hydrogen, drawing from a large central core that makes up about a tenth of the star's mass. In less massive stars, whose centers are not as hot, this convection of fresh fuel is less important; instead the hydrogen-burning zone eats its way out like a slowly moving grass fire. In either case, the balances of pressure and of energy flow can maintain themselves in their usual form only until about a tenth of the star's fuel is converted from hydrogen to helium. Thereafter the structure of the star changes. The over-all rules of pressssure and energy balance are the same, but the local ground rules have changed. The star now has a central core that consists almost completely of helium, at a nearly

340

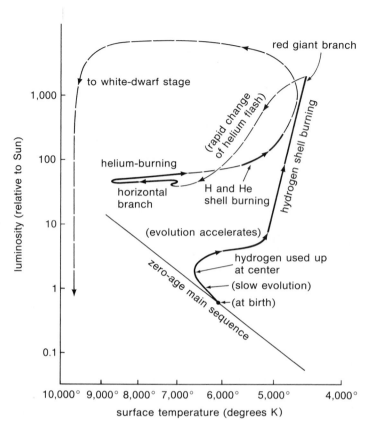

FIGURE 18.6.
The evolution of a globular-cluster star, in an HR diagram that is
marked according to physical rather than observational quantities.
(After I. Iben, *Publications of the Astronomical Society of the
Pacific,* **83** (1971), 708.)

uniform temperature because no new energy is being generated within
it. The energy generation is now taking place in the thin shell where the
high temperature of the core meets the fresh hydrogen of the envelope.
As the shell source of energy eats its way outward, the result of the various balances is such that the surface of the star gets larger while the
core becomes denser and hotter. The increasing surface size means that
the star moves to the right in the HR diagram, away from the main
sequence.

As the temperature rises at the star's center, we must ask whether other
nuclear reactions will begin. The material there is mostly helium; can
helium nuclei fuse with each other to release more energy? The answer is
that they can, but only with difficulty, because the product of fusing two
He⁴ nuclei, beryllium-8, is unstable and remains together for only the

tiniest fraction of a second. A stable result can be achieved only if this Be^8 nucleus again fuses—during its all-too-brief lifetime—with a third He^4 nucleus, to form a stable nucleus of C^{12}. This succession of steps is called the "triple-alpha process," according to the old nickname "alpha particle" that the He^4 nucleus has borne ever since it was first encountered among the products of radioactive decay.

The triple-alpha process produces energy in appreciable amounts when the temperature at the center of a star reaches 100 million degrees or more. It produces only about a tenth as much energy per gram as hydrogen fusion did, but this is still a lot of energy in a region where none was being generated, and the onset of helium burning causes further changes in the structure of the star.

In a massive star helium burning begins in a gradual way, but once established it of course makes an additional contribution to the star's evolution. Stars of low mass, however, are subject to a different phenomenon, called the helium flash, in which helium burning begins with extreme suddenness and changes the entire structure of the star in a short time. The reason for the difference in behavior is the difference in the nature of the gas at its center.

Ordinarily, when a gas is subjected to a higher pressure it becomes denser. At a fixed temperature, in fact, the density is proportional to the pressure. Even under the high-density conditions at the centers of most stars, this "perfect-gas" law continues to hold. Under really extreme conditions, however, the electrons of a gas strongly resist being pushed any closer together. This phenomenon, which is a consequence of the quantum nature of matter, is called degeneracy. Strangely, in a degenerate gas the density depends on temperature in a backward way, in the sense that higher temperature allows more compression! This reverses the usual behavior, in which heating *causes* a gas to *expand*; if a gas is degenerate, heating *allows* it to *contract*. As a result, the thermostat effect of gas expansion on nuclear-energy production, described on page 338, now has its wires crossed. Thus if a nuclear reaction begins in a degenerate gas, it heats the gas and allows it to be compressed further by the surrounding pressure. The compression heats the gas further, this makes the nuclear reaction go faster, and the whole process runs away, until the gas finally reaches a temperature at which it is no longer degenerate. Then the usual thermostat process is again effective.

This is what happens in the late stages of evolution of a low-mass star. Theoretical calculations have been followed in special detail for Population II stars in globular clusters. When hydrogen is exhausted in the star's core, the star continues to burn hydrogen in a shell surrounding the core. The surface begins to swell, and the star leaves the main sequence. As the star moves up the subgiant branch, its core gets hotter and denser. When the star evolves up the red-giant branch, the gas in its core is degenerate but still getting hotter. Finally, when the temperature

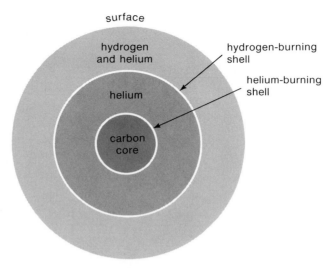

FIGURE 18.7.
Cross section of a star at a late stage of helium burning.

at the center reaches about 100 million degrees, helium begins to burn. Quite suddenly the degenerate gas turns into a normal gas of higher temperature. Helium then continues to burn, but the whole structure of the star readjusts as rapidly as it can. The helium flash at the center takes only seconds, but the restoration of balance to the whole star occupies the length of time that it takes radiation to flow through the star, or thousands of years. For stellar evolution this is still a short time, however, and the star can be said to make a sudden jump in the HR diagram— probably from the red giant branch to the horizontal branch.

These late stages of stellar evolution are still not well-understood theoretically, nor is there a clear enough guide from observational details of HR diagrams of clusters. What we can be sure of is that as evolution proceeds, the center of the star can heat up further, and a bewildering range of nuclear reactions becomes possible. Which ones will occur depends not only on the temperature and density but also on the composition that has been established by all the previous reactions.

The final end comes when previous nuclear processes have exhausted their fuel, and the temperature fails to rise high enough to ignite a new energy-generating reaction. The star can then only cool slowly, radiating away its stored-up internal heat. How far a star will go before this happens, we are still unable to predict, but it is likely that the average star carries considerably beyond the mere burning of hydrogen to helium and then of helium to carbon.

It is also likely that the late stages of a star's evolution involve the ejection of a good part of its material into interstellar space. On the deductive side, we can examine end-product stars, in which the generation of nuclear energy has ceased; their masses tend to be low, as if their progenitors had thrown off most of their mass before reaching this final stage. On the observational side, we observe many processes by which stars of various types are losing mass at the present time. Some stars have a slow, steady outflow, like an exaggerated solar wind. Some throw out repeated shells of material. The objects called planetary nebulae throw out one large shell and appear to be subsiding toward a white-dwarf-like object. And the supernovae blast out most of their material in one great explosion, leaving behind only a small, condensed core.

WHITE DWARFS

Even though we have only a hazy understanding of how a star makes the transition from nuclear activity to its final quiescent stage of evolution, it is clear that many stars have reached that stage, and some things about their structure are also quite clear.

The best-known final-stage stars are the white dwarfs. It is easy to see, just from their position in the HR diagram, that these stars must have very high densities. As was explained in Chapter 16, the color of a star indicates its surface temperature, and from the star's luminosity and Stefan's law we can then tell how large the radius of the star is. When combined with the star's mass—which can be calculated for some white dwarfs that are members of binaries—this small radius implies a tremendous density. Matter that is so highly compressed must be degenerate, and this implies that no nuclear reactions are going on. As was explained in the preceding section, nuclear reactions in degenerate matter are unstable and lead to a rapid change of structure. Since this does not happen, it must be that there is no nuclear energy generation inside a white dwarf.

What then keeps a white dwarf star shining? It is not nuclear energy, and it cannot be gravitational energy either, since the degenerate gas refuses to be compressed any further and this makes contraction impossible. The answer is that here, finally, is a star in which the energy radiated from the surface is *not* compensated for by an internal energy source. The energy that a white dwarf radiates is never replaced. The star is "living on capital," radiating away slowly but irrevocably all the internal energy that it has. Eventually it will cool off completely and have nothing left to radiate. In a sense, the only thing that keeps a white dwarf shining is the great length of time that it will take to cool off.

The lifetime of a white dwarf is long because it is a star that has a low luminosity but a relatively large store of internal energy. Furthermore, it continually becomes cooler and radiates less, so that the cooling-off process slows with time. The star evolves more and more slowly down the white-dwarf sequence in the HR diagram, which is simply the path that a star follows when its surface cools while the star keeps a constant radius.

If the star could continue in this way, it would take an infinite length of time to cool completely. Two changes intervene, however. One is the solidification of the interior. Like other processes of freezing, this releases energy. (We are more familiar with the reverse process, in which an ice cube *absorbs* energy when *melting*; this is what gives ice cubes their great cooling power.) Now the star has, temporarily, an internal energy source again. As it loses energy by radiation, it no longer compensates by changing its structure; instead it replaces the energy by solidifying more of its interior. It is in balance again, like a normal star. But this is disastrous, because the star now stays at a constant luminosity, instead of continuing to dim gradually. It continues to radiate at a rate that it can afford less and less as its interior continues to solidify. Furthermore, the solid state that it attains is one from which no more heat energy can be extracted. Eventually, when solidification is complete, the last available energy leaks out, and the star is finished as a star.

The second unfortunate thing that happens to a white dwarf is that as it cools, its atmosphere develops currents of warm gas rising and cool gas sinking. This process of convection transports the internal heat upward through the atmosphere much more efficiently than radiation can, and it further hastens the loss of the star's internal energy.

Modern theories of the structure and evolution of white dwarfs are by no means certain, but they predict that a white dwarf will last less than a billion years. The number of white dwarfs that we see in the region of space surrounding the sun is indeed consistent with this picture. The ones that we see are the residues of the local main-sequence stars that have ended their energy-burning lives in the last few hundred million years. This picture also suggests that there are many invisible, burned-out stellar remnants in the space around us, not numerous enough to constitute a large part of the material present, but hiding there nevertheless.

NEUTRON STARS

The theory of white-dwarf structure predicts one interesting physical relationship, which has an even more interesting consequence. The theory predicts that the radius of a white dwarf will be determined only by its mass and its composition, regardless of its temperature. Thus a typical white dwarf, with a slightly smaller mass than the Sun, should have a radius only a little larger than that of the Earth; and white dwarfs are indeed observed to be of this general size. The really interesting thing about the relationship, though, is that the greater the mass, the smaller the radius—and there is a limit! The exact limit has not been reliably enough calculated, but at somewhere between 1 and 1.5 solar masses the radius of white-dwarf configuration becomes zero. In other words, no white dwarf is possible at all above this limit; a larger mass would simply collapse to an infinitely small size under its own self-gravitation.

But a complete collapse does not take place; here nuclear physics comes to the rescue. Under the very high compression that takes place, the electrons of the ionized gas are forced into the nuclei, where they combine with protons to form neutrons. Since the star began with an equal number of positively charged protons and negatively charged electrons, the result consists totally of neutrons. By the time the star has reached this degree of compression, the whole inside of the star is as dense as an atomic nucleus. At this stage the nuclei blend with each other and lose their separate identity, and the interior of the star becomes simply a dense gas of neutrons. Their degenerate pressure is enough to support the weight of the star, and the result is a stable neutron star. The compression is now so great that a solar mass has a diameter of only a few miles.

Theory predicts that neutron stars can exist, and observation shows that they do exist. They are the pulsars. The nature of pulsars is clear from the rapidity and regularity of their flashes. The repetition is so regular, at so constant a period, that what we see must be a period of rotation. But it is very rapid: typically a second, and sometimes much shorter. It is easy to show theoretically that such a body would be torn apart by its rotation, unless it were held together by a very strong gravitational field. Here the required mass and size imply such a high density that only a neutron star will do.

Because of the extreme states of matter that are involved, the structure of neutron stars is poorly understood. In the few years that pulsars have been known, observational facts have been accumulating rapidly, however; and understanding of these exciting objects is sure to grow each year.

BLACK HOLES

Like white dwarfs, neutron stars also have an upper mass limit, beyond which no density, regardless of how great, will produce a pressure large enough to bear the weight of the material. At the limit, which is a little greater than 2 solar masses, the theory of neutron-star structure predicts that the radius of the star becomes zero. For any greater mass there is no possible equilibrium at all, and as with an over-massive white dwarf, a collapse takes place. This time, however, there is no salvation. The neutrons cannot blend into more compact particles that will respond with enough pressure to support their now quite fantastic weight. For such a star there appears to be no alternative to a complete and unrestrained collapse. The material falls together into a smaller and higher-density region, without any limit to its smallness. That is, it becomes infinitely dense and infinitely small.

This strangest of stellar destinies is called a black hole. It is a hole because nothing can ever escape from it again, and black because there is a level at which the gravitational field is so strong that not even light can get out.

A black hole is a confusing object to contemplate, in a quite literal sense. It looks different, depending on whether one looks at it from outside or from inside. This is a consequence of the very strong gravitational field; the theory of relativity says that under such circumstances the rate at which time appears to flow depends on the location of the observer. To an observer on the outside, a clock in a gravitational field appears to run slow. For a gravitational field as weak as that of the Earth, only a very cleverly designed experiment can detect the difference; but near a black hole the gravitation is so strong that the time effects are very large. If someone were to fall into a black hole, his own experience would be mercifully brief. In an extremely short time he would "hit bottom," with all the consequent disaster. To us, however, watching from the outside, the events of his demise would appear to run more and more slowly, so that for us he would appear to take an infinite length of time to reach the point of no return.

All these strange effects happen only within a very small region, however. The critical surface, from which light rays fail to escape, has a radius of only a few kilometers. Farther away, the gravitational field is much weaker and behaves more and more like the simple Newtonian gravitation. From a distance, the mass appears to be just that of the object that collapsed to form the black hole.

Thus when a massive star collapses to form a black hole, it stops radiating but continues to exert the same gravitational force as before. The exotic effects associated with its immediate neighborhood affect only the tiniest volume of space. For all practical purposes a black hole is just a mass that does not radiate. This makes it extremely difficult to find a black hole, even if they exist in appreciable numbers. Although theories of stellar evolution suggest that some stellar remnants should indeed form black holes, and several claims have been made for their discovery, there is no compelling evidence, at this writing, of the identification of a black hole anywhere.

There are several possible ways in which black holes might be found. First, and best, if a black hole is a member of a binary pair, it should make its presence known by its gravitational effect on its companion. We would see the bright companion as a member of a spectroscopic binary whose other member contributes no light. If the invisible star can be established to have too large a mass to be a white dwarf or a neutron star, then it must be presumed to be a black hole.

The most plausible case yet presented for the identification of a black hole is the strange object known as Cygnus X-1. First discovered as a strong source of X-rays, this object was then identified as a star. The spectrum is that of a B star, and the radial velocity is variable, so that the system must contain an unseen companion, which is presumably the source of the X-rays. All the evidence points to the companion being a highly condensed object, either a neutron star or a black hole. The choice hinges on its mass, whose correct value depends on various observed quantities in a complicated way. Many astronomers believe that Cygnus X-1 contains a black hole, but others regard the case as not yet proven.

A second possible way of detecting black holes would be to find a stellar system whose structure indicated the presence of a great deal of mass in a region that gives very little light. If there were other reasons to expect the formation of black holes there, the extra mass might be taken as evidence of their presence. It would be very difficult to be sure of such a conclusion, however, since the missing mass could easily be in the form of stars of very low luminosity or planet-sized dark bodies.

Still another possibility is that a black hole may exist in a very dense region, like the center of a galaxy, where material is constantly drawn into it. Matter that is close to the critical surface would not be conspicuous, since the apparent slowing of its clock makes its light very dim; but the material that is just starting down might funnel together violently enough to produce some conspicuous light. When quasars were first discovered (see Chapter 22), models of this sort were suggested, to explain their very high luminosities; but the models have not stood the test of closer examination, and black holes at the centers of galaxies are still an unfounded speculation.

Finally, we might hope that a black hole could be recognized by the way in which it strongly perturbs light that passes close to it. The trouble here is that the region in which this happens is so small that our chances of seeing a distorted-looking object behind a black hole are almost nil.

All in all, the idea of black holes is tantalizing. Even if we are unable to find them, there is good reason for believing that black holes should exist. If a star is unable to generate any more nuclear energy, it *must* contract to a condensed state; and if it is too massive to be either a white dwarf or a neutron star, it can only become a black hole. Unless all massive stars are able to throw off nearly all their mass, the final state of many of them must be black holes. It is indeed frustrating that these objects are so hard to find.

NUCLEOSYNTHESIS

Throughout the ages, stars have continued to shine, burning hydrogen to helium, helium to carbon, and then starting up reactions that build up even heavier nuclei. This is a one-way process; these larger nuclei are not broken down again. Thus stellar energy processes have a permanent effect on the chemical composition of the material of the universe. How serious is this effect? Is it possible, indeed, that such processes in the past have helped to determine the present composition of the material? Modern astophysicists incline very much to this view, and it seems possible that the abundances of the different chemical elements are determined almost completely by what has happened inside past generations of stars.

The strength of this opinion in based largely on the *un*likeliness that the chemical elements were made in any other way. Discussion of this question involves cosmology, and it will be taken up in Chapter 23. At this point we note briefly that the history of the universe began with a period when all of its material was at a high temperature. At that stage it was a mixture of what could only have been simple particles—neutrons, protons, electrons, etc.—together with radiation. When this material expanded and cooled, the neutrons either decayed into protons and electrons, or else combined with protons in reactions whose result is almost completely He^4. Thus, in the era prior to the formation of stars and galaxies, the expanding universe became a mixture of hydrogen and helium. It seems quite impossible for elements heavier than helium to have been formed in this initial condensation of the material into atomic nuclei. The only possible conclusion, then, is that the heavy nuclei were formed at a later date, and the only places with high-enough temperature and pressure are the insides of stars.

The basic problem of manufacturing the chemical elements inside the stars has received a great deal of attention from nuclear astrophysicists. Many processes have been studied by which heavy nuclei can be built up under suitable conditions of temperature and pressure. In understanding the setting of this work, it is important to appreciate the nature of the scientific predictions that are made: once the temperature, pressure, and initial composition of the gas are specified, the astrophysicist calculates how the nuclei *will necessarily* behave. There is no "might" or "maybe" about it; if our understanding of the nature of atomic nuclei and of the rules of nuclear structure are correct, the nuclei *must* behave that way. The only guesswork is in choosing the gross conditions of temperature and pressure and in predicting whether they will really ever occur in some type of star.

At temperatures and pressures higher than those that build He^4 into C^{12}, these nuclei continue to react with each other. It is relatively easy for successive He^4 nuclei to be captured; this "alpha process" builds up oxygen-16, neon-20, magnesium-24, etc. Along with these reactions, however, some neutrons and protons can be released, and these in turn react to form the intermediate nuclei of masses 17, 18, 19, etc. Under extreme enough conditions, however, reactions of innumerable kinds occur freely, and an equilibrium is set up in which energy is randomly distributed between the various possible kinds of nuclei. As normally happens in any such energy-sharing process, the lowest-energy nuclei—that is, the most tightly bound ones—end up being most numerous. Thus there is a preponderance of nuclei with masses near that of the most tightly bound nucleus of all, iron-56. In astrophysical terms, the *e*-process builds up the iron-peak elements.

Under suitable conditions, all the preceding reactions will go spontaneously, since they give off energy rather than requiring energy to be put in. Figuratively speaking, the production of elements up to the iron peak can be spontaneous, but production of the heavier elements must be forced. Yet we know that it must have happened somewhere, somehow, because the heavier elements do exist.

The most plausible way to force the build-up of heavy nuclei is to take iron-peak nuclei and bombard them with neutrons. A nucleus can gradually increase its mass by absorbing one neutron after another. Whenever it becomes too neutron-rich for stability, one of its neutrons ejects a negatively charged electron and thus turns itself into a positively charged proton. Calculations of this process, and the rate at which it will go for each kind of nucleus, have been successful in accounting for many of the detailed abundances of heavy nuclei. It is called the *s*-process; *s* is for "slow," to signify that the build-up is slow enough to allow a a neutron-rich nucleus to decay, by emitting an electron, before it absorbs another neutron. A different set of abundances will result, however, if the flux of neutron bombardment is so high that a neutron-rich nucleus absorbs even more neutrons before it has time to decay into a stable form. This is called the *r*-process (*r* for "rapid"). The heavy elements that we observe in the universe appear to be a mixture of products of the *s*-process and the *r*-process.

For either the *s*-process or the *r*-process, there must be very high temperature and density *and* a reaction that produces a lot of neutrons. The details of how this happens are not at all clear, but the presumption is very strong that it happens during the explosion of a supernova, which then throws this "enriched" material out into interstellar space. There it will become a part of the stars that are later formed out of that interstellar material.

The details are complicated and uncertain, and even more complications must be added in order to get all the abundances to fit; but this over-all picture is almost certainly right. All the elements except hydrogen and helium have been created in the insides of stars, which then spewed their material into space for other stars to be made from. The Sun is at least a second-generation star and may even belong to a third or later generation. The Earth that we live on, the book that you are holding, and even the atoms of your body itself are second-hand material, which owes its structure to a process of nuclear cookery in the distant past, deep in the center of some forgotten star.

EPILOGUE

The study presented in this chapter illustrates vividly the major points of this book: that the bodies of the universe follow the same laws of physics that hold here on the Earth, that they are as they are because the workings of those laws would not allow them to be otherwise, and that we probe the nature of the universe simply by applying those laws to the conditions that apply in stars.

We have seen how the requirements of equilibrium determine the structure of a star, how the initial contraction of a star heats it until nuclear reactions begin to generate energy, how its subsequent development depends on how the internal temperature develops and on what nuclear reactions are able to take place, and how the fate of the burned-out remnant depends on the way in which the nuclei are able to pack themselves together.

Much of this depends on the structure of atomic nuclei. The way in which particles join together in stable nuclei governs the kind of nuclei that exist, how they can be transformed, and the energy that is produced when they are. Finally, it governs the way that material has changed its composition in the interiors of stars, and the chemical composition that has resulted. Much of the nature of the universe is just a consequence of the way in which its smallest particles behave.

REVIEW QUESTIONS

1. Outline the principles by which astronomers can calculate the internal structure of a star. What simple physical balances are involved?

2. How do the principles of stellar structure account for the existence of the main sequence in the HR diagram? How is it possible, then, for a star to be a red giant?

3. Nuclei of oxygen each have 8 protons. What is the difference between oxygen-16, oxygen-17, and oxygen-18? Why is it reasonable to call all of them oxygen? How many electrons does a neutral oxygen atom have? Why?

4. Distinguish clearly between fusion and fission, in terms of
   (a) the physical process,
   (b) typical elements that are involved,
   (c) the occurrence of each process in nature,
   (d) nuclear explosions,
   (e) nuclear power plants.

5. Outline the processes that are involved in the proton-proton chain and the CNO cycle. What do they have in common, and in what basic way do they differ? Why do they go on readily only at high temperature?

6. In the birth of a star, what makes the nuclear energy-burning begin?

7. What is degenerate matter? Why are we sure that there are no nuclear reactions going on inside a white-dwarf star?

8. What is a black hole? Why is it hard to detect in any way?

9. Outline the processes by which we believe that the chemical elements of the universe were formed.

19 | INTERSTELLAR MATERIAL

Although the stars are the conspicuous objects of the sky, they do not by any means contain all the substance of the universe. Spread through the space between the stars is a large quantity of gas and dust. This interstellar material is a very thin scattering—imagine a thimbleful of air spread over a cubic mile—but over the tremendous interstellar distances, it adds up to almost as much material as there is in the stars.

Not only is interstellar material an important component of the universe in its own right, but it also contains the debris of stars gone and the stuff of which future stars will be made. Perhaps even more interesting, it is responsible for the star births of our own time. For that process, from the dimmest dusty seed to the glowing protostars, it is to the interstellar material that we must look.

The interstellar medium, as we call the general spread of this material, is a mixture of two components: gas and dust. The gas consists of separate atoms, in a mixture whose composition is very similar to that of the Sun and the other stars of Population I: about three parts of hydrogen to one of helium by weight, with all the heavier chemical elements adding up to only a few per cent of the total. This gas has a very low density—about 1 particle per cubic centimeter, compared with 3×10^{19} for the air we breathe—and it is also quite cool. In most parts of interstellar space the temperature of the gas is about 50° K. (Remember that on this absolute scale the temperature of the Earth's surface is around 300°.) The gas does not freeze, however, because at such low density the atoms have too little opportunity to meet and join each other.

FIGURE 19.1.
The bright nebula Messier 8, also known as the Lagoon Nebula, shows clearly that the interstellar medium is a mixture of gas and dust. (Kitt Peak National Observatory.)

Interstellar dust consists of small grains of solid material, each a tiny fraction of a centimeter in size. We are not at all sure of its composition, but it is probably made largely out of the elements heavier than hydrogen and helium. The grains, which are also very cold, make up only 1 per cent or less of the total interstellar material, but they have an important effect both on astronomical observation and on the development of the universe.

In a few places the interstellar medium makes itself evident as a glowing cloud or as an obscuring dark cloud, but for the most part it lies hidden. The astronomer who wishes to study interstellar material must rely on observations that are specifically designed to reveal it. To make the problem even more complicated, observations often give separate infor-

353

mation about the different components of what is really a single mixture. There is no neat single picture in which we can point to one part or another of the structure, nor is there a single neat spectrum in which we can pick out the lines of each element. We shall thus approach the interstellar problem as the astronomer does, beginning with the broad distinction between gas and dust, and introducing the various kinds of observations that bear on the nature and distribution of each.

EMISSION NEBULAE

The most conspicuous parts of the interstellar medium are the regions surrounding hot stars, where the gas glows brightly as a result of the energy that it absorbs from those stars. Such a region is called an *emission nebula,* because its spectrum consists almost completely of emission lines. ("Nebula" is Latin for "cloudlet.")

We can understand emission nebulae by first noting that they occur only around the hottest stars—spectral class B2 or hotter—and by then applying some straightforward reasoning about the behavior of gas and radiation. The key is in the very different transparency of the gas at different wavelengths. Over most of the spectrum it is almost completely transparent; but at wavelengths shorter than a certain "cutoff" in the ultraviolet, the gas is opaque. The cutoff point is the wavelength at which a photon has just enough energy to ionize a hydrogen atom that is in its lowest energy level. Photons of longer wavelength have too little energy to do this; in this energy range the hydrogen can absorb only those very few photons whose energy happens to coincide with a difference between some pair of the particular energy levels of hydrogen. Only those wavelengths, which are of course the spectral lines, can be absorbed, and other wavelengths pass through freely. By contrast, below the cutoff wavelength (which is called the Lyman limit), any photon can be absorbed, throughout a whole continuum of wavelengths, and any hydrogen atom that absorbs such a photon loses its electron in the process. As in all such photoionization processes, the required minimum energy detaches the electron from the atom, and the remainder of the energy of the photon sets the elctron in motion away from the atom.

Hydrogen gas is an avid absorber of photons whose wavelength is shorter than the Lyman limit; thus all the energy in that part of the spectrum goes into the first gas that the radiation encounters. This makes little difference around a cool star, most of whose radiation is at the longer wavelengths, for which the hydrogen is transparent. A very hot star, however, gives out much of its radiation shortward of the Lyman limit, and this is soaked up completely by whatever interstellar gas happens to surround the star. The gas is ionized by this absorption, but each absorption by an atom in turn uses up a photon. As a result, the hot

FIGURE 19.2.
One of the best-known of emission nebulae is the Orion Nebula, which is dimly visible to the naked eye as the hazy middle star in Orion's sword. (Lick Observatory; see also Plate 9.)

FIGURE 19.3a.
A short-exposure photograph of the
Pleiades star cluster shows only a
grouping of stars. (Yerkes Observatory.)

FIGURE 19.3b.
A longer exposure shows the Pleiades stars to be imbedded in clouds
of wispy dust, which reflects the light of the stars. (Kitt Peak National
Observatory.)

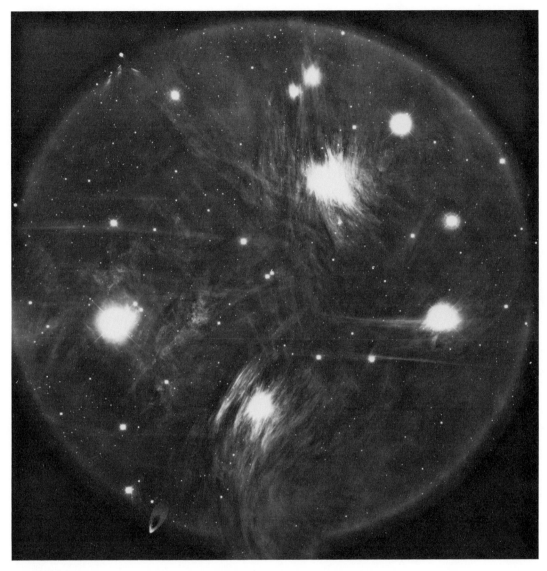

FIGURE 19.3c.
In a very deep photograph, the dust is visible everywhere, and its structure has bewildering complexities. (Kitt Peak National Observatory.)

star (or often a group of stars) keeps the gas around it ionized, out to the distance where the last ionizing photons are used up. Beyond this rather sharp boundary, there is no more ionizing radiation, and the gas is not heated up. Thus every hot star is surrounded by a nearly spherical region in which the interstellar gas is ionized and hot, but elsewhere the gas is neutral and cool.

The amount of ionizing radiation that comes from a star depends very sensitively on its temperature. O stars are surrounded by large ionized regions, early B stars by smaller ones, and beyond spectral type B2 the region of ionization is too small to be at all conspicuous. Many cooler stars do have bright nebulae surrounding them, but these are not spheres of hot gas that emit light. Instead they are "reflection nebulae," in which we simply see the surrounding interstellar dust illuminated by the star, just like the bright patch around a street light seen in a fog. The starlight is simply reflected by the dust grains, as we can tell from the spectrum of the nebula, which is identical to the spectrum of the star, dark lines and all, just as the spectrum of the Moon is that of the sunlight that it reflects.

The sphere of ionized gas surrounding a hot star is called an "HII region," in accordance with the spectroscopic custom of referring to an ionized atom by the Roman numeral II. The neutral-hydrogen regions, which are the major part of interstellar space, are called HI regions.

Because of the energy that is absorbed from the ionizing radiation, an HII region is hot, being typically around 10,000 degrees. But what is most striking is its spectral emission. This is a consequence of the recombination of atoms and electrons. Each ion, with its positive electric charge, attracts the negatively charged free electrons that are running about in the gas; and if an electron comes close enough, there is a chance that the atom will capture it. The atom is then neutral again, with the electron bound in one of the allowed energy levels. The binding energy gets radiated as a photon; but the photon's energy also includes the original energy of motion of the captured electron, and this may have any of a continuous range of values. Thus the radiation coming from recaptures is spread over a continuum of wavelengths and is usually quite inconspicuous, except possibly at ultraviolet wavelengths that have not yet been studied.

What *is* conspicuous is the radiation that the atoms emit *after* recapturing an electron. If the electron, immediately after recapture, is in any but the lowest energy level, it will drop to lower energy levels; and the atom will emit a photon at each downward jump in energy. There may be a single jump down to the ground level, or the atom may cascade down through several intermediate energy levels, emitting a photon at each step. Thus the recombining atoms, between them, will emit the whole line spectrum of the atom.

Although it is the hydrogen atoms that absorb most of the ionizing photons, collisions occur between all types of atoms in the hot gas, and

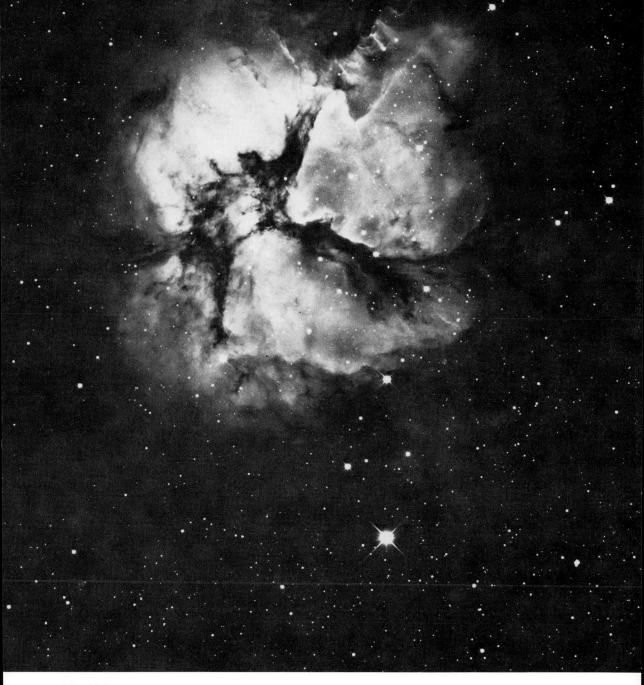

FIGURE 19.4.
Messier 20, the Trifid Nebula, is a typical H$_{II}$ region. (Lick Observatory.)

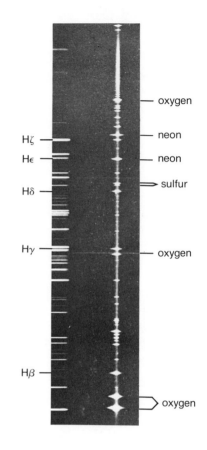

FIGURE 19.5.
This spectrum of the planetary nebula NGC 7027 is typical
of the spectrum of gas surrounding a hot star. Recombination
lines of hydrogen are marked on the left, and some of the
more prominent forbidden lines are marked on the right, by
element. (Lick Observatory.)

atoms of all types get ionized. They too recapture electrons and cascade
down through their energy levels, so that elements other than hydrogen
also emit their line spectra.

Included in the spectrum of some elements are strong lines that for a
long time defied identification. These correspond to transitions between
well-recognized energy levels, but they are missing from laboratory
spectra. The reason is that they can be emitted strongly only under the
very low-density conditions of interstellar space. Ordinarily these are
very improbable transitions for an atom to make; in fact, under the simplest picture of how electrons change from one orbit to another, they are
forbidden. More accurate theories show that these transitions are not
utterly impossible, but just very unlikely; nevertheless, "forbidden lines"
has become the standard name for them.

If an atom is in an excited energy level, it will normally make some
downward transition in energy, emitting a photon, in a millionth of a
second or less. This is a typical waiting time before a "permitted" transition takes place. Forbidden transitions, by contrast, are so unlikely that

atom must typically be left undisturbed for a whole second or so before it will make such a transition. Clearly, the atom is overwhelmingly more likely instead to make a permitted transition downward—unless it is in an energy level that *has no* permitted downward transitions. Such energy levels exist; they are called *metastable.* An atom in a metastable energy level can radiate only by emitting a forbidden line. At ordinary gas densities, however, the atom is very unlikely to do this, because it is rarely left undisturbed for a long enough time. Collisions with other atoms are much too frequent, and a collision can de-excite the atom without regard to the rules that restrict spontaneous emission of radiation. At the very low density of interstellar gas, however, collisions are rare. An atom that has recombined and has cascaded down, by chance, into a metastable energy level will often be left undisturbed for so long that it finally takes the otherwise unlikely step of emitting a "forbidden" photon.

Forbidden lines of many elements, particularly oxygen, are prominent in the spectra of emission nebulae. They are also important in other astronomical contexts, just because they give such a clear clue about the physical conditions: whenever we see forbidden lines, they must come from a gas of low density, either in interstellar space, or in a very tenuous envelope surrounding a star.

INTERSTELLAR ABSORPTION LINES

The cooler gas of the interstellar medium also produces spectral lines, but as absorption lines printed on the spectra of distant stars. Just as the atoms of a star's own atmosphere absorb their characteristic wavelengths, and produce dark lines in the spectrum of the star, so also do the atoms of the interstellar gas impress their spectral fingerprints. The interstellar contribution can be separated from the star's own spectrum by its different Doppler shift, which corresponds to the motion of the interstellar gas rather than to that of the star. In practice, however, interstellar lines are easily observed only in the spectra of hot stars; the spectra of cooler stars are inherently too complicated to allow interstellar lines to be disentangled from the numerous stellar features. Also, interstellar lines tend to be weak; to study them requires careful, accurate analysis of the spectrum of the background star. Such work can be done only on bright stars. As a result, there are rather few directions—the directions of hot, distant stars that are also bright—in which this method of probing can be used.

The results are particularly interesting, though, because they not only detect the presence of the interstellar gas, but also measure its motion. In the spectra of many stars, each interstellar line is multiple—i.e., has several components, each with a different radial velocity. These multiple lines show that there are several interstellar clouds along the line between us and the star. We cannot tell where they are, along the line of

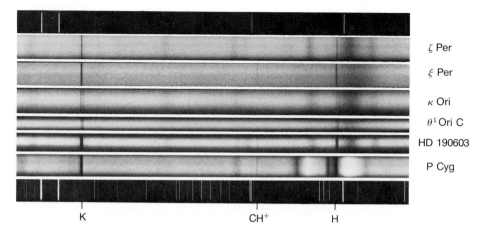

ζ Per

ξ Per

κ Ori

θ¹ Ori C

HD 190603

P Cyg

K   CH⁺   H

FIGURE 19.6.
The sharp absorption lines marked H and K are produced by interstellar gas. In some stars they have multiple components, showing the velocities of different clouds along the line of sight. Note also the line marked CH⁺, which comes from an interstellar molecule. (Lick Observatory.)

sight, but the difference in velocity shows their separate identity. A typical line of sight shows five or ten clouds per thousand parsecs. From other observations, which will be discussed below, it is probably not accurate to speak of these as separate, discrete clouds; but this figure nevertheless shows us the scale of the major structural features in the arrangement of the interstellar gas.

The strengths of the interstellar absorption lines also tell us something about the composition of the gas. These lines complement in a useful way the information that we get from emission lines in the hot HII regions. Just as in the atmospheres of the stars, the visible range of wavelengths singles out the excited energy levels for nonmetallic elements, which are seen only in a hot gas. In the cooler gas, again as in stars, the lines that we see arise from the ground level, and it is for the metallic elements that these lines fall in the visible part of the spectrum. Thus, by looking at both the hot and the cool regions, we can put together information on the abundances of nearly all the chemical elements.

Our knowledge of interstellar matter is being immeasurably increased now that observations are being made from Earth satellites. Outside the atmosphere, the astronomer can make measurements at the ultraviolet wavelengths that do not reach the Earth's surface. In this part of the spectrum are lines that arise from the lowest energy levels of nearly all the most common chemical elements. As in ground-based observations, the technique is to observe the spectrum of a distant star, using Doppler shift to distinguish those lines that have an interstellar origin. The engineering effort of placing a high-quality, remotely controlled telescope in

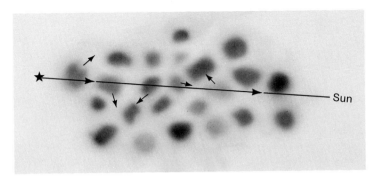

FIGURE 19.7.
The light from a distant star may pass through several interstellar
clouds, each with a different Doppler shift. If so, each cloud will
produce interstellar lines with slightly different wavelengths, as in
some of the examples in Figure 19.6.

orbit is immense; but there is hardly a field of astronomy where the
results are as rich as those derived from ultraviolet measurements of
stellar spectra.

THE RADIO LINES OF HYDROGEN

Much of our information about the interstellar medium has come from
the long-wavelength end of the spectrum, in the radio domain. A radio
telescope receives this radiation just as an antenna receives a radio message or program; the difference is that an intestellar gas cloud does not
emit a regular signal. If we put interstellar radio radiation into a speaker,
it would just give noise, a steady hiss and crackle. What the radio astronomer measures is the "loudness" of this noise, that is, the strength of the
radiation.

The most useful feature in this whole spectral region is a line of the hydrogen atom, at a wavelength of 21 centimeters.* As always in atomic
processes, this line must represent a transition between two different energy levels of the atom; and such a long wavelength—half a million times
as long as that of visible light—must correspond to an exceedingly small
jump in energy. The two levels concerned are, in fact, very narrowly
separated energy states within what for all other purposes is simply called
the ground level of hydrogen.

*So important is this line to astrophysics that international telecommunications treaties
have set aside its frequency band—1420 megaHertz, in the more common radio terminology—as a world-wide "protected" band, barred to communications or broadcasting use,
and reserved exclusively for radio astronomy.

excited levels

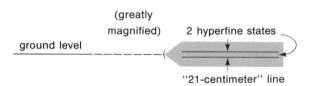

FIGURE 19.8.
The ground energy level of hydrogen actually consists of 2 "hyperfine" energy states. The transition between them leads to emission or absorption of the 21-centimeter line.

The distinction between these states is connected with a property of the electron that we have not needed previously to consider, although it is important to the details of atomic spectra. Along with its orbital motion around the atomic nucleus, an electron spins on its own axis. The spinning charge has a magnetic field, and in a complicated atom the interaction of the electron spins has a lot to do with the spacing of the energy levels and the probabilities of transitions between them. In hydrogen, however, with only a single electron, the interaction that we now have to consider is the one between the electron spin and the nucleus, which also has a charge and a spin. If their spin is opposite in direction, then these two little magnets attract—as oppositely directed magnets always do—and the atom is a little bit more tightly bound than if the electron and nucleus had their spins in the same direction.

Thus what we had previously called the ground level is really made up of two possible energy states: an upper state in which the spins are parallel, and a lower state in which they are anti-parallel. A hydrogen atom in the upper state can spontaneously jump to the lower state, emitting the energy difference as a photon. Because the force due to the spins is very weak, the energy difference is very small, and this is why the photon has such a long wavelength.

As a simple jump in energy, the transition that gives rise to the 21-centimeter line sounds very reasonable. Physically, however, it involves an act of atomic acrobatics—flipping the electron spin over—that an atom is very unlikely to carry out. In fact, a hydrogen atom would wait in the upper state for 10 million years, on the average, before making such a transition. It is clearly ridiculous, because of collisions, to imagine an individual atom remaining undisturbed for that long; what happens instead is that, out of all the atoms in the upper state, about $1/10,000,000$ will make the transition each year. Correspondingly in each second, then, out of every 3×10^{14} hydrogen atoms in the upper state, one atom will emit a photon of 21-cm radiation. This is a very small fraction, but there are very many hydrogen atoms in interstellar space, and in the cool gas nearly all of them are in one state or the other of the ground level. Statistical calculations predict that three quarters of these will be in the upper state; so at any given moment there are plenty of potential emitters. Enough of them do indeed emit, and modern radio techniques can easily measure the 21-centimeter line.

Indeed, it is a fortunate thing for astronomy that the 21-centimeter line has such a small transition probability. If hydrogen atoms in the upper state were able to emit this line readily, atoms in the lower state would also *absorb* it avidly. Interstellar space would then be practically opaque at this wavelength, and we would be unable to use it to detect the remote parts of our Galaxy.

The 21-centimeter line of hydrogen is the only line of this sort that is emitted by the interstellar gas. All the other abundant nuclei—He^4, C^{12}, N^{14}, O^{16}, etc.—have an even number of charge units. In such nuclei the spins of the protons arrange themselves in opposite directions, so that the nucleus has no net spin, and hence no spin interaction.

The 21-centimeter line is emitted by hydrogen gas in all cool regions, where the atoms are nearly all in the ground level. Hot hydrogen also emits radio-frequency lines, but they come instead from its highest energy levels. Up close to the maximum energy that a hydrogen atom can hold without losing its electron completely, very many permitted energy levels are crowded very close together. Their energy differences are so small that transitions between successive levels give photons with radio wavelengths. The transition between 167th and 166th levels, for instance, gives a line whose wavelength is within the interference-free band that is reserved for radio astronomy near 21 centimeters. Other such transitions fall in other parts of the radio spectrum. Lines of this sort are emitted by hydrogen atoms in HII regions, as they combine and cascade down through their various energy levels. The physical process is the same as the one that produces the visible-wavelength emission lines in an HII region; the "radio recombination lines" that we are considering here are just the very-long-wavelength part of that same spectrum.

Observing the radio-frequency lines of hydrogen takes special equipment and special effort; but astronomers have been very willing to expend the effort, because of the high value of the results. The value arises from several factors. First, most of the interstellar medium *is* hydrogen, and it pervades the whole Milky Way. Second, these radio observations are unimpeded by the interstellar dust that makes it impossible to see the distant parts of the Milky Way at optical wavelengths, so we see the whole Milky Way. Third, this radiation is in the form of lines, which have predictable wavelengths and allow Doppler shifts to be measured; from them comes nearly all our knowledge of the rotation of the Milky Way and its over-all structure. Radio astronomy will play a prominent role in our discussion of the Milky Way as a galaxy, in Chapter 20.

INTERSTELLAR DUST

It takes only a glance to see that the Milky Way is full of dust clouds that obscure the light of the stars beyond them. Careful study shows, moreover, that the full problem is even worse; in addition to the obvious dark clouds, thinner obscuration is everywhere, so widespread as to be unavoidable and so chaotic as to be unpredictable. Typically, a star 1,000 parsecs away is dimmed a whole stellar magnitude by the intervening dust clouds. One such star, however, may lie behind only half a magnitude of obscuration; another at the same distance may happen to lie behind denser clouds that dim it by three magnitudes.

Like the stars, the interstellar dust is concentrated mainly in the extended, flat disc of the Milky Way, whose central layer we live close to. We can see *out* of the disc without great difficulty, and we can thus study the sparse stars of the galactic halo and, beyond them, other galaxies. But *along* the direction of the disc, where most stars lie, the dust lies distressingly in our way.

Interstellar absorption, as this general problem is called, is the curse of stellar astronomy. At its worst, it totally wipes out our view of the distant parts of our Galaxy. And even at best, it dims the light of stars by an amount that is difficult to determine. This dimming plays havoc with the measurement of stellar distances. on which so much else depends. Beyond 100 parsecs or so, nearly all our distance measurements are photometric—that is, they come from comparing the apparent magnitude of a star, as it looks to us, with the absolute magnitude that we

FIGURE 19.9.
A large region of the Milky Way in the general direction of the Galactic center. It is clear that dark clouds are numerous, and in fact even the brighter areas are actually obscured by clouds that are less obvious. (Hale Observatories.)

know that star to have. But this calculation is based on the assumption that the faintness of the star is due to its distance. What happens to our argument when part of the star's faintness is due instead to absorption by interstellar dust clouds? The answer is that a distance *cannot* be measured photometrically *unless* the interstellar absorption is first evaluated and allowed for.

Here the astronomer is saved by a fortunate circumstance: the dust dims the light of some wavelengths more than others, and as a result the apparent color of a star is changed. Like the setting Sun, it looks redder than it really is. On the face of it, this would appear to be a further disaster; not only is a distant star dimmed by some unknown amount, but its color is not really what it seems to be. But this interstellar reddening is in fact the observer's salvation; if he can somehow determine what the *true* color of the star should be, then its observed color will show how much it has been reddened. The amount of reddening shows, in turn, how much dust is in the line of sight and therefore how much the star has been dimmed by interstellar absorption. Thus the key to absorption is reddening. By carefully studying, over a broad range of wavelengths, stars that have large and small amounts of absorption in front of them, astronomers have determined the so-called "ratio of absorption to reddening." That is, we know how much absorption corresponds to any given amount of reddening.

This still leaves the question, how do we determine the amount by which a particular star is reddened? One answer is simple and straightforward: the spectral types tell us what the true color of each star should be. Physically this is perfectly clear; both the spectral type and the color index are determined by the temperature of the star's surface, so the one should correspond closely to the other. Interstellar reddening changes the color index, of course, but not the spectral type; it reddens by reducing the blue brightness more than it reduces the red, but this does not in any way change the spectral lines. Thus we can measure interstellar absorption by comparing the observed color of a star with the true color that we deduce from its spectral type.

More commonly, however, astronomers use a method that, like the color index itself, depends on magnitude measurements alone and does not require observing the spectrum. Not only is this less time-consuming, but magnitude measurements can also reach stars that are too faint for spectroscopic observation. This purely photometric method of determining reddening uses observations made at *three* different wavelengths, comparison of which allows the astronomer to determine separately the star's true color and its reddening. Whatever method is used, however, the absorption must be allowed for before the distance of the star can be correctly calculated.

THE NATURE OF INTERSTELLAR DUST

Just as the different dimming at different wavelengths allows us to determine how much starlight the dust hides, so this same selectivity tells us something about the nature of the dust grains themselves. The way in which they affect light of different wavelengths is, in fact, our best clue to the size and composition of the grains. Unlike a gas, a solid material does not reveal its composition by producing spectral lines. Because each atom is bound to its neighbors (this is what distinguishes a solid from a gas), the electrons do not have the neat, discrete energy levels that they have in a free atom. As a result, absorption is spread rather widely over the spectrum and depends relatively little on the composition of the material.

What does matter very much, however, is the size of the particles. At the one extreme, a large chunk of material will simply act like an opaque screen, blocking every wavelength in the same way. When the particles are small, however, with a size comparable to the wavelength of light, the light waves are able to bend around each particle, and the shadowing becomes less effective. What matters, then, is the size of the particle relative to the wavelength of the light that it is trying to block. When the particle is very small, it blocks light quite inefficiently, compared with the shadowing that its actual size would lead us to expect. For extremely small particles, this inefficiency goes as the fourth power of the ratio of particle size to wavelength. In addition, the light tends not to be absorbed by the particle, but instead is scattered in all directions.

It is atmospheric particles in this size range that produce the light of the blue daylight sky. Since the particles are even smaller relative to long-wavelength red light than they are relative to the shorter-wavelength blue light, they scatter the red light much less than the blue, and the scattered light thus looks blue.

The interstellar particles discriminate much less than this between red light and blue; hence their size must be only a few times smaller than the wavelength of visible light. By also assuming a modest spread in particle sizes, theoreticians are thus able to explain the observed "extinction curve" over the range of near-ultraviolet, visible, and near-infrared wavelengths at which observations are most commonly made.

These theoretical interpretations tell us the *size* of the particles in a clear way, but they say very little about what the particles are made of. In one sense this is fortunate; we are certainly happy to know the sizes unequivocally, without having to face the confusing questions of composition. But in another sense it is unfortunate, because it means that the optical extinction curve cannot tell us what substances the interstellar grains are made of.

370

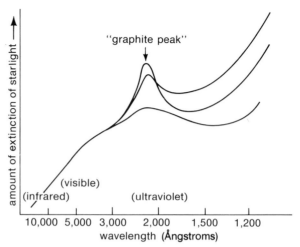

FIGURE 19.10.
Extinction curves of three different stars, from
observations by the University of Wisconsin. The
curves have been scaled to agree at visible
wavelengths; it is clear how much their ultraviolet
extinctions differ. (Adapted from *The Scientific
Results from the Orbiting Astronomical Observatory,
OAO-2*, National Aeronautics and Space
Administration, 1972.)

Here the recent "opening up" of the electromagnetic spectrum has made major contributions. One of the most important results of observations by the first Orbiting Astronomical Observatory, which operated from 1968 to 1972, was the extension of the interstellar extinction curve to ultraviolet wavelengths that cannot be observed from the ground. The most striking feature in the curve is a "bump" of sharply increased absorption within a small range of ultraviolet wavelengths. From studies of the absorption spectra of various substances, it seems almost certain that the bump is due to small particles of graphite, which is one of the several possible solid forms that carbon can take. If all the grains were graphite, however, the bump would be very much stronger than it is. We are thus forced to conclude that interstellar dust is a mixture of grains of different kinds. This inhomogeneity is further emphasized by the way the extinction curve behaves in the extreme ultraviolet; it is observed to be different for different directions in space. It may well be that we have to deal with three kinds of grains: one that produces the extinction of visible light, one for the extreme ultraviolet, and in addition the graphite, which leaves a clear "signature" at one wavelength but cannot be responsible for what happens elsewhere.

Ever since the existence of interstellar dust was first recognized, astronomers have wondered not only what it is made of, but also *how* it is made. Interstellar space seems a singularly poor place to build up solid particles, because at such low densities the atoms collide with each other relatively rarely. Even a small grain is made up of a large number of atoms, and there will just not be enough collisions to build it up. Once the grain reaches some size and presents a good-sized target, it may have a chance to grow; but getting it started seems too difficult under interstellar conditions.

An alternative would be for the grains to be formed in the dense atmospheres of stars and then ejected into interstellar space. Indeed, infrared observations of some stars suggest that this is exactly what happens. With the recent development of better infrared detectors, astronomers are now able to observe at wavelengths many times as long as that of visible light, and they find many stars to have "infrared excesses." That is, some stars emit far more infrared radiation than objects of their size and temperature ought to. This radiation seems to arise from a cool shell of dust surrounding the star; the dust shell absorbs a large part of the star's radiation and reradiates it. Being cool because of its distance from the star, the dust shell radiates predominantly in the infrared part of the spectrum, and this is what is observed.

We observe more, however, than the mere existence of a lot of infrared radiation. Within certain ranges of wavelength, the radiation is enhanced; and these wavelengths indicate something about the composition of the solid grains that make up the circumstellar dust shell. Although such broad spectral features do not label the composition of a solid nearly as sharply as spectral lines do for a gas, two substances seem likely identifications for features in the spectra of some dust shells: silicates, and silicon carbide.

Both of these substances could plausibly be formed in the atmosphere of a cool star. Which of them occurs is controlled by a quirk in the compositions of some stars, depending on whether carbon or oxygen is more abundant. What happens in the atmosphere of a cool star is that these two atoms combine very readily into the molecule CO (carbon monoxide), so readily, in fact, that the less abundant of the two elements is completely used up in making CO, and only the excess amount of the more abundant element is left free. In most stars, oxygen is somewhat more abundant than carbon; so in a cool atmosphere all the carbon goes into CO and there is free oxygen but no free carbon. If such a star has a dust shell, it contains silicates, which are compounds of silicon and oxygen with metallic atoms. In the atmospheres of a few stars, however, as a result of some evolutionary process that has not yet been identified, there is more carbon than oxygen. A cool atmosphere of this composition contains free carbon, but all the oxygen is bound up in CO. In a dust shell surrounding such a "carbon star," we see silicon carbide, which is the molecule SiC.

In a star that has an excess of carbon and is surrounded by a dust shell, it is chemically plausible that the shell will also contain grains of graphite—which unfortunately does not have infrared spectral features that are easily observable. Thus it seems quite possible that cool stars in the late stages of their evolution produce solid grains of at least three different types, which are spread through interstellar space when these stars shed their outer envelopes.

A further problem is durability and possible growth of the grains once they are in interstellar space. Many astronomers have suggested that the grains might collect icy coverings, made up of molecules of the more common elements in interstellar space. Such coatings would be unlikely to survive heating, however, either in collisions between clouds or in the formation of H II regions. The grain materials suggested above would be much more durable, however. Even graphite, known to us as the soft "lead" in a pencil, is difficult to vaporize. The other two substances are even tougher: silicates are the main constituent of rock, and silicon carbide is commonly known as carborundum, a very hard abrasive.

This discussion began with the observed properties of interstellar dust, and we then applied general theories of light-scattering to find the general properties of the grains; but we have ended up with a picture that is only informed speculation. Such speculation plays an important role, however, in directing both observation and theory toward specific studies that may be profitable. From the infinite number of substances that can be assembled out of atomic building blocks, the universe has made a few choices for its grains of interstellar dust. Research is now directed toward narrowing the possibilities, and eventually finding out what those substances are. Then, somewhere in the future of astronomy, it will become clear why those specific particles are the ones that form. Foresight would be preferable, but today there are far too many directions to follow. We shall have to settle for eventual hindsight; but we have the faith that when we are finally told the answer to the puzzle, we will see enough about it to realize that it was the only possible answer all along.

THE STRUCTURE OF THE INTERSTELLAR MEDIUM

The interstellar dust is a chaos of clouds, and emission nebulae also look like cosmic puffs and swirls, so it should not be surprising to find that the cool hydrogen has an irregular structure. Seeing it directly is impossible, however; the observations of 21-centimeter radiation do not give pictures but only a measurement of the intensity of the radiation at one point at a time. It is only recently that enough point-observations have been collected to make possible the creation of a synthetic picture, which shows what the sky would look like if we could "see" at 21 centimeters. This sky is cloudy indeed, but the clouds are wispy rather than round.

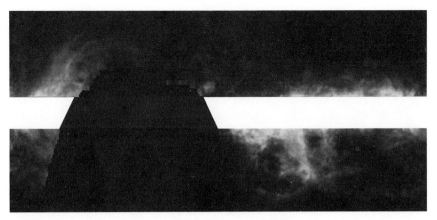

FIGURE 19.11.
A "picture" of the sky in the radiation of the 21-centimeter line of hydrogen.
Generated by a computer from radio measurements at half a million points in the
sky, this picture shows what the sky would look like if our eyes were sensitive to
21-centimeter radiation. The picture is laid out in galactic coordinates; the white
strip in the middle is close to the Galactic circle, where we see a large amount of
hydrogen everywhere. The large gap in the left side of the picture is the extreme
southern sky, which cannot be observed from the University of California's Hat
Creek Radio Observatory, where these observations were made. (Courtesy of
C. E. Heiles; see also Plate 8.)

Along the galactic plane we see little in such a picture except a con-
confused blackening, as we pile layer upon layer in the distance. Looking
away from the plane, however, we see the clear, graceful wisps of our
local hydrogen; in these directions there is no confusing background,
because our line of sight rapidly passes out of the thin layer of gas.

The picture of neutral hydrogen is a fascinating one, but it is not at
all clear just what it really shows us. Is the hydrogen really spread in this
wispy pattern, or are these just the regions where we *see* it? The 21-
centimeter line can be emitted only by neutral hydrogen in single-atom
form. If the atoms are joined into molecules, or if they are ionized, they
will not emit at 21 centimeters and we will not see them. Are the blank
regions in our picture really empty, or are they just places where the
hydrogen is either ionized or molecular? We do not know the answer,
but many astronomers believe that the latter is the correct interpretation,
at least for a large region on either side of the central dense layer of gas.

There has been, of course, a great deal of theoretical work on the behavior of gas in interstellar space, and some of it gives reasons for sus-
pecting that hydrogen should show just this sort of three-way behavior:
that in regions of low density it will be ionized, whereas in dense dust
clouds the atoms will pair into molecules. Only further observations
will tell.

The structures that we have been discussing here range from wisps
a fraction of a parsec thick to streamers hundreds of parsecs long. For

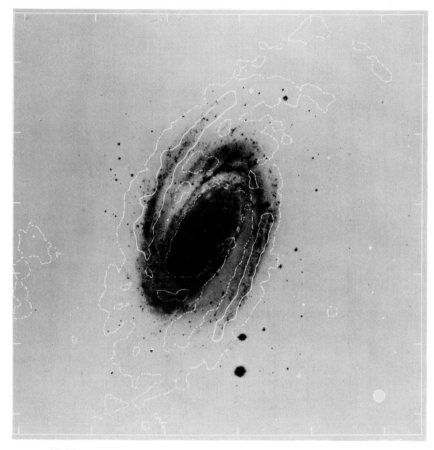

FIGURE 19.12.
In a spiral galaxy, the interstellar material is very much concentrated into the
spiral arms. In this map of the Galaxy M81, the contour lines show the distribution
of hydrogen gas, as observed in the 21-centimeter line. The spiral arms
themselves, in the underlying Hale Observatories photograph, are most clearly
outlined by young supergiant stars and by HII regions. (Courtesy of A. H. Rots
and W. W. Shane. Adapted from *Astronomy and Astrophysics,* **31** (1974), 247.)

structures of a much smaller scale, observations of pulsars have made an
unexpected contribution, not to knowledge of the pulsars, but to knowledge of the interstellar medium. The beam of radiation from a pulsar to
us is a very narrow beam, coming not from a normal star a million miles
across, but from a neutron star ten miles across. As such, it is an almost
needle-like probe of the interstellar medium. Slight variations in the
properties of the gas affect the passage of the beam of radiation; and
as the interstellar gas drifts through our line of sight, the pulsar "twin-
kles" just as stars twinkle when we look at them through the Earth's
unsteady air. This interstellar scintillation of pulsars, as it is called, is the
finest probe that we have of the interstellar medium.

On a much larger scale, also, the interstellar material shows a pro-nounced structure, as the spiral arms of our Milky Way. As in other galaxies, the gas and dust show up most prominently along these great pinwheel-type regions that wind erratically outward. In the following chapter we shall see how observations of interstellar gas are used to plot out the locations of the spiral arms in the map of our Galaxy.

THE INTERSTELLAR MAGNETIC FIELD

Along with the gas and dust, interstellar space contains another com-ponent that is insubstantial but nevertheless important: a magnetic field. We still know very little about this magnetic field, but what we do know is tantalizing, because the field may well have a strong effect on the motion of the interstellar gas. It may also play an important role in the condensation of dust clouds into stars. And we are sure that it has much to do with the motions of the high-energy particles that we call cosmic rays.

Nearly all observations of interstellar magnetic fields depend on a characteristic of light called *polarization.* Although most of our discussion of light has relied on the particle behavior of light as photons, here we must consider the wave aspect of light. A ray of light can be considered to have electric and magnetic fields oscillating in directions perpendic-ular to its direction of travel. These oscillations cover a whole circle of orientations; and in ordinary, or unpolarized, light the electric fields oscillate equally in all these directions. If, however, the oscillation of the electric field is mostly or completely along one direction rather than others, then we call the light partially or completely polarized.

We most commonly encounter polarized light after an oblique reflec-tion from a surface. Sunlight reflected off a road is highly polarized horizontally, for instance. Polarizing sunglasses are constructed to block light that is polarized horizontally, so they reject this glare quite effec-tively. Reflections off a vertical shop window are vertically polarized, however, and the same glasses do not block these at all.

Just like the layer of tiny crystals that causes the sunglasses to polarize, so the interstellar grains are able to polarize light that passes through

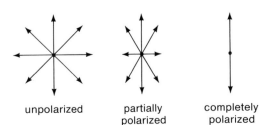

unpolarized   partially   completely
                 polarized   polarized

FIGURE 19.13.
Electrical oscillations in a ray of light. In every case the ray is coming toward you.

them. To do this, however, they cannot be round, because all directions would then look the same and there would be no favored direction of polarization. Furthermore, they must also be lined up, because with a random orientation there would again be no favored direction for the polarization of light.

In fact, the light of distant stars is often observed to be somewhat polarized, and the tendency toward polarization is indeed greater when a greater amount of interstellar absorption indicates the presence of more dust. What is more interesting, however, is that interstellar polarization is different in different directions in space. This tells us the direction of the interstellar magnetic field that aligns the grains—as soon as we figure out how the alignment is related to the direction of the field.

The key question is, how do the elongated interstellar grains line up in the field? The answer seems to be that they do not line up *along* the field, as an iron magnet would, but instead they tend to be perpendicular to the field. As a result, we see some alignment when we look across the field, but not when we look along it, as examination of Figure 19.14 will show; and the polarization that we see is parallel to the direction of the field.

When the observations of interstellar polarization are entered on a map of the Milky Way, it appears that the magnetic field lies approximately along the spiral arms. We can also get an idea of its strength, from calculations of the ability of the field to realign particles after collisions have set them spinning and have thus disturbed their orientation.

Other magnetic-field observations come from another polarization phenomenon. Some radio sources outside our Galaxy emit radiation that is polarized when it leaves the source. If such waves pass through an ionized gas in a magnetic field, the direction of polarization is rotated. Ionized gas in the Milky Way does this. Because the amount of rotation depends on wavelength, we can calculate the amount of rotation by comparing different wavelengths, even though the original direction of polarization is unknown.

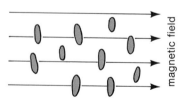
looking across the field

looking along the field

FIGURE 19.14.
Grains aligned perpendicular to a magnetic field. As seen across the field, they look partially aligned; as seen along the field, they are randomly oriented.

The result, like so many observed quantities, does not give us a single answer; instead it gives a quantity that depends on the strength of the magnetic field, on its direction, and on the number of ionized atoms. When combined with other observational results that are related to these quantities in different ways, it can lead to a set of equations that have a single solution for all the unknowns. The final answers in such a solution are much less obvious than if they could be directly observed, but they may be reliable enough nevertheless. Most important, these are answers that are difficult or impossible to reach in any other way.

The interstellar magnetic field also shows us its broad structure, in another kind of radio phenomenon. This is the emission that we see from the whole disc of the Milky Way, throughout a broad continuous range of wavelengths. From its distribution over the radio spectrum, and its polarization, astronomers are convinced that this radiation is emitted by electrons moving with very high speed, whirling about in the Galactic magnetic field. These electrons presumably have an origin connected with that of cosmic rays, and they are thus clues to the violent processes—such as supernovae—that might be capable of producing such high-energy particles.

Our picture of the interstellar magnetic field is being assembled slowly, and present knowledge is still far from satisfactory. Least satisfactory of all, however, is our understanding of the behavior of magnetic fields in a gas. Electromagnetic theory is difficult enough, and so is the hydrodynamics of gas flow; but the combination of the two—magnetohydrodynamics—is far worse. The basic equations are easy to formulate, but specific problems are far from easy to solve. Much effort has gone into magnetohydrodynamic theory, largely because one of the best approaches to controlled nuclear fusion is to confine a hot gas by means of electromagnetic fields, but results are disappointingly few. Interstellar gas dynamics is similarly frustrating; we suspect that magnetic fields are important, but we are unable to solve significant problems in which they are included.

COSMIC RAYS

The continuous-spectrum radio emission coming from the Galaxy introduced us to a new component of the interstellar medium: high-energy electrons. There are also rapidly moving positive ions in interstellar space, that is, nuclei of atoms. Physicists have long been familiar with these ions, some of which strike the Earth; because of their extraterrestrial origin, they are called cosmic rays.

These particles are moving with energies that are very high compared to the energies of ordinary particles or photons. Physicists commonly express particle or photon energies, or energies stored within atoms, in a unit called the *electron-volt* (eV). The air molecules bombarding your skin at this moment have energies of about a thirtieth of an eV. In the Sun's atmosphere the atoms bounce about with energies around 1 eV, while typical energies of motion of atoms at the Sun's center are around 2,000 eV. Photons of visible light have energies around 2 eV, and a 13.6-eV ultraviolet photon can ionize a hydrogen atom. Medical X-rays use photons of around 100,000 eV, which are able to pass completely through the human body. Fusion of hydrogen to helium releases 6 million eV per hydrogen atom.

Cosmic rays are particles with energies in a much higher range, from 10^6 eV all the way up to 10^{20} eV. The most energetic particles are very rare, however, and most of the total energy seems to be carried by particles with individual energies of around 10^9 eV. The importance of cosmic rays is that these particles, rare as they are, carry in them about as much energy as all the rest of the interstellar gas put together. Because of their high speed, they interact very little with other material; but even so, they may be responsible for much of the heating and ionization of the interstellar gas.

Like other charged particles, cosmic rays follow curved paths in a magnetic field; the more energetic particles curve less, however. Thus the cosmic rays of lowest energy do not penetrate the Earth's magnetic field, and even when observed from high-altitude spacecraft they are seen only after being deviated by other magnetic fields in the Solar System. Above about 10^9 eV, cosmic rays are affected very little by local magnetic fields, but they are held into the Milky Way by the Galactic magnetic field. As weak as this field is, distances in the Galaxy are so large that the particles curve round and round and never escape, except possibly for the rare particles that have the very highest energy.

This bending of particle paths completely hides the origin of cosmic rays; they approach the Earth equally from all directions. We can tell something about their origin from their composition, however. Most are protons and electrons, but heavier nuclei also occur; and over-all the cosmic rays look like an ordinary sample of the material of the universe, but one whose atoms have somehow been accelerated to fantastic speeds. There are some composition differences, however, and these seem to indicate the extent to which the cosmic rays have interacted with the interstellar gas through which they pass. There is some depletion of heavy nuclei, which tend to interact more and would thus be slowed down and lost. There is also an excess of some light nuclei of kinds that are otherwise very rare in nature; these must be the product of nuclear reactions between cosmic-ray particles and other nuclei with which they have collided. In both cases the abundance changes are such as would be produced by travel through the interstellar gas for a few million years. The cosmic rays may well be older than that, if they have spent much of their time in the Galactic halo, where there is little or no gas; but they are certainly young compared with the Galaxy. Furthermore, the cosmic rays of low energy show no signs of having spent any appreciable time traveling through interstellar gas, so they may be very young. This would suggest that some cosmic rays have been very recently created, whereas others persist for millions of years.

The origin of cosmic rays is another of our unsolved problems, however. There have been ample suggestions—supernova explosions, acceleration by the galactic magnetic field, etc.—but as yet there is no reason to believe that any one explanation is the correct one.

DUST CLOUDS AND STAR FORMATION

Stars are born from the interstellar gas and dust, but how this happens is another of the intriguing mysteries of astronomy. Since the process must be one of condensation, it is to the denser clouds that we should look for clues. Many of these are visible in absorption, as dark dust clouds. It is natural to suspect that a dense cloud of dust should also contain gas, since gas and dust are found together elsewhere and are hard to separate.

Gas is indeed found in dust clouds. Because the clouds are cool, the gas atoms are not excited enough to emit lines in the visible spectrum; but again radio astronomy comes to the rescue. As we would expect, the 21-centimeter line of hydrogen is often observable. But even more interesting, other lines are found in the radio spectra of these dense regions. These lines come not from atoms but from molecules, structures in which two or more atoms are bound together.

Just as a pure chemical element is made up of a large number of identical atoms, so a chemical compound is made up of a large number of identical molecules, each containing the same kind of atoms bound together in the same way. The formula for water, H_2O, reminds us that each molecule consists of two hydrogen atoms bound to one oxygen atom. (The arrangement is actually $H-O-H$, with a bend in the middle.) The forces that hold the atoms together are weaker, however, than those that hold the electrons into an atom. Very few molecules survive the buffeting that they get in the hot atmosphere of a star, so they figure very little in stellar spectra. In interstellar space, also, there are enough ultraviolet photons with energy enough to disrupt a molecule, and the molecules have little chance to survive. Only in the cool, dark interior of a dust cloud are the conditions favorable. The surfaces of the dust grains actually aid the formation of molecules, and the cloud then protects them from disruption.

Like an atom, a molecule can hold energy by means of distortions of the forces that hold it together. Because of its more complicated structure, however, it can store energy in many different ways, and its spec-

FIGURE 19.15.
The Horsehead Nebula in Orion is typical of a large, dense region of dust. We see it so well only because it is silhouetted against glowing gas. (Kitt Peak National Observatory.)

trum is more complicated than that of an atom. Among the very many possible energy levels, most molecules are able to make some transitions with the tiny energy jumps that correspond to radio wavelengths. Radio lines of more than two dozen different molecules have been observed.

The molecules in interstellar clouds are just the ones we would expect to find if the atoms of the common chemical elements were to join together in whatever ways they can. All of them are compounds of hydrogen, carbon, nitrogen, oxygen, and sulfur. Only helium is conspicuously absent, but this is fully understandable, because its two electrons bind themselves to the helium nucleus in a tight, closed shell that keeps helium completely aloof from chemical relationships.

What is surprising, however, is the complexity of the molecules, some of which have as many as seven atoms joined together. Many of them are representatives of the broad class called "organic molecules." This name dates from a past era when chemists believed that only living organisms possessed the "vital spark" needed to form these molecules. It has long been suspected (and much more recently demonstrated in the laboratory) that this distinction was quite false; the interstellar parade of methyl alcohol, acetaldehyde, and others of their clan shows even more clearly that "organic" substances can appear under circumstances that are far from any possible abode of life. The reasoning thus goes in quite the opposite direction: not only is organic material not unique to living material, but it is created so easily in nature that the genesis of life may also be an easy process, in places where conditions are less rigorous than those in an interstellar dust cloud.

Missing from the radio spectrum, however, is the most abundant of all molecules, H_2. The reason is quite simple: the hydrogen molecule has no radio spectrum. This is true of all molecules that consist of two identical atoms; although they do have energy levels with a suitable spacing, the symmetry of the molecule makes transitions between those levels almost impossibly unlikely. The only strong transitions of cool hydrogen molecules are in the far-ultraviolet part of the spectrum; these can be observed only from space vehicles. Hydrogen molecules have indeed been thus detected. The first observations indicate that the distribution of molecular hydrogen is very cloudy: where it exists, it exists in great quantities, and elsewhere there is very little. This all-or-nothing behavior is easily understood. Hydrogen molecules are easily dissociated—broken into separate atoms—by absorbing ultraviolet radiation, which is so prevalent in interstellar space that an isolated hydrogen molecule has little chance of survival. In a dense cloud, however, only at the outside are molecules dissociated; the damaging radiation is unable to penetrate to the interior. Meanwhile, inside the high-density cloud, separate hydrogen atoms have a better chance to meet each other and combine into molecules, especially on the hospitable surfaces of dust grains.

Where hydrogen molecules are observed, they are so abundant that it seems probable that all the hydrogen is in molecular rather than separate-atom form. This is true especially in dusty regions, where there is often a lack of 21-centimeter radiation, just as we would expect if none of the hydrogen was left in atomic form. There are still very few direct observations of H_2, but if all dust clouds contain comparable amounts of it, then there must be almost as much molecular hydrogen as there is atomic hydrogen in the Milky Way.

The dust clouds that we observe are the densest samples of the interstellar medium, but they are still a far cry from the much higher density and much smaller size of a protostar. Even the black "globules" that are seen silhouetted against some emission nebulae are many times the size of the Solar System. It is very likely, however, that these globules are already dense enough that their own gravitation will now cause them to contract further. Eventually the gravitational energy that is released will heat the cloud up, and it will radiate enough to be detected in the infrared; but between now and that time, the cloud will pass through an elusive stage in which it is too small to be seen as a dark cloud, but still too cool to be detected as an infrared source.

This invisible stage is long in terms of the amount of change, but it is short in years. During this stage the cloud is already opaque to starlight, but the energy released in its contraction is still able to leak out readily at longer wavelengths, and the cloud contracts as fast as it can, in what is called "free fall" in its own gravitation. From the sizes and estimated masses of globules, we calculate that the free-fall stage will take only a few million years. This stage ends when the cloud becomes so dense that it is opaque to its own radiation. Its behavior then resembles that of a star in many ways. The gas stops falling together and is supported by its pressure, and heat flows out of the interior to be radiated at the surface. The configuration is now properly called a protostar. Further contraction will inevitably heat up its interior until, finally, a star is born.

There is much in this picture that is satisfying, but there are also serious gaps. At the protostar end of the development, we are beginning to understand the processes; and newly developed techniques of infrared observation are accumulating facts about these objects, often faster than we can interpret them. The earliest stages of star formation, however, are still lost in the darkness of interstellar space. Ordinary interstellar material is far too chaotic and tenuous to contract under its own gravitation; what then makes the dark clouds, and what sets them contracting? A good bet is to look for some process of physical instability, whereby any excess of density will make itself even denser—but what *is* this process? And what triggers it in one part of the Galaxy but not another? Only future study can tell, and no doubt it will.

REVIEW QUESTIONS

1. What is the physical difference between gas and dust?

2. What is the relation of interstellar material to stars? Answer this question in as many different ways as you can.

3. What is an H II region? What makes it shine?

4. What are "forbidden lines"? When we see forbidden lines, how do we know that they originate in low-density gas?

5. What do we learn from interstellar absorption lines in stellar spectra?

6. What is the 21-centimeter line? What is its importance?

7. Discuss in detail one aspect of the interstellar grains: either their scattering of starlight, their size and composition, or their origin.

8. Describe the effects of interstellar magnetic fields on radiation, on dust grains, and on cosmic rays.

9. Why do interstellar molecules tend to be found in dust clouds? What is the importance of dust clouds to an understanding of the processes of the universe?

20 | THE MILKY WAY, OUR HOME GALAXY

In studying the Milky Way as a galaxy, we look at the universe on a new, larger scale. The questions are now different. We are interested not in the nature of the stars but in their numbers, not in their locations but in their over-all arrangement, and not in their individual lives but rather in the history of the whole population to which they belong. The interstellar material now appears in a changed role too: instead of its local structure, we look for its grand design, and we now regard it not as a separate set of problems but rather as one ingredient in the total melting pot.

The focus of our attention is a galaxy, one of the countless islands that make up the universe. It has lived in isolation for ten billion years, building whatever structures the laws of physics create out of the materials with which our Galaxy began. These structures are the stars, gas, and dust that we now see around us. We, as curious inhabitants of this scene, ask what things are there, and how they became what they are. And as always, we look with some awe on our own role: how, endowed only with our telescopes and our wits, can we presume to find answers to such questions?

AN OVERVIEW

The galaxy in which we live has about 200 billion stars. Most of these are spread through a thin, circular disc that is at least 30,000 parsecs in

FIGURE 20.1.
NGC 628 (M74) is a spiral galaxy that is similar in structure to
the Milky Way. We see it nearly face-on. (Hale Observatories.)

diameter. They are the stars of Population I. Most of them are old; but a
few stars continue to be born from the small remaining amount of interstellar material, which is largely concentrated in spiral arms that wind
outward through the disc.

The stars of Population II, by contrast, are spread through a halo that
is centered on the disc but is much less flattened and probably extends
farther out than the disc. All the stars of the halo are old. They interpenetrate the disc, in which they are a small minority; for the whole
Galaxy, however, the halo may well make up a good fraction of the total
number of stars.

Both disc and halo are densest at their common center—that is, the
stars are closest together there—and the density falls off outward. The
Galaxy has no definite boundaries, however; any limits we quote are
merely distances at which we can no longer detect appreciable numbers
of stars with any ease.

The Galaxy rotates on its axis, the disc relatively rapidly and the halo
more slowly. A star takes hundreds of millions of years to go around;
it rarely passes close to any other star, and stars never collide. This
smooth rotation has gone on for most of the age of the Galaxy, which
is nearly as old as the universe itself.

FIGURE 20.2.
NGC 4565, which we see nearly edgewise, is also similar in structure to the Milky Way. We see only the disc; the halo is too faint to be detected except by very sensitive measurements. (Hale Observatories.)

The Sun is an old star, but not one of the oldest. We are located in the central layer of the disc, but we are far from the dense central hub of the Galaxy, some 10,000 parsecs away (or 10 kiloparsecs, as we commonly express it). The stars that we study as individuals are our near neighbors, some of them only a few parsecs away, some 100 parsecs, and very few more than a kiloparsec away from us.

We see very little of the structure of our Galaxy directly, because of the disastrous effect of absorption by the interstellar dust. Our position is like that of a general on a fog-bound battlefield, who grasps at intelligence reports of all sorts and tries to plot out the locations and movements of the armies that surround him. Like him, we are often puzzled by conflicting reports, and sometimes deceived by them, and we constantly seek better and more reliable sources of information.

PLOTTING THE OVER-ALL PICTURE

We can build up a picture of our Galaxy by beginning with what we see and then seeking clearer information by whatever methods we can devise. To start, consider the stars once again. This time, however, look

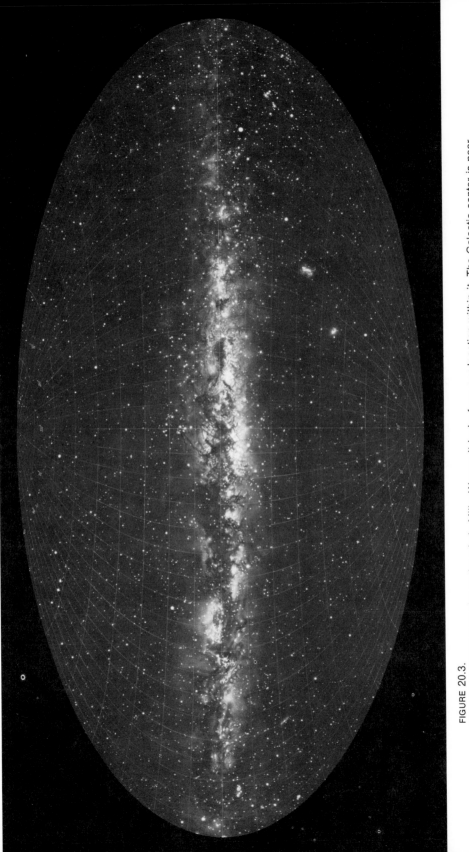

FIGURE 20.3.
This hand-drawn map shows the whole Milky Way as it looks from our location within it. The Galactic center is near the center of the picture, which represents the entire circle around the sky as one continuous map. (Lund Observatory, Sweden.)

at their arrangement over the sky. Some parts of the sky are richer in stars than others; and, over-all, there are clearly more stars near a certain circle that stretches around the sky. Furthermore, if the Moon is down and we are away from city lights, we can see this circle marked by a continuous band of faint light. This band around the sky is the Milky Way, from which our Galaxy takes its name.* The light that we see in it is that of countless stars, too faint to see individually but adding up collectively to a visible glow. This circle marks the directions in which we look through the disc of the Galaxy edgewise.

The Milky Way that we see in the sky is just a pale glimmer of what it would be if it were free of dust. The light that we see comes almost completely from the first thousand parsecs all around us. Most of the Galaxy lies hidden in the distance, and even the part that we do see is cut up by the nearer dust clouds. One part of the over-all structure does show through, however. The part of the Milky Way that we see in the night skies of summer is markedly brighter than the winter part of the Milky Way. (Seasonal differences in our view of the sky were discussed in Chapter 4.) The brightest part of all is low in the summer sky for inhabitants of northern latitudes. This is, in fact, the direction of the Galactic center. Not that we see the center itself; it is just that the Galactic disc is denser toward the center and less dense away from the center, and even within the first thousand parsecs the difference is enough to be noticeable.

To get a better view of our Galaxy, we need to observe at wavelengths that are less obscured by the interstellar dust. We do considerably better even in the near infrared, wavelengths a little longer than those that we can see. Infrared photographs do indeed show the over-all outline of our Galaxy, with the bright central bulge that is so characteristic of a spiral galaxy seen edge-on. At even longer wavelengths we can "see" even better, but unfortunately there are no materials that can produce photographs directly at those wavelengths. Maps have to be built up by means of point-by-point measurement. Such maps, using far-infrared wavelengths, clearly show the radiation that comes from the stars in the densest central region of our Galaxy. Then by comparing the brightness at different infrared wavelengths—in effect, by watching the Galactic center fade out as we shift our observation back to shorter wavelengths— we judge that the visible light of the Galactic center must be dimmed by 25 magnitudes. That is, only one ten-billionth of the visible light gets through the obscuration; small wonder that we fail to see it!

---

*"Milky Way" is the old folkname for the bright band of Galactic light around the sky. "Galaxy" comes from the Greek word for milk; it originally referred to the Milky Way only, but it has now become the standard term for an independent stellar system. It is usually capitalized when referring to our own system. We often call our whole Galaxy the Milky Way, especially when distinguishing it from other galaxies. In purely local views, however, the term "Milky Way" is used for the visible band of Galactic light around the sky. In either case, the adjective corresponding to "Milky Way" is "Galactic."

FIGURE 20.4.
In infrared photographs, the Milky Way actually looks like a spiral galaxy. Top: Infrared photograph of the central part of the Milky Way, taken with a special camera whose field of view reaches almost from horizon to horizon. The three dark lines are the shadows of the camera supports. (University of Bochum, Germany.) Bottom: Conventional photograph of the galaxy NGC 891, a spiral seen edge-on. (Kitt Peak National Observatory.)

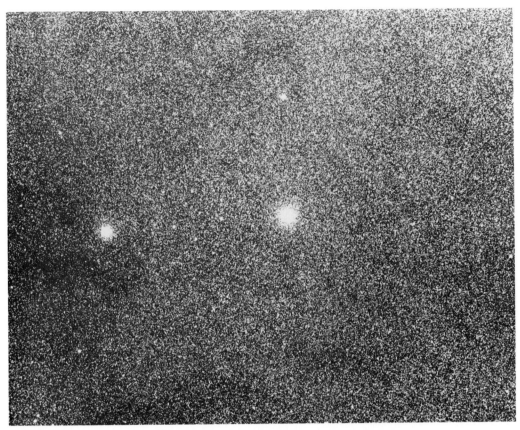

FIGURE 20.5a.
The clear "window" 4 degrees below the Galactic center. Note the apparent absence of dark clouds. The two globular clusters lie beyond the absorbing material, and their reddening allows the absorption to be calculated. (Kitt Peak National Observatory.)

Although infrared observations are best for penetrating the interstellar haze, the measuring techniques are difficult and expensive, and we prefer to work with visible light where we can. Penetrating great distances through the Galactic plane is quite hopeless; but since the absorbing layer is only a few hundred parsecs thick, we can try oblique lines of sight and hope to avoid most of the obscuration. In particular, we search carefully for directions in which we have the good fortune to peep between the major dust clouds of the foreground. One such "window" near the Galactic center has been studied extensively; it has only two magnitudes of absorption, even though the direction is only 4 degrees from that of the Galactic center, and the line of sight passes only 600 parsecs from the Galactic center itself. In this field large num-

FIGURE 20.5b.
Schematic drawing indicating the direction of the field shown in Figure 20.5a.

bers of RR Lyrae stars have been found, with a peak of density that shows where our line of sight actually passes through the central bulge of the Galaxy. To calculate the distance of these stars, we are forced to use a somewhat uncertain measure of the amount of interstellar absorption, but their distance is certainly around 8 to 10 kiloparsecs.

Another way of looking for the Galactic center is even more indirect: study the halo and find its center, which must be the same as that of the disc, since they are both held together by the same over-all gravitational field. Although individual stars of the halo are hard to pick out in the distance, globular clusters are quite conspicuous. Their distances can be found from observations of the RR Lyrae stars that they contain (as explained in Chapter 15), and we can then calculate the center of the whole system of globular clusters. This is the method by which the center—and the extent—of the Milky Way were first discovered, more than fifty years ago.

The center of the system of globular clusters is of course in the same direction as we find for the Galactic center by other methods. The scale of the whole picture of the globular clusters depends, however, on the absolute magnitude that we choose for the RR Lyrae stars, and this is still somewhat uncertain. Again we find the distance of the Galactic center to be 8 or 10 kiloparsecs. As a round number, we shall use 10 kiloparsecs in further discussions.

THE WHIRLING GALAXY

As one might expect from its flattened and extended shape, the whole disc of the Milky Way is in rapid rotation. The Sun and all its immediate surroundings are traveling around the Galactic center at a speed of about 250 kilometers per second. Our orbit is a large one, however—

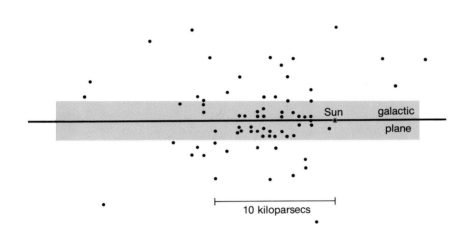

FIGURE 20.6.
The distribution of globular clusters, projected on a plane that is perpendicular to the Galactic plane and passes through the Galactic center and the Sun. It was from a diagram like this that the great extent of the Milky Way and the distant location of the Galactic center were first discovered. This version of the diagram, which includes the best data available in mid-1975, illustrates the two most serious observational problems of galactic structure.
(1) Interstellar absorption hides distant objects that are near the Galactic plane. Note how the known clusters within 2 kiloparsecs of the plane (shaded zone) are nearly all within a few kiloparsecs of the Sun. One can presume that another dozen or two clusters lurk undetected in the lefthand part of the shaded zone; interstellar absorption hides them from our view.
(2) The whole scale of distance is uncertain, since it depends on the absolute magnitude that we assume for the RR Lyrae stars, from which we determine the distances of the globular clusters. This diagram suggests that the distance to the center of the system of globular clusters (excluding the very incomplete shaded zone) is about 8 kiloparsecs—rather less than the standard working value of 10 kiloparsecs that was adopted by most astronomers in the mid-1960's. If the RR Lyraes are actually half a magnitude more luminous than is assumed here, then the scale of the diagram would expand enough to place the center 10 kiloparsecs from the Sun. On the whole, however, astronomical opinion seems to be moving toward a value of 8 or 9 kiloparsecs for the distance to the Galactic center. Uncertainties like this are typical of modern astronomy.

10,000 parsecs in radius—and a full revolution takes us about 250 million years.*

This motion of galactic rotation is normally ignored when we study the motions of local stars, since for purely local relationships it makes no difference that all of us have this additional motion. In just the same way, passengers can walk past each other in the aisle of an airplane, without having to take into account its high speed over the Earth's surface—or, for that matter, the rotation or other motions of the Earth. Locally, stars move past each other with speeds that are typically about 20 kilometers per second. The revolution of the whole neighborhood is additional to this.

Thus the "rotation" of the Galaxy is really just the collective motion of all of its stars, as each revolves around the center of the Milky Way in its own individual orbit. The average of the motions of all the stars in the solar neighborhood is a circular path around the Galactic center; but each individual star has its own orbit, which differs from the average circle according to the star's individual motion. Because the "peculiar" motions (as the individual local motions are called) are only a fraction of the total speed around the Galactic center, the individual orbits are not very far from circles.

Just as a star's peculiar motion causes its path to be different from the circle that the average of its neighbors follows, so the part of its motion that is perpendicular to the Galactic plane causes it to oscillate up and down through the disc. The total effect is like a ride on a merry-go-round, where the horses go up and down while the whole structure goes around.

The Sun, for example, follows an orbit around the Galactic center that deviates about 600 parsecs on either side of a circular orbit. Meanwhile it bobs up and down about 75 parsecs on either side of the central plane, making three or four of these up-and-down oscillations per quarter-billion-year tour around the Galaxy.

From this picture of motions of individual stars, it should be clear that the Galaxy does not rotate like a rigid disc. As in the Solar System, the inner parts go around fastest, and the outer parts lag behind. This "differential rotation" is quite evident in the solar neighborhood, as soon as we look at the radial velocities of stars that are 500 or 1,000 parsecs away. Because we are overtaking objects that are farther from the Galactic center, whereas the inner regions outrun us, the radial velocities that we see vary systematically around the sky.

---

*This is easy to see, if we note that each kilometer per second carries a star about a parsec in a million years. The circumference of the orbit is 2π times its radius, or about 63,000 parsecs; so the period of revolution is about 63,000 parsecs divided by 250 km/sec, or 250 million years.

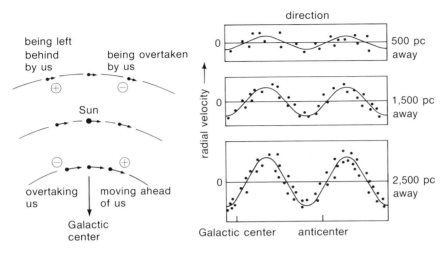

FIGURE 20.7.
The effect of differential rotation on the apparent velocities of stars. The lefthand figure shows how differing speeds of galactic rotation cause stars in some directions to appear to be receding from us (⊕ symbols) and stars in other directions to be approaching (⊖ symbols). The righthand figure shows how the observed radial velocities of stars depend on direction. Note how the magnitude of the effect increases with distance.

Using luminous objects, such as cepheid variables or star clusters, we can study Galactic rotation in our small neighborhood and a little beyond; but as usual, interstellar absorption prevents us from seeing as far as we would like to. Again we resort to radio observations, and here is where the 21-centimeter line of hydrogen plays its most important role. Not only can we use it to see all over the Galaxy, but from this spectral line, as distinct from a continuous spectrum, we can measure radial velocities.

The 21-centimeter observations are full of information, but they are difficult to interpret. The trouble is that the radio telescope records the radiation coming from all the hydrogen along the line of sight, regardless of its distance. These contributions coming from the various points along the line of sight must somehow be disentangled. To do this, we must use the radial velocities. Because the rotation of the Galaxy has a different speed at each distance from the center, the material at each point along the line of sight will have a velocity that depends on where it is. As a result, the 21-centimeter line that we observe in a particular direction will have a profile that is spread over some range of wavelength, because of the different Doppler shifts. Thus if we know just how the Galaxy rotates—that is, the speed at each distance from the center—we can then

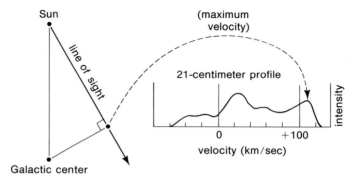

FIGURE 20.8.
The largest velocity seen in a given direction corresponds to the
point where the line of sight passes closest to the Galactic center.
It is from observations such as this that the rotation curve of the
Milky Way is calculated.

tell the location of each separate hydrogen cloud, from its wavelength
in the line profile.

The 21-centimeter observations play a double role in the study of Galactic structure. Not only do they let us map the hydrogen, but they are
also the source of the Galactic rotation curve that we use in the analysis.
Our ability to use them this way is due to a lucky accident of the observational geometry. Along any direction that is less than 90 degrees from the
Galactic center, there is a maximum velocity, which occurs at the point
where the line of sight passes closest to the Galactic center. Measuring
this tells us the speed of Galactic rotation at that point, and a series of
such measurements allows us to determine the "rotation curve" for the
whole region of the Galaxy interior to the Sun. Then from dynamical
studies (which will be described in the following section), we can calculate a fairly reliable rotation curve for regions beyond the Sun.

FIGURE 20.9.
The velocity of rotation of the Milky Way,
at different distances from the Galactic
center. (Data from M. Schmidt, in A.
Blaauw and M. Schmidt, eds., *Galactic
Structure.* University of Chicago Press,
1965.)

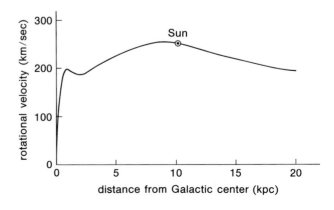

GALACTIC DYNAMICS

Much of our knowledge of the Milky Way comes not just from seeing where the stars are but from studying how they move. As usual in astronomy, the key is gravitation. From observing how stars move, we deduce the strength of the gravitational force that controls their motion, and this tells us in turn how great a mass is responsible for that force.

Thus a study of the rotation curve of the Milky Way shows us how its mass is distributed. If a star moves in a circular orbit around the Galactic center, there must be a force of a certain strength to keep it curving along that circle rather than running off in a straight line. From the size of the orbit and the period of time that the star takes to go around, we can calculate the strength of the force.

This calculation is analogous to the use of Kepler's third law, where we use the size and period of an orbit to find the mass of a central attracting body. The situation here is more complicated, however, since the mass is distributed throughout the Milky Way rather than concentrated at a single point in the center. Fortunately, however, it turns out that a star is affected largely by the matter interior to it, which all pulls in similar directions. The exterior material, being on all sides, pulls at cross purposes, and its total effect nearly cancels out. Thus the speed of Galactic rotation at each distance from the center tells us in an approximate way how much mass our Galaxy has interior to that point. And of course a more refined calculation gives more accurate results.

It is from studies of Galactic rotation that we find out how great the total mass of the galaxy is. It is about 200 billion times the mass of the Sun. Since the mass of a typical star is of the order of that of the Sun, we judge that the Milky Way also contains about 200 billion stars.

At the same time as the 21-centimeter observations allow us to map the velocities from which we deduce the total mass of the Galaxy, the amount of this radiation tells us how much hydrogen there is: simply enough, the more hydrogen, the more radiation. Since most of the interstellar material is hydrogen, it is an easy step to find the total amount of interstellar material. It makes up a few per cent of the total mass of the Milky Way. The gas seems to favor the outer parts of the Galactic disc, however; in our outlying neighborhood it probably makes up some 20 per cent of the local material.

From the analysis of Galactic rotation, we also find that the material of the Milky Way is crowded toward the center, with the density dropping off very much through our region and beyond. This is of course very much what we would expect to find, by analogy with other galaxies. Our own location in our Galaxy is really quite far out. We can see this by adding up the light of all the stars in the solar neighborhood and asking how bright our region would look if we saw it from the distance of an-

396

FIGURE 20.10.
NGC 2903, a spiral galaxy. The arrow marks a region whose star density is comparable to that in the solar neighborhood. (Lick Observatory.)

other galaxy. The answer is that our neighborhood corresponds to what we would certainly call the outer fringes of a galaxy.

Yet we know that there is plenty of material out here; sparseness is only a relative thing. Just how much material is in our neighborhood can be found, in fact, from another kind of dynamical study. In this case we examine the motions of stars up and down through the Galactic plane. By seeing how far they go above and below the central plane, and how long they take to move up and down, we can tell how much material is holding the local part of the Galactic disc together. We cannot of course follow the oscillation of a single star for 50 or 100 million years; instead we compare the thickness of the disc—which is about 500 parsecs—with the speeds of stars up and down through it. If there were less local mass, the stars would range farther from the central plane before being pulled

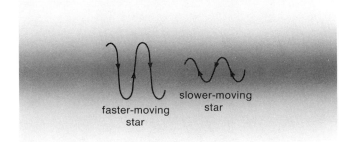

faster-moving
star

slower-moving
star

FIGURE 20.11.
Individual stars move up and down through the Galactic plane.
The higher a star's speed, the farther from the central plane
it goes. It is the amount of mass in the local region of the disc
that determines the relation between a star's up-and-down
speed and how far it moves from the central plane.

back; from the distances they go, we find the total density of material in
the solar neighborhood.

This local density is an important figure, because it includes all the
mass that is present in our part of space, whether we see it or not. We do
see most of it, in fact. From the local luminosity function—the number of
stars of each luminosity—and the mass-luminosity relation, we can add
up the masses of all the visible stars; this accounts for about half the local
mass density. Another 10 or 20 per cent is in interstellar gas of the sort
that emits the 21-centimeter hydrogen line. Of the rest, some is undoubtedly interstellar hydrogen in molecular form, which space observations
are just beginning to detect in quantity; some may consist of inconspicuous dwarf stars that are not yet detected; and some part of the discrep-
ancy may just reflect an uncertainty in our dynamical determination of
the total density. But it is reassuring that our accounts balance even this
well; at least the space around us is not filled to distraction with invisible
objects—burned-out stars, black holes, or cold lumps.

SPIRAL ARMS

It is obvious, from the appearance of other galaxies whose over-all structure is similar to ours, that the Milky Way must also have spiral arms.
They have been annoyingly difficult to find and to map, however. The
problem, as usual in Galactic structure, is that we sit in the midst of the
picture and are beset by interstellar absorption besides. Two methods of
searching for the spiral structure have paid off: one maps them as con-

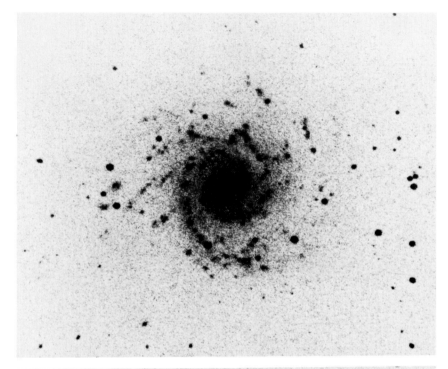

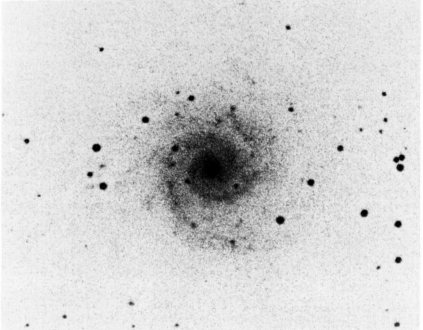

FIGURE 20.12.
Two photographs of the same spiral galaxy, taken in different wavelength ranges. The upper photograph is particularly sensitive to wavelengths that include strong emission lines from Hɪɪ regions, while the lower photograph is not. Note how the Hɪɪ regions stand out. (Yerkes Observatory.)

centrations of interstellar gas; the other focuses on their role as the birth-place of stars.

Examination of the spiral arms of other galaxies shows that the arms are most strikingly outlined by the bright O and B stars of the upper main sequence and the HII regions that surround them. This suggests that in the Milky Way we should also look for these conspicuous objects. Fortunately they are, among distant objects, almost the easiest to find—or rather, to state it more realistically, the least difficult to find. HII regions can be made to stand out, by taking photographs through a color filter that isolates a narrow spectral interval around the red line of hydro-gen, in which much of the emission radiation is concentrated. Spectral surveys of the region can then identify the hot stars that are responsible for ionizing the HII region. From their spectral classification their abso-lute magnitudes can be estimated, and comparison of the spectral types with the observed colors tells us the amount of interstellar absorption between us and these stars. The apparent magnitudes and absolute mag-nitudes then tell us the distance.

This procedure is typical of the way in which astronomy weaves its fabric. In order to plot a point on the spiral-arm map, we have drawn upon our knowledge of other galaxies (location of HII regions in spiral arms), interstellar gas (HII regions created by hot stars), stellar properties (absolute magnitudes, as calculated for these stellar types by statistical analysis of proper motions, and intrinsic colors), interstellar dust (ratio of absorption to reddening), and finally the basic photometric method of measuring distance. Some of this knowledge, it is interesting to note, went only into our initial strategy and did not figure in the measurement itself. The distance was finally measured by means of stars alone; the role of the HII region was merely to find stars of the right type.

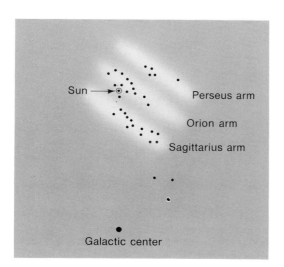

FIGURE 20.13.
Map of HII regions in the region of the Milky Way surrounding the Sun. Their distribution crudely outlines the local parts of spiral arms.

When many such points are plotted on a schematic map, they cluster along lines that represent the local pieces of the spiral structure of our Galaxy. On early maps of the spiral structure, astronomers named these pieces with their customary talent for bad terminology; each arm was named for the constellation (that is, direction in the sky) that contained its most prominent objects. Thus the outer arm was called the Perseus arm, the feature that passes near the Sun was called the Orion arm, and the inner arm was called the Sagittarius arm. The names have persisted, even though more complete maps now show these arms winding around large parts of the sky.

For a more extensive picture of the spiral structure, we turn to the 21-centimeter observations. Here we "see" radiation from the hydrogen all over the Milky Way, and must disentangle the contributions coming from material at different distances. This is done, as we have seen, by using the Doppler shifts that result from Galactic rotation. Typically, the 21-centimeter radiation from a specific direction in the Galactic plane will have a profile with several peaks, each coming from a gas cloud that has a different velocity. These clouds are at the points where the line of sight intersects the spiral arms. From a knowledge of just how the Galaxy rotates, we can assign each cloud to the proper distance; and by putting together the various lines of sight we build up a map of the spiral arms. There are observational uncertainties, and a few ambiguities have to be resolved, but a clear over-all picture emerges.

The picture has one unfortunate lack, however: the directions toward and away from the Galactic center have unavoidable gaps. Along those directions, Galactic rotation is across the line of sight; hence there is only a small component of velocity along the line of sight, and the Doppler shifts are too small for us to separate the clouds at different distances.

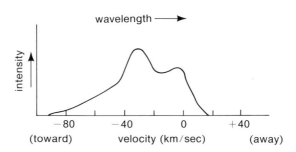

FIGURE 20.14.
The profile of the 21-centimeter hydrogen line, as measured in a particular direction. The peak around zero velocity is produced by local hydrogen, whereas the peak at −35 km/sec comes from hydrogen in the next spiral arm. (See Figure 20.15.)

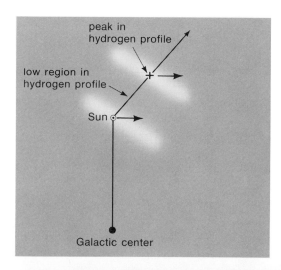

FIGURE 20.15.
Where the line of sight crosses a spiral arm, the hydrogen gas in that arm contributes radiation; we see it at a Doppler shift that is caused by its velocity with respect to us. This line of sight corresponds to the profile shown in Figure 20.14.

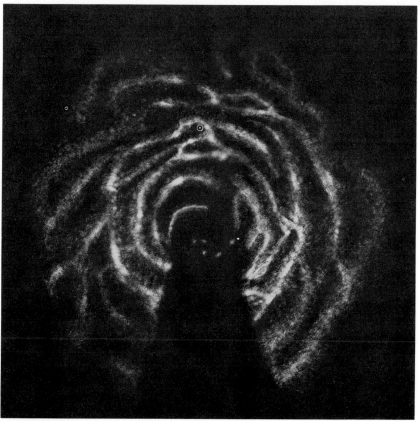

FIGURE 20.16.
A map of the spiral structure of the Milky Way, compiled from 21-centimeter radio observations. (Courtesy of G. Westerhout, University of Maryland.)

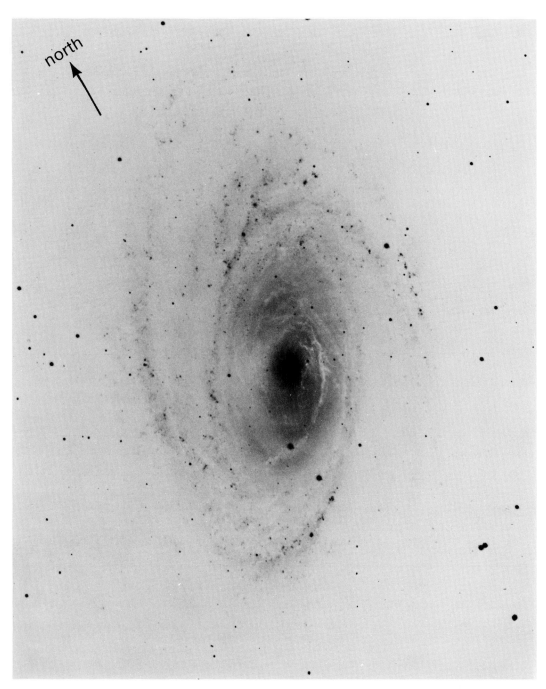

FIGURE 20.17.
The spiral galaxy M81 has a more symmetric two-armed structure than most. This photograph is more lightly exposed than the one in the frontispiece, and it allows detail to be more easily seen. It is printed as a negative, the form in which photographs are studied by astronomers. (Kitt Peak National Observatory.)

FIGURE 20.18.
The spiral structure of M101 is typical. Spirality is everywhere, but one cannot trace a regular two-armed structure. (Kitt Peak National Observatory.)

The hydrogen map undoubtedly shows a spiral galaxy, but the picture is not nearly so neat as that of "typical" spiral galaxies. Part of the problem is probably due to a real raggedness in the structure, such as is common in galaxies; regular, two-armed structures are almost the exception among spiral galaxies rather than the rule. It is very likely, however, that a part of the sloppiness is in our drawing of the picture. We have assumed that all the gas moves in perfect, neat circular orbits around the galactic center. If a cloud has a peculiar motion besides, we will miscalculate its distance by a corresponding amount.

In the region within 2 or 3 kiloparsecs of the Sun, we have optical maps of spiral structure, which can be compared with the radio map. There is some general agreement; but there are also some disagreements,

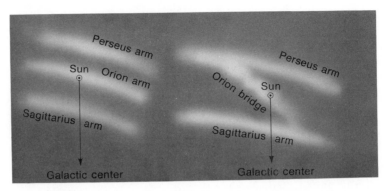

FIGURE 20.19.
In the earliest pictures of the local spiral structure, the Orion feature
was assumed to be one of the major spiral arms (left). It now seems
more likely that it is a bridge between two spiral arms (right).

which are probably due to the difficulties just referred to in the interpretation of the radio observations. One interesting conclusion does emerge, however. The "Orion arm," which passes through the solar neighborhood and runs more sharply outward than most features, appears not to be a part of the major spiral structure. It is more likely a spur, or a bridge between arms. It appears that the major arm features that are nearest to us are the Perseus arm and the Sagittarius arm.

We see spiral structure in the Milky Way, with some uncertainty, and in other galaxies more clearly. Spiral structure appears to be a general property of disc galaxies that contain quantities of interstellar gas. But why? What makes the gas concentrate in these winding arms, and why are the new stars born there? This has been a puzzle ever since the existence of galaxies was first recognized, and much of the answer is still obscure, but an answer is now beginning to take shape.

The starting point is to recognize that a spiral arm cannot consist of a fixed set of gas clouds, moving along with the stars. The rotation of the Galaxy makes this impossible; the inner parts of the Galaxy go around so much faster than the outer parts that such an arm would soon be wound up like a clockspring. Yet we know that the gas clouds themselves do move in this way, with a strong differential rotation; how can an arm then maintain itself? The only plausible answer seems to be that the arm does *not* continue to contain the same material; it is rather a progressive wave of compression that passes through the gas. Where the density of gas is highest at the moment, that is where we see the arm. At a later time, however, the density pattern will have shifted; the gas will have expanded locally where the arm formerly was and contracted at neighboring locations, so that the peak density continually moves along in the gas. Theories of the motion of such a wave can be made to produce pictures that look very much like spiral arms.

This does not, however, answer the more fundamental question: why does the density wave exist at all? Here our understanding is still rather poor. It is likely, however, that the cause is connected in some way with the phenomenon called *gravitational instability.* This is the tendency of material to contract under its own gravitation unless internal motions balance the gravitational forces. In a star the internal motions are those of the individual atoms; it is these motions that constitute the pressure of the gas. In the interstellar medium, each cloud or wisp of gas has its own motion—but collisions between clouds can dissipate this motion. It is possible that gravitational instability will then set in, drawing the clouds together into large clumps. These in turn are stretched out, by the differential rotation of the Galaxy, into the familiar spiral shapes.

This general picture of the origin of spiral structure probably has basic truth in it, but we cannot claim a real understanding. Not only are many of the details vague, but the existing theories do not include the effects of the Galactic magnetic field, which may turn out to be important in how the arms form.

Whatever the basic cause may be, however, it does seem very likely that spiral arms are waves of compression traveling through the interstellar gas. This phenomenon suggests an interesting possible connection with the process of star formation, which we know happens in the spiral arms. Some astronomers have suggested that the compression of the gas actually triggers the condensation of small interstellar clouds into new stars. To establish whether this is so, much more detailed work will be needed, both in theory and in observation; but it is working hypotheses such as this one that suggest what problems should be examined further, out of the infinite range of directions that a scientist might follow.

THE GALACTIC HALO

Extending beyond the Galactic disc in every direction are the more thinly spread stars of the halo. The outline of the halo is clearly shown by the distribution of globular clusters. They are strongly concentrated toward the Galactic center, but they also extend a long distance out in every direction. The most distant globular clusters are about 100 kiloparsecs from the Galactic center; some astronomers have called these "intergalactic globular clusters," but in fact they probably mark the outer fringes of the Galactic halo. The other good tracers of the halo, the RR Lyrae stars, confirm this picture, although they have not been followed to such great distances.

It is fortunate that we have these good tracer-objects, the globular clusters and RR Lyrae stars, because the Galactic halo, important though it is, is otherwise quite inconspicuous. The problem is our position in the middle of the disc. We see the halo only through a relatively dense foreground of disc stars, so that it contributes almost nothing to the total appearance of the sky around us.

We can, however, pick out those members of the halo population that are in our immediate neighborhood, because their motions are so completely different from those of disc stars. For both populations the motions correspond to the way the stars are distributed in space. The disc stars follow orbits that are nearly circular and are close to a single plane; so the space that they occupy is a thin circular disc. The halo stars, on the other hand, have orbits that are both elongated and inclined, and they thus fill a much more widely spread region. Their motions carry them through the disc, of course, and at the present time there are quite naturally some halo stars passing through our immediate neighborhood. Because their motions are so different from those of disc stars like the Sun, they move past us with high velocities; and our surveys of stars of large proper motion are able to pick out a number of stars that turn out, on closer examination, to be members of the halo population. From the number that have been found, along with an assessment of our own ability to find them, we judge that about 1 per cent of the stars in our neighborhood are halo stars.

The Galactic halo is very sparse, but it is also very large; how much material does it actually add up to? We can give only a poor answer to this question, by combining our local count of halo stars with our overall picture of relative star densities all over the halo. It appears from reasoning of this sort that the halo has perhaps 10 per cent as much mass as the disc. But our knowledge is so uncertain that, in fact, the halo could be only 1 per cent, or it could be the dominant constituent of the Galaxy.

As we saw on first encountering stellar populations, the stars in globular clusters are of a quite different type from those in the open clusters of the Galactic disc. This distinction extends also to the "field" stars, as we call the stars that are scattered individually through space. The stars of the disc are Population I, having "normal" metal abundance, with a complete range of ages; the halo stars are Population II, having low metal abundance, and all old.

Our knowledge of the halo population is still incomplete, but it is intriguing. Color-magnitude arrays have been measured for a number of globular clusters, and from these we study the brighter parts of their HR diagrams. Because these clusters are all many kiloparsecs away, however, their main-sequence stars are too faint to study; so for the main sequence of Population II we must rely on the local-neighborhood halo stars, which are recognized by their high velocities with respect to the Sun. These stars do indeed have low metal abundances; as a result their spectra look somewhat different from those of Population I stars, a difference which has earned them the name *subdwarfs*. This term would seem to imply that they lie below the normal main sequence of the HR diagram; but in fact they appear, on more careful study, to have temperatures

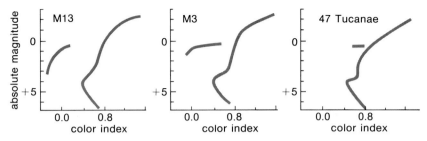

FIGURE 20.20.
HR diagrams of 3 globular clusters whose horizontal branches have very different shapes.

and luminosities that are very similar to those of the main-sequence stars of Population I.

The intriguing aspect of the Type II population is that it is by no means homogeneous in its characteristics. Its metal abundances seem to range from about a hundredth of that of Population I up to levels that are barely—if at all—distinguishable from the metals in Population I stars. Along with these abundance differences, which show up in the strengths of the spectral lines, there are large differences between HR diagrams of individual globular clusters. Indeed, it is likely that there are more differences than can be accounted for by varying a single "metal abundance." Perhaps different elements differ in separate proportions.

One general trend within the halo population is clear, however: the objects of greater metal abundance tend to have a flatter distribution, closer to that of the disc. Some schemes of classifying globular clusters distinguish, in fact, between "halo globulars" and "disc globulars." This latter term is something of an exaggeration, since these strong-lined clusters are more widely spread above and below the Galactic plane than are the stars of the disc, but the terminology makes its point nevertheless.

This population trend within the halo, converging toward the disc, raises the question of whether the "two" populations can really be sharply separated—or do they merge into each other continuously? Again, only future studies will tell.

STELLAR GROUPS

Star clusters exist within all components of our Galaxy—the halo, the disc, and the spiral arms. They have been invaluable in helping us to understand the nature of stars in general, but the clusters are also interesting for their own properties.

FIGURE 20.21.
These two globular clusters, which happen to lie in nearly the same direction, have very different properties. Messier 53 (upper right) is typical of rich, dense globular clusters, while NGC 5053 (lower left) is poor in stars and loosely bound. (Hale Observatories photograph by the author.)

The clusters in the halo are nearly all very rich, from tens of thousands up to millions of stars. In recognition of their striking symmetry, they are called *globular clusters.* In a photograph of a globular cluster the stars in the central region—a few parsecs in diameter—look utterly jammed together. In reality, however, it is only the picture that is crowded; the stars are much tinier than their images on the photograph, and they are very far from brushing against each other.

In the disc (and in the spiral arms, which are of course just the young-star regions of the disc) the star clusters do not have nearly so many stars. A typical cluster has a few hundred stars, and they do not look so densely packed on photographs; hence they have acquired the name *open clusters*. Since most open clusters are seen near the plane of the Milky Way, they are often called "galactic clusters."

Our catalogs of star clusters are of uneven completeness. Globular clusters are so conspicuous that we can see them all over the Galaxy; and because most of them are well away from the Galactic plane, absorption hides relatively few of them. We know of about 125 globular clusters, and there may be a few dozen more hidden behind the dust clouds near the plane. The open clusters, however, lie near the plane of the Milky Way, and our knowledge of them is limited very much by interstellar absorption. The 700 or so open clusters in our catalogs are nearly all within 2 kiloparsecs of the Sun; they must be only a small fraction of the open clusters in the whole Milky Way.

Each star cluster contains stars that are characteristic of the component of the Galaxy within which it is found, but the stars in each cluster are a narrow selection. The simplicity of the individual HR diagrams shows that within each cluster we have a set of stars that have the same age and the same chemical composition. Apparently these stars were born together at a single time, and the cluster has since then neither brought

FIGURE 20.22.
This photograph of a small region of the Milky Way shows several star clusters. NGC 654, 659, and 663 are all young clusters, recently born in the Perseus spiral arm. IC 166 is a much older cluster, which belongs to the galactic disc. (Hale Observatories photograph by the author.)

new stars to birth nor allowed interlopers to join its society from outside.

This behavior is easy to understand, if we examine how a cluster holds together. The force, as usual, is gravitation; and the structure of the cluster is a balance between the motions of the stars, which tend to make them move farther apart, and their gravitational attraction for each other, which tends to draw them together. Each star moves as an individual; but any time that its motion is directed away from the cluster, the attraction of all the other stars slows it down and draws it back into another plunge through the group. So the cluster maintains itself. At any given moment some stars are looping outward and some inward, but each star maintains the energy of its in-and-out swing, bound by the joint attraction of all its companions.

Only rarely do two stars pass so close together that they influence each other's motion as individuals. Thus a foreign star that happens to pass through the cluster does so with no interference to its motion; and it passes on, with no chance of being captured. For stars inside the cluster, however, over a long period of time quite the contrary happens. The member stars, swinging repeatedly through, do occasionally perturb each other's motions during individual encounters. The effect of these random changes is to slow some stars down and to speed others up. Once in a long while a star acquires such a large speed that it leaves forever. It cannot be replaced, and so the cluster slowly dwindles.

Meanwhile the cluster is also buffeted by outside forces—also gravitational—from concentrations of mass in the Galactic plane and particularly in spiral arms, and from the steady tidal force due to the main concentration of mass in the center of the Galaxy. Between the cluster's own tendency to eject stars and the effect of external forces, its lifetime as a cluster is limited. The stars can go on living their individual lives, but they cannot remain together forever as a star cluster.

Calculations of all these dynamical effects show that the expected lifetime of a cluster depends on its size. A rich globular cluster is almost impervious, during the 10 billion years since it was born. The open clusters of the Milky Way, however, have many fewer stars, and the erosive processes work faster on them. A typical cluster should be broken up in half a billion years. This is indeed what we find; young clusters are plentiful, but old clusters are rare—and, in fact, the few old open clusters that have been found are the massive ones that are best equipped to survive.

Is it possible, then, that all stars begin their life in clusters? It is very easy to answer no. Most young stars are in the field rather than in clusters, and the time since their birth is too short for them to have come from disrupted clusters, because clusters just do not break up that fast. From a deeper point of view, however, this is a hasty answer. To be sure, most stars are not formed in real clusters, in the sense of groups that will remain gravitationally bound together. But it does appear that they are *born* in groups nevertheless. These groups are called *stellar associations*. Their stars have too little gravitational attraction to counteract the velocities with which they are born, and they separate from each other. After a short time—a few million years—the stars of an association have spread unrecognizably into the general field of stars.

The number of young stellar associations that we see is enough to support the hypothesis that all stars are born in associations. It may well be that the star clusters are just those few groups that happened to be born in such a tight clump that they were able to remain together.

Some of the waves of star formation must have indeed been prolific. In the solar neighborhood are some sets of field stars whose motions are so similar that they seem to have come from a common event of star birth in the past. The best-recognized such "moving group" also contains two widely separated star clusters, whose identical HR diagrams allow us to date this "super-association" as having been born several hundred million years ago.

From evidence like this—and, indeed, from current activity in the spiral arms—it appears that star formation in the disc of the Milky Way has taken place in great local bursts. What happened in the halo is much harder to say, since we can examine the evidence only 10 billion years later. Certainly each globular cluster represents a single event of star formation, creating up to a million stars at once. As in the disc, however, only a small fraction of the halo stars are members of clusters, 1 per cent or fewer. Whether the halo once had its great stellar associations is much harder to say; but there are some faint indications, from similarities in the motions of high-velocity stars, that this may have been so.

THE MILKY WAY AS A GALAXY

Throughout this chapter we have been attempting to probe the nature of our Galaxy, hampered by our parochial position, embedded in the outer part of its dusty disc. Because of these difficulties, there are many ways in which we know our own Milky Way less well than other galaxies. But there are obviously ways in which we know it infinitely better. We can study individual stars in great detail, and we have a wealth of information about the interstellar material and the way in which it forms stars. In studying other galaxies we shall make use of much of this information. Other galaxies are so distant that at best we can see as individuals only their most luminous stars—and, indeed, in only a few of the nearest galaxies can we see individual stars at all. But as always, we shall count on an orderly cosmos, in which properties that we have found in our own stellar system can be presumed to apply in a corresponding way in remoter places.

FIGURE 20.23.
Facing page: a part of the great spiral galaxy in Andromeda, Messier 31, enlarged from a photograph made with a large telescope. (The relation to the smaller-scale photograph below is shown by the white line. The cut-off corner is the edge of the field of the telescope.) Note, along the spiral arms, the prominent regions of star formation and the rich associations that they contain. (Kitt Peak National Observatory.)

REVIEW QUESTIONS

1. Contrast the extent, the mass, and the composition of the disc of the Milky Way, the halo, and the spiral arms.

2. Why does our position in the Milky Way make it difficult for us to see the over-all structure of our Galaxy? How do we deduce this structure nevertheless?

3. How do stars move in the disc of the Milky Way? Distinguish between Galactic rotation and peculiar motions.

4. What do we mean by the "rotation curve" of the Milky Way? How do we determine it? What do we deduce from it?

5. By what methods can the spiral arms of the Milky Way be mapped out? Describe the structure in the few kiloparsecs around the Sun.

6. Discuss the processes by which spiral arms are formed. What do we feel reasonably sure about, and what do we not know?

7. Why do we believe that the Galactic halo is a mixture of subpopulations? What could this signify?

8. What holds a star cluster together, and what are the processes that tend to break it apart?

21 | GALAXIES:

THE STELLAR CONTINENTS

Our Milky Way is just one of the countless galaxies that populate the universe. There are single galaxies, doubles, groups, and clusters of galaxies, scattered in all directions as far as our telescopes can see. These are the units into which the material of the universe divided itself, early in its history. Since then, each galaxy has developed in its own independent way, usually with little or no interaction with its neighbors.*

GALAXY TYPES

Galaxies have developed many different structures, and our first approach is therefore to classify them. The forms of most galaxies can be arranged in a fairly neat sequence, ranging from *elliptical* galaxies, which

---

*Our terminology has developed in an awkward way as our knowledge of the universe has broadened and deepened. The word "galaxy" is now used for any independent stellar system, but originally it referred only to the Milky Way. When capitalized, however, "the Galaxy" still refers to the Milky Way. Unfortunately, the adjective "galactic" is understood to apply only to the Milky Way, even when it is spelled with a small letter. There is no satisfactory adjective corresponding to "galaxy." Some astronomers use "galaxian," but others gag on this word and refer instead to "galaxy distances," "galaxy magnitudes," etc. Worst of all, galaxies have not always been called by that name. The earliest cataloguings of nonstellar objects did not distinguish galaxies from gaseous nebulae, so that when the difference was realized, the former became "extragalactic nebulae." The recognition that other galaxies are comparable to the Milky Way led to the term "external galaxies," and finally the word "external" was dropped.

FIGURE 21.1.
The giant elliptical galaxy M87. (Hale Observatories.)

FIGURE 21.2.
NGC 205 is an E5 galaxy of intermediate luminosity. It is one of the
companions of the Andromeda spiral, M31 (at the upper right in Figure
1.3 and on page 412). In this photograph the brightest individual
stars of Population II can be seen in large numbers. (Hale
Observatories.)

show no structure other than a smooth concentration of light to their
centers, through *spirals*, which have a centrally concentrated but flat disc
with spiral arms imbedded in it, to *irregulars*, which have a more or less
chaotic form. These classes grade into each other, and there is a progression within each class.

An elliptical galaxy has an extremely smooth and symmetric distribution of brightness. What we see is the combined light of a large number
of stars, too faint to detect individually except in our closest neighbors
among the galaxies. There are no outstandingly bright stars, which
means that there is no very young population. It appears, in fact, that an
elliptical galaxy contains only old stars. But it is not at all surprising that

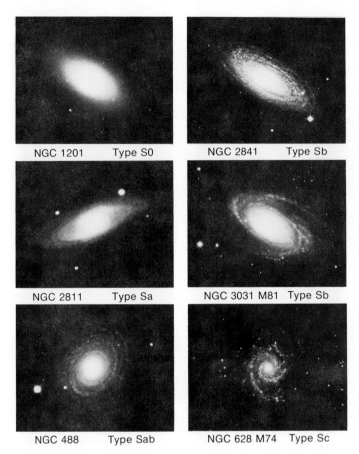

NGC 1201   Type S0   NGC 2841   Type Sb

NGC 2811   Type Sa   NGC 3031 M81 Type Sb

NGC 488   Type Sab   NGC 628 M74 Type Sc

FIGURE 21.3.
An S0 galaxy, and various types of spirals. For better photographs of M81, see the frontispiece and Figure 20.17; for M74, see Figure 20.1. (Hale Observatories.)

there should be no young stars in an elliptical galaxy, because it contains very little interstellar material; so it has nothing to make stars from.

Elliptical galaxies have a range of shapes that differ in the degree of flattening, from spheres through pumpkins, down almost to discs. The difference is in the relative amount of rotation. In all cases the motions of the stars are a mixture of randomly directed motions and over-all rotation around the central axis; when the rotation predominates more, the galaxy is more flattened. Astronomers classify ellipticals according to their apparent flattening, from E0 for a round system to E7 for a very flattened one. The relation between the apparent shape of an individual elliptical galaxy and its *true* shape is unknown, however, because we do not know from what sort of angle we view it. Thus an E2 may be a galaxy that we see at an oblique angle, whose real shape is E6, for instance. But even though we can never tell the true degree of flattening of an individual galaxy, we can be sure, from the statistics of the apparent shapes and the assumption that we see them from random directions, that the true shapes do indeed range through all degrees of flattening.

There is no sharp dividing line between elliptical and spiral galaxies, and indeed astronomers recognize a transitional class called S0. This is a disc, of old stars only, without spiral arms. The center of the disc bulges out thickly, and the edges are thin. Dust is sometimes seen, but there is very little gas.

There is a continuous progression of properties among spiral galaxies, which astronomers classify by the letters Sa, Sb, Sc, and in a few cases Sd. An Sa has a large, prominent central bulge, outside which are thin, tightly wound spiral arms. In an Sb, the central bulge is less prominent and the arms more extensive. In an Sc the arms wind all the way in to the center, and there is little or no bright, smooth bulge at the center.

Between spiral galaxies and irregulars there is again no sharp dividing line. An Sd has barely enough symmetry to be called a spiral at all, whereas an irregular galaxy looks like a star cloud that has no particular form.

As with spectral types of stars, astronomers sometimes use the words "early" and "late" to designate positions in this classification sequence; thus an Sc galaxy is "later" than an Sb. Here again, however, these terms have no evolutionary significance.

In addition to this variation in the relative prominence of central bulge and spiral arms, spiral galaxies differ in one other way. In some cases the spiral arms wind smoothly outward from a round central region, but in other cases they begin at the ends of a straight bar that extends across the central region. The "barred spirals" are designated SB0, SBa, etc.

Along with the progression in the optical appearance of galaxies goes a corresponding progression in the amount of interstellar material, which is easily seen from radio measurements of the total amount of 21-centimeter hydrogen radiation coming from a galaxy. Relative to the total mass of the galaxy, the amount of gas rises from practically nothing, at type S0 and earlier, to 10 per cent or more in irregulars. This whole picture makes very good sense; the key is the amount of interstellar material and the resulting prominence of the young stars that form within it. The sequence of galaxy types just described is, after all, just a sequence of prominence of the young population. In the E's and S0's the young population is absent, whereas it dominates the picture of an irregular galaxy. This goes hand in hand with the amount of interstellar material; when there is a lot of interstellar material, there is a lot of star formation.

It might seem, then, that the irregular galaxies, still full of interstellar material and young stars, are themselves young galaxies. This is almost certainly not the case, however. There is convincing evidence that gal-

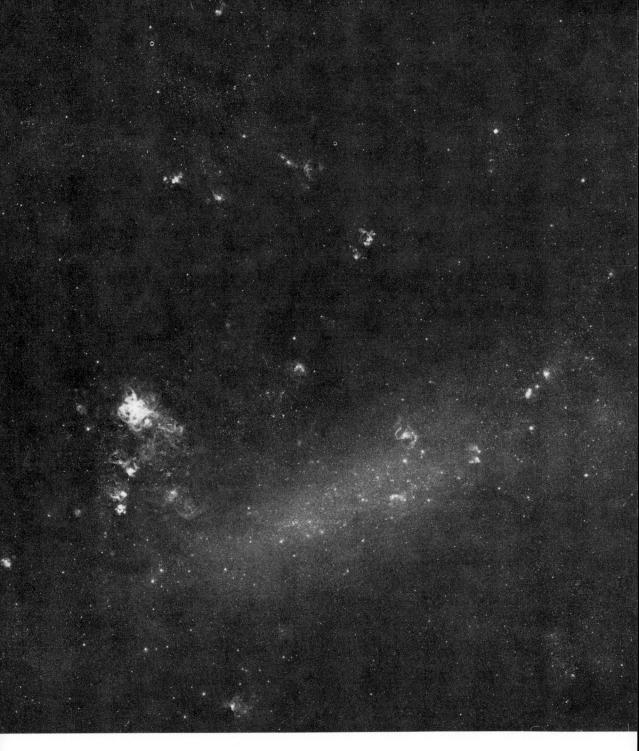

FIGURE 21.4.
The Large Magellanic Cloud is an irregular galaxy that gives some hints of barred-spiral structure.
(Cerro Tololo Inter-American Observatory.)

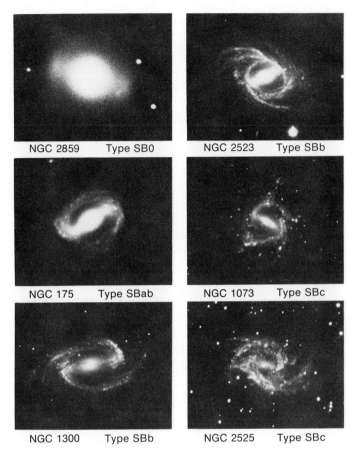

NGC 2859   Type SB0   NGC 2523   Type SBb

NGC 175   Type SBab   NGC 1073   Type SBc

NGC 1300   Type SBb   NGC 2525   Type SBc

FIGURE 21.5.
An SB0 (a galaxy with a bar but no spiral structure) and barred
spirals of various types. (Hale Observatories.)

axies of all types are old. The centers of spirals contain an old population,
and even the irregular galaxies contain a Type II population like that of
the Galactic halo. We clearly detect its individual stars in those irregulars
that are close enough for us to observe stars of such a luminosity. It
would appear, then, that irregular galaxies are as old as the Milky Way.
If they are still full of interstellar material, it is because they have kept it
in interstellar form, rather than turning it all into stars.

This seems paradoxical, that the galaxies that seem to be making stars
most rapidly are really those that have been most conservative. But if we
follow the arithmetic through, it really works. An irregular galaxy is
indeed making many stars at the present time; but it has so much more
interstellar material, relative to the rate at which the material is being
used up, that its depletion time is still very long. Thus the basic difference

between galaxies is in the rate at which they have turned their material into stars.

Why galaxies should differ in this respect, we unfortunately do not know. This is the way in which scientific understanding develops; we collect information and always attempt to understand many facts in terms of a few principles. At some point we encounter questions to which we have no answer; this is today's frontier. The depth of our penetration into the problem is measured not by the state of things on the frontier, but rather by how far away the frontier lies.

THE DISTANCE SCALE

We can classify galaxies just from looking at their appearance in photographs. But to study them in detail, and particularly to measure their sizes, we must know how far away from us they are. This requires identifying in other galaxies stellar types whose absolute magnitudes we know, from having studied similar stars in the Milky Way. Only highly luminous stars will do, because the distances are very great; even for nearby galaxies, distances are commonly counted in megaparsecs (a megaparsec is a million parsecs).

The most fundamental yardsticks for the distances of galaxies are the cepheid variables. Their absolute magnitudes can be found from the period-luminosity relation; and because they are such luminous stars, our

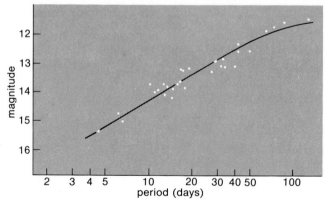

FIGURE 21.6.
Periods and magnitudes of cepheids in the Small Magellanic Cloud. The curve is a standard period-luminosity relation, slid fainter by 18.85 magnitudes. This shows that the distance modulus of the Small Magellanic Cloud is 18.85 magnitudes, which corresponds to a distance of 59 kiloparsecs. (Adapted from A. Sandage and G. A. Tammann, *Astrophysical J.,* **151** (1968), 541.)

large telescopes can observe them in galaxies as far away as 4 or 5 megaparsecs. It is a laborious task, first to discover them by blinking photographs, and then to plot their light curves by measuring many time-consuming long-exposure photographs; but the study of cepheids in a dozen or so nearby galaxies is the cornerstone of the extragalactic distance scale.

Once the distances of a few galaxies are known, we can measure the characteristics of other objects that are both easier to observe than cepheids and more luminous. Thus the cepheids served as calibrators, but the other objects now take over as secondary yardsticks. The best of these other distance criteria appear to be the absolute magnitudes of the brightest stars and the diameters of the largest HII regions. These can be used to measure distances of galaxies out to about 10 megaparsecs. At that distance, apparent magnitudes are fainter than absolute by 30 magnitudes, so that a star of absolute magnitude –9 has an apparent magnitude of 21, about as faint as can be measured accurately with our large telescopes.

All these distance-measuring objects are members of a young population. For the galaxy types that contain no young stars at all—the ellipticals and S0's—these particular distance criteria are not available. The brightest stars are the red giants of the Type II population, and these are hard to detect even at a distance of 1 megaparsec. There are brighter individual objects, however: the globular clusters. The total luminosity of a globular cluster is equivalent to that of a very luminous young star, and the light of the cluster is sufficiently spread out to make it look like a fuzzy patch on a photograph, even at a distance as great as 10 megaparsecs.

In galaxies at greater distances we cannot see individual objects at all, except for rare supernovae. The over-all structure of the galaxy is quite apparent; but in estimating its distance, all we have to go on is its total brightness and its angular diameter. Diameters are a poor distance-measuring criterion, both because galaxies have ill-defined edges and because we know from study of nearby galaxies that there is a tremendous range of sizes. Thus for the greater distances of the universe we are thrown back on the magnitudes of galaxies.

These too have a very large range; nearby galaxies are known whose absolute magnitudes are as faint as –10 or as bright as –23. Among faint, distant galaxies, how are we to know which is which? The answer is to take a large sample at once, all at the same distance, so that we can tell which galaxies are more and less luminous. This opportunity occurs in clusters of galaxies; in practice what we do is to take the brightest galaxies in a cluster and assume that they have a standard absolute magnitude, which is appropriate for the most luminous galaxies in any large sample. Thus clusters of galaxies are the most distant objects for which

FIGURE 21.7.
M33 is a nearby Sc galaxy. Its largest HII region is marked here with an arrow. (Lick Observatory.)

we can measure photometric distances. The faintest that have been studied so far are a billion parsecs away.

This leaves us with a serious problem, however. How are we to estimate distances of remote individual galaxies, most of which are not members of rich clusters? Here we can make use of one of the striking properties of the universe itself: its expansion. As we shall see in more detail in Chapter 23, all the galaxies are moving away from each other; as a result, every distant galaxy appears to be moving away from us with a velocity that is proportional to its distance. This velocity shows up as a Doppler shift in the spectra of distant galaxies—the famous

FIGURE 21.8.
A faint cluster of galaxies in the constellation Hydra, more than a billion parsecs away. (Hale Observatories.)

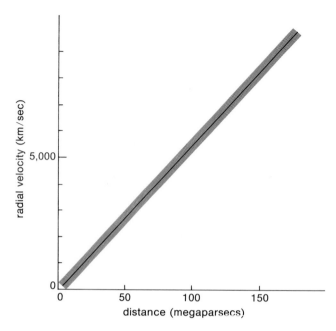

FIGURE 21.9.
The velocity-distance relation for galaxies. The width of the band indicates the general size of the deviations of individual velocities from the mean line.

redshift. In Chapter 23 we shall attempt to understand the redshift as one of the puzzles of cosmology, but here we can accept it gratefully as an indicator of the distance of each remote galaxy. From nearby galaxies with photometric distance measurements, we can calculate the scale of the redshift; it is about 50 kilometers per second per million parsecs of distance from us.

From these nearby galaxies, we also see that galaxies have individual motions of 100 or 200 kilometers per second, in addition to the average velocity of the redshift. This means that the velocity of an individual nearby galaxy may deviate by a serious fraction from the average red-shift rule, so that for nearby galaxies the velocities are not reliable indicators of individual distances. For more distant galaxies, however, the extra individual velocity makes very little difference. If a galaxy has a velocity of 10,000 kilometers per second, an additional uncertainty of 200 kilometers per second hardly matters. For distant galaxies, then, the redshift is an excellent distance indicator.

If the redshift tells us distances, why then do we need to bother with measuring photometric distances of clusters of galaxies? The answer is that redshift distances are indeed adequate when our problem is only to study galaxies. But when we come to study the redshift itself, as a problem of cosmology, we need an independent way of measuring large distances. Again, we shall return to these problems in Chapter 23.

POPULATIONS IN GALAXIES

As soon as we measure the distances of galaxies, it becomes clear that they have a wide range of luminosities. The most luminous galaxies are giant ellipticals, with absolute magnitudes around –23; but the ellipticals also span the full range of luminosity, down to dwarfs with absolute magnitudes as faint as –10. This last is no brighter than the most luminous globular clusters; the difference is that a globular cluster is part of a larger galaxy, while a dwarf elliptical is an independent system and also tends to have a much larger diameter than any star cluster—a kiloparsec rather than tens of parsecs.

The most luminous spiral galaxies do not quite match the most luminous ellipticals. The spirals also have a more restricted range of luminosities and do not include any extreme dwarf galaxies. Even so, they range through a factor of 100 in luminosity. Irregulars, in turn, do not include

FIGURE 21.10.
NGC 4594, the "Sombrero" galaxy, is a giant Sa. Its central region has an old population and very little interstellar material, whereas dust and young stars can be seen closer to its edge. (Kitt Peak National Observatory.)

FIGURE 21.11.
A dwarf irregular galaxy in the constellation Sextans. (Hale Observatories.)

any galaxies as bright as the giant spirals, but some irregular galaxies are extreme dwarfs.

Our general knowledge of dwarf galaxies is still quite poor. Just as with stars, we know only a few nearby dwarfs, because the more distant ones are so inconspicuous, whereas we survey giant galaxies throughout a much larger volume of space. When we allow for this volume effect, it appears that the universe contains far more dwarf galaxies than giants. On the other hand, when we take sizes into account, it turns out that the giants contain most of the total material.

The stellar population types in other galaxies, in so far as we can study them, are like those that we find in the Milky Way; but they often occur in quite different proportions. The dominant stellar population of the universe is an old Population I. It is responsible for most of the light of the giant elliptical galaxies, and the central regions of spirals are quite similar to elliptical galaxies. Even in the region covered by spiral arms, a large part of the light comes from an underlying old Population I. And there are indications that this is also an underlying population in irregular galaxies, even though it is even less conspicuous there.

The other population that probably occurs everywhere is Population II—all old, of course. In most galaxies it is relatively inconspicuous, but the smoothly spread red giants of Population II show up in every galaxy that is close enough for us to see such stars, in elliptical galaxies, spirals, and irregulars.

The most conspicuous population variable is, of course, the amount of young Population I. As we have seen, its prevalence and its distribution account for the major differences in galaxy types.

One other population characteristic varies among galaxies in a systematic and intriguing way: the metal abundance, which is, of course, the basic distinction between Populations I and II. The differences are particularly striking among the elliptical galaxies. Giant ellipticals have a dominant population that is Type I, with "normal" metal abundance; whereas dwarf ellipticals appear to be pure Type II, with low metal abundance. Within many galaxies there are also some suggestions of a radial change of population, in the sense of a higher metal abundance near the center. The central few hundred parsecs of giant ellipticals and spirals have very strong heavy-element spectral lines, and the stars in those regions may well have a higher metal abundance than even the conventional Population I. Also, the emission lines of the interstellar gas in spirals suggest a similar effect: unusually high heavy-element abundance at the center, and a decrease outward that persists far beyond the small central region. Thus there is a general tendency for metal abundances to drop off radially within a galaxy and also to be a lower in dwarf galaxies. The latter tendency is well-established for elliptical galaxies. For spirals and irregulars, the information is fragmentary, but it does tend in that direction, raising the question of whether some dwarf irregular galaxy will eventually be found to contain a young Population II!

What does it all mean? The stellar populations that we see in a galaxy are indicators of its history and its development. Written in the populations, in a code which we are still not able to decipher at all well, is the way in which gas has formed into stars, and the way in which successive generations of stars have changed the chemical composition of the material. Future astronomers will read this code, and the messages will change the face of astronomy.

THE DYNAMICS OF GALAXIES

There is much to be learned about galaxies by studying their dynamics. As usual, the key to masses is gravitation, and the gravitational forces are indicated by the motions of the stars and gas within a galaxy. In

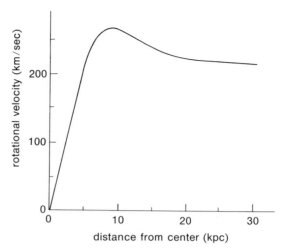

FIGURE 21.12.
A typical rotation curve of a giant spiral galaxy.
The continuing high velocity far from the center
suggests that much of the galaxy's mass is widely
spread.

spiral galaxies, with their flattened disc-like forms, the motions are circular, and all that we need to study is the rotational velocity at various points within the galaxy. Elliptical galaxies have a mixture of random and rotational motions; for them the analysis is more complicated, but again the speeds of the stars and the size of the galaxy can be combined to give an estimate of its mass.

The masses of galaxies have a large range, as might be expected from the large range in luminosities. Unlike stars, however, where masses range through a factor of only 100, but luminosities differ by factors of 10^{10} or more, galaxies have masses that go very much as their luminosities. The reason is simple: a galaxy is made up of stars, and a large galaxy simply has more of them. Thus it is natural that the mass should go up in proportion to the luminosity. To state this in another way, the *ratio* of mass to luminosity should be nearly constant, regardless of the size of the system, as long as we are dealing with galaxies that are made up of similar mixtures of stars. But for a galaxy with a different population mixture, M/L (as it is commonly called) should be different; thus the M/L of a galaxy tells us something about its population content.

Specifically, M/L tells us the relative proportion of dwarf and giant stars, since most of the light comes from giants, but most of the mass is in dwarfs. Furthermore, we can find out a great deal about the giant stars in a galaxy from an analysis of its spectrum, so that the M/L can be used as an indication of the number of dwarf stars, about which we would otherwise know nothing.

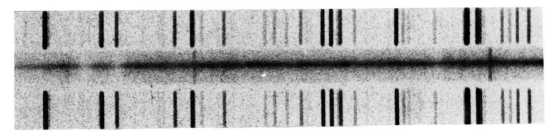

FIGURE 21.13.
The spectrum of the giant elliptical galaxy NGC 4406. Because its individual stars are moving toward and away from us with speeds of several hundred kilometers per second, the lines in the integrated spectrum are so broadened that they are almost invisible. This spectrum also illustrates light pollution; the bright lines across the galaxy spectrum come from mercury street lighting in San Jose. (Lick Observatory spectrum by the author.)

We normally express M/L in units of the mass and luminosity of the Sun, so that for the Sun itself $M/L = 1$. A young population has an M/L much less than 1, because of the large amount of light that is contributed by its luminous stars. For the mixed population of the solar neighborhood, we can add up all the masses and luminosities, and we find an M/L of about 4. Among old populations the value of M/L seems to vary from one type to another, indicating that different populations have different numbers of dwarf stars. For globular clusters, dynamical studies indicate an M/L of about 1. Similarly, the weak indications that we have of the M/L in dwarf elliptical galaxies suggest a comparably small value. These systems are all rather poor in dwarf stars. Whether all Type II populations have few dwarfs, we do not know, because the proportion of dwarf stars in the Galactic halo is still very uncertain.

Giant elliptical galaxies are quite different in this respect; they have M/L's of around 30. This dwarf-rich population seems also to exist in the central bulges of giant spirals. In more fundamental terms, star formation in these systems, long ago, put their material predominantly into small stars. Why this should be, we do not know; this whole problem area is one in which population studies have given us only these few first clues.

There is one aspect of galaxies that does appear to have some straightforward dynamical explanations, however. The general form of each type of galaxy seems to be a consequence of the rate at which it turned its material into stars. The dynamical picture has two parts: the behavior of gas in the earliest history of a galaxy, and the subsequent behavior of a mixture of gas and stars.

In its earliest form, before it had made any stars, a galaxy must have consisted of a large cloud of gas. Interestingly, we can draw some strong conclusions about how such a protogalaxy will behave. First, it must be held together by its own gravitation; otherwise it would fly apart and never become a galaxy. As with any other self-gravitating system, there is a contest between gravitation, pulling it together, and internal motions, keeping the system extended. Unless the internal motions are large enough to support the gas cloud, then, it will collapse under its own gravitation. The motions can be of two sorts. They can be motions of the individual atoms, the helter-skelter movement whose amount is an index of the temperature of the gas; or they can be motions of separate clouds of gas within the galaxy, in orbits that are like those of stars in a cluster. But the motion must be there in one form or the other, or else gravitation will rapidly draw the whole mass together.

The trouble is that a protogalaxy made of gas cannot maintain motions of *either* of these kinds for very long. In either case collisions will rapidly cause the energy of motion to be dissipated. For thermal motions it is the individual atoms that collide. Some collisions excite one of the atoms involved, and the two colliding atoms consequently rebound less rapidly. The excited atom then drops to its ground level by radiating a photon, whose energy is lost into space. Thus the gas continually loses the energy of motion that is needed to support it against its own gravitation. Calculations of these energy losses show that they should be very rapid, so rapid, in fact, that a gaseous galaxy cannot support itself at all. Gravitation simply pulls it together as fast as it is able to fall. The "free-fall" time for a cloud with the mass of a galaxy, starting at several times its present dimensions, is a billion years or less.

Nor can cloud motions support a protogalaxy. Here the trouble is that the clouds very soon collide with each other. A collision dissipates the motion of the clouds, and the energy goes into heating them up. The hot gas then radiates away its energy, in the manner that we have just seen. Thus in either case the energy of motion is dissipated so fast that the protogalaxy collapses.

The next question to ask is, what does it collapse *to*? Here one of the fundamental rules of dynamics determines the answer. According to the principle of *conservation of angular momentum,* the amount of rotation in the protogalaxy must remain constant. This is measured by its speed of rotation times its size; thus as the protogalaxy contracts, the product can remain unchanged only if the galaxy spins faster. Its ultimate state is a completely flattened disc in which all the remaining motion is rotation. All the gas particles then move in circular orbits; they no longer collide with each other, so that no more energy is lost by excitation and radiation.

This final picture is strikingly similar to that of the disc of gas in a galaxy like the Milky Way. The only difference is that the gas layer in a real galaxy is not infinitely thin, nor are all motions exactly circular. The reason is that the real galaxy has processes that keep putting a small amount of energy back into the gas. The thickness of the disc is then determined by how fast this energy is put in and how fast the gas clouds dissipate it by collisions. We do not know specifically what the energy source is, but there are two processes that seem capable of stirring up the interstellar gas sufficiently. One is the explosion of supernovae; the other is the creation of HII regions around newly born stars. The latter process is one of sudden heating, which causes the gas to expand and blow its surroundings away. Within the uncertainties of the birth rate and the death rate of stars, we do not yet know which process stirs up the gas more.

Thus dynamical arguments show us why the present star-forming layer is a thin disc. More important, our analysis of this process tells us a great deal more about the early history of the Galaxy. The old stars of the Galactic disc formed only after the gas of the protogalaxy collapsed into a disc, but the much rounder Galactic halo must have formed its stars before the collapse was complete. To understand this distinction one must recognize the basic difference between the dynamics of stars and those of gas: in a gas the atoms collide and dissipate energy; but stars do not collide, and therefore they maintain their energies. Thus a gaseous system will collapse but a stellar system will maintain its size forever.

In the Milky Way, then, the halo stars were all born during the first few hundred millions of years, when the Galaxy was much rounder than it is today. Most of the material did not form into stars at that time, however. Left in gaseous form, it necessarily dissipated its energy and collapsed into the inevitable disc, where star formation has gone on at a slower rate ever since.

Presumably, the same argument applies equally to other galaxies. An elliptical galaxy must be a system that turned all its gas into stars before any dissipative collapse could take place. The flatter ellipticals, however, may indicate the occurrence of some dissipation. The S0 galaxies were not quite so quick; there the gas did collapse into a flattened disc, but the gas then formed itself into stars immediately. In all these systems we see only old stars. In the ellipticals, in particular, it could not be otherwise; any material capable of forming stars at a later date would quickly flatten into a disc, and that is where the later stars would form, just what we indeed see in spiral galaxies.

Thus the early dynamical behavior of gas explains much about the structure of galaxies. By another line of reasoning, the present behavior of gas may be the basic reason for another structural difference: between spiral galaxies and irregulars. A very likely key here is the basically different behavior of stars and gas. In a word, stars mix, but gas clumps. Because stars never collide with each other, the various parts of a stellar system interpenetrate each other freely until the whole distribution of stars is exceedingly smooth. No astronomical object is as featureless as an elliptical galaxy. In a spiral galaxy too, the older stars in the disc also have a very smooth distribution. Not so the gas, however; for it, chaos is the rule. Its early collisions were responsible for the original formation of the disc, and even today the collisions go on. Between the shock of the collisions themselves and the gravitational imbalance that is left by their energy dissipation, the gas is full of turbulent condensations.

In the dynamical behavior of these gas condensations lies a likely explanation of the difference between spiral and irregular galaxies. In both the gas forms condensations under its own gravitation. In a spiral galaxy, however, the gas is only a tiny fraction of the total. The dominant gravitational force is that of the symmetric, centrally condensed disc of stars. In this force field, the gas follows the general rotational pattern, in which the edges lag behind the rapidly rotating central parts. As a result, condensations of the gas are twisted into trailing spiral structures, just like cream that is stirred into a cup of coffee. In an irregular galaxy, on the other hand, the gas makes up a sizeable fraction of the total mass; and the force of the stellar component is not great enough to force the clumping gas into a regular pattern.

This picture may or may not be right. It is more than a wild guess, but it is much less than an explanation. It does invoke physical processes and predict their behavior in a plausible way; but it lacks, at this stage, a quantitative working out of detail. Yet this too is a stage of science: the sketching of an outline of how a theory might be developed. For this problem, unfortunately, we can do nothing more solid at the present time.

THE LOCAL GROUP OF GALAXIES

Few galaxies live alone. Most are members of doubles or multiples, and many galaxies belong to larger aggregates, ranging from small groups up to rich clusters that contain a thousand galaxies or more.

Our own Milky Way is part of a loose clustering of nearly two dozen galaxies that we call simply the Local Group. Within the group, many of the galaxies belong to two prominent subclusters, one centered on each of the two largest galaxies, the Milky Way and the Andromeda spiral, which are about 0.7 megaparsec apart. Close to the Milky Way are two irregular galaxies, the Magellanic Clouds, which have already figured in our story of how cepheid variable stars are studied. Accompanying the Andromeda spiral (which astronomers usually call M31, after its earliest catalog designation) are four small ellipticals. The Local Group also contains one other smaller spiral galaxy, two more irregulars, and a number of very faint dwarf ellipticals.

These extreme dwarfs are still poorly known, because they are so hard to find. Six have been found in the space surrounding the Milky Way; and three have been found in the immediate neighborhood of M31, which is the only distant region in which a search for such inconspicuous objects has been made. It may be that dwarfs of this type occur only as

FIGURE 21.14.
The nearby dwarf elliptical galaxy in the constellation Sculptor.
(Boyden Observatory photograph, courtesy of P. W. Hodge.)

companions to giant galaxies, in which case our catalog is reasonably complete. If, however, such systems are equally likely everywhere in space, then those that we have discovered are only a tiny fraction of the total, even within the modest confines of the Local Group. Under that hypothesis, the Local Group would contain hundreds of dwarf elliptical galaxies, and more than 90 per cent of the galaxies of the universe would be of this virtually undetectable type.

This possibility, that most of the galaxies of our neighborhood may be missing from our catalogs, does not disturb astronomers seriously. The function of astronomy is not to catalogue the universe but to understand it. What matters is not the individual galaxies but rather the clues they give us about how the universe has developed and why it has become what it is. In this respect the dwarf ellipticals are one of the galaxy types, and as such they represent one of the ways in which a stellar system can form. Not much of the material of the universe has taken that course, in fact. The mass of each dwarf elliptical is so small that even when we add them all up, they constitute only a tiny fraction of the material of the universe.

Taken as a whole, the Local Group is a very modest grouping of galaxies. If we viewed it from a distance of 20 megaparsecs, as we do many other groups of galaxies, we would see two rather handsome spirals, the

Milky Way and M31, about two degrees apart (or closer, depending on the oblique viewing angle). Each would have a diameter of several minutes of arc. The Magellanic Clouds would show up clearly near the Milky Way, as would the two brighter elliptical companions of M31 near it. We would also see the other spiral and the two other irregulars quite easily, and we might even guess that all were members of the same physical group. The less luminous galaxies of the Local Group, however, would be indistinguishable from the host of faint background galaxies that appear in every region.

CLUSTERS OF GALAXIES

Small groups of galaxies, such as the Local Group, are very common. In addition we find, scattered through the distant scenery, much richer groupings that are called clusters of galaxies. These are systems that are held together by the gravitational attraction of the galaxies for each other, just as stars are held together in a star cluster. The dynamical principles are the same in both cases, even though the scale is very different. The diameter of a cluster of galaxies is a million parsecs or more, and the galaxies move through it at speeds around 1000 kilometers per second, taking a billion years or more to cross the cluster.

In some clusters, indeed, the time that an individual galaxy takes to move across the cluster—the "crossing time"—is even longer and amounts to a considerable fraction of the age of the universe. In such a cluster the galaxies have had relatively little opportunity to smooth out their arrangement in space, and it is probably this dynamical "youth" that accounts for the ragged appearance of many of the looser clusters of galaxies.

Loose clusters of galaxies differ in another striking way from dense clusters. They contain comparable numbers of spiral and elliptical galaxies; the relative numbers are similar to those for the general field, outside clusters. Dense clusters, by contrast, have almost no spiral galaxies (or irregulars either); nearly all their members are eliptical galaxies or S0's. Physically the distinction is easy to state: in a dense cluster almost none of the galaxies have any appreciable content of interstellar material.

It is very tempting to blame this condition on the crowded circumstances in a dense cluster. Collisions between galaxies must happen occasionally. In such a "collision," the stellar parts of the galaxies pass through each other like insubstantial ghosts; there is unlikely to be a single collision between stars. The interstellar gas behaves quite differently, however. Gas clouds do not penetrate each other; so the interstellar component will undergo a real collision. The result is to sweep

FIGURE 21.15.
The central part of a loose cluster of galaxies in the constellation Hercules. Note the large proportion of spiral galaxies. (Hale Observatories.)

FIGURE 21.16.
The center of the best-studied dense cluster of galaxies, in the constellation Coma Berenices. Nearly all the galaxies are ellipticals or S0's. X-rays have been detected coming from a region about the size of that shown in this picture. (Kitt Peak National Observatory.)

the gas out of both galaxies and to leave it hanging in the space between them.

We do not know enough of the details of how this process works, however. It may be, for instance, that after a few galaxies have been swept clean by such collisions, the space around them is so filled with gas that this intergalactic gas itself can sweep out other galaxies that pass through the region. It is certainly true that many dense clusters of galaxies contain a large amount of widely spread intergalactic gas. We observe it by detecting the X-rays that are emitted on account of its high temperature, which results from the rapid motions of the galaxies from which the gas came.

If a cluster of galaxies has a short enough crossing time to have settled into an equilibrium by now, we can expect that the speeds of the galaxies through the cluster, which would tend to make them fly away from each other, are just balanced by the gravitational forces that hold the cluster together. Thus by measuring the radial velocities of individual galaxies, as we can easily do from their spectra, we can deduce how much gravitation is needed to hold the cluster together; and from that, we can calculate the total mass of the cluster of galaxies.

The results are surprising. In a typical rich cluster of galaxies, the ratio of mass to luminosity is around 200 solar units, almost ten times as high as the M/L that we measure for the galaxies that appear to make it up. Or, to put it another way, there seems to be almost ten times as much mass present as we can account for by adding up the estimated masses of the galaxies.

The problem of the "missing mass" is one of puzzles of modern astronomy. A few astronomers believe that it is not there at all, and that clusters of galaxies are flying apart. They are unable to explain, however, how a cluster of old galaxies could have persisted until the present era and then suddenly decided to fly apart. Most astronomers accept the existence of the missing mass and wonder what form it can take. Is it spread through the space between the galaxies? As we have seen, dense clusters contain intergalactic gas. Its amount can be considerable, perhaps totaling as much mass as there is in the galaxies; but this is not nearly enough to account for the missing mass. Intergalactic stars are another possibility. They would have to be very faint dwarf stars, however, in order to have enough mass without showing a large amount of intergalactic light, which we do not see. Still another possibility, of course, is that the missing mass consists of more exotic objects that are unobservable except for their gravitational force: rocks, or perhaps black holes.

Or it may be, on the other hand, that the missing mass is in the galaxies, which would mean that we have grossly underrated their masses. This could happen if every galaxy were surrounded by an extended, massive halo of objects that emit very little light. Our dynamical studies of galaxies are largely confined to their central regions; we can be sure that the missing mass is not hidden there. We have very few observations of motions in the outer parts of galaxies, however, which would indicate how much mass lies out there. It may be significant that the few existing observations of this type do indeed indicate a widely spread mass, out where the galaxy has very little light. Again, we can only speculate at this stage about what the form of such mass could be: dwarf stars, black holes, or what?

It is perhaps a healthy thing to end this chapter on such a note of uncertainty. The structure that we have built up in understanding the universe is often brilliantly successful, but parts of it are very flimsy. There may be places in the picture that will need major revisions. Only time can tell—and it will. Not a single page of this chapter could have been written sixty years ago; it is awesome to wonder what will be written sixty years hence.

REVIEW QUESTIONS

1. Describe the chief types of galaxy. What kinds of stars does each contain, and how much interstellar material?

2. How is the stellar population of a galaxy related to the amount of interstellar material that it contains? Show how this relationship is a natural consequence of the way in which stars are formed.

3. How are the distances of galaxies determined? Trace the chain, from stars in the Milky Way to nearby galaxies to distant galaxies.

4. Where do we find populations of high metal abundance, and where do we find populations of low metal abundance?

5. Describe the collapse process through which we believe galaxies to have passed during their formation. How does this process explain the difference between elliptical galaxies and discs?

6. How do we determine the masses of galaxies of various types? What does the ratio of mass to luminosity tell us?

7. What evidence is there of unobserved mass, in individual galaxies or in groups of galaxies?

22 | COSMIC VIOLENCE: FROM RADIO GALAXIES TO QUASARS

Although the normal inhabitants of the universe are galaxies, not all galaxies are normal. Some have violent, explosive events at their centers, and in a few the violence seems to affect the whole galaxy. The most violent objects of this whole class, the quasars, are the most luminous things in the universe; and they are therefore the most distant objects that we can observe. Yet the reason for all these cosmic explosions is still a complete mystery, and even the events themselves are poorly understood.

The story of this branch of astronomy began after the end of the Second World War, when some of the scientists who had developed radar turned their new-found radio skills toward the observation of the universe. Listening for cosmic noise at short wavelengths for which sensitive radio receivers had never existed before, they began mapping new things in the sky. The first maps were very crude, however, because a radio antenna—called in this context a radio telescope—has only a poor ability to determine accurate directions. As we saw in Chapter 2, the sharpness of the map depends on the size of the antenna relative to the wavelength of the radio waves; to achieve a sharpness of one degree, an antenna must be about 70 wavelengths across. For one-meter waves, typical of the wavelengths used, no radio telescope existed in 1950 that was large enough to map with even one-degree accuracy. Fortunately, however, radio astronomers have a technique of tying together two separated antennas; this arrangement, called an interferometer, has a resolving power that depends on the distance between the antennas rather than their individual sizes. With interferometers it soon became possible to

FIGURE 22.1.
NGC 5128 is one of the most peculiar galaxies in the sky. It is a strong source of radio waves and
X-rays. (Cerro Tololo Inter-American Observatory photograph, courtesy of S. van den Bergh.)

measure the "positions" (directions in the sky, that is) of radio sources
with a precision of a small fraction of a degree.

Radio maps showed many "point sources," and with good interfer-
ometric positions to pinpoint them, their locations could be examined on
optical photographs, to see just what sort of objects might be responsible
for the radio waves. Some of the results were surprising: many of the
radio sources were galaxies. If we detect them at such a great distance,
they must be emitting radio radiation at an intensity that was completely
unexpected. Furthermore, although some of the "radio galaxies" had
a very peculiar optical appearance, others looked like perfectly normal
galaxies.

Often, however, a search around the radio position revealed nothing,
just the scattering of random stars that appear on any photograph. It

FIGURE 22.2.
The interferometer array at Westerbork, in the Netherlands, can make detailed radio maps of small regions of the sky. (Aerophoto Eelde.)

was only when radio positions became even more accurate that identifications could be made, and again the result was a complete surprise: these radio sources looked just like stars. The astronomers who identified them called them at first by the cumbersome title "quasistellar radio sources," which usage has since shortened to "quasars."

FIGURE 22.3.
M87 is one of the few elliptical-galaxy radio sources that show any optical peculiarity. The jet coming out of its center is over 1,000 parsecs long. In Figure 21.1, which covers a much larger area of the same galaxy, the jet is lost in the overexposed central region. (Lick Observatory.)

Astronomers could be sure from the beginning that quasars were not ordinary stars, because their spectra show bright lines, always a sign of peculiarity. The quasars were particularly puzzling, however, because the wavelengths of their spectral lines made no sense at all. The puzzle was finally solved when it was realized that all the lines could be easily identified if it was assumed that the spectrum of each quasar had a sizeable redshift. Thus the quasars must all be rushing away from us with high velocities.

The pattern of large velocities, all receding, would make sense if the quasars were very distant objects, taking part in the expansion of the universe just like the distant galaxies, and visible to us so far away only because they are so luminous. The luminosities required are astonishing, however. If quasars are at "cosmological" distances, as their large redshifts would indicate, then they must be several magnitudes more luminous than the most spectacular giant galaxies. Yet quasars are much smaller in size than galaxies; in fact, as we shall see, they are as small as the tiny centers of galaxies. Although some astronomers have found this picture too fantastic to believe and have sought alternative explanations, the conventional interpretation is almost certainly correct: the quasars are the most luminous and most distant objects that we see in the universe.

445

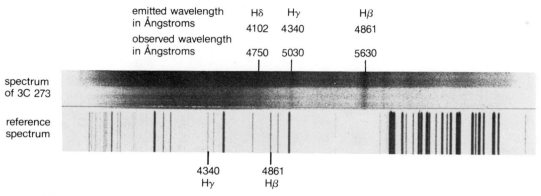

FIGURE 22.4.
The spectrum of 3C273, one of the nearest and brightest quasars. (Hale Observatories spectrum, courtesy of M. Schmidt.)

The bestiary of violent galaxies is strange indeed. It ranges from small radio sources at the centers of otherwise normal galaxies, through strongly disturbed galaxies, up to the stupendous quasars that dwarf the rest of the galaxy in which they occur—if, indeed, there is a galaxy there at all. Although we do not understand these objects in any basic way, there are family relationships between them that give us a few clues to their physical nature. We shall see some of these in looking at each class in some detail.

RADIO GALAXIES

Deep in their cores, some galaxies carry an unknown disease that can flare up as the release of a large amount of energy. This appears in the form of a cloud of highly energetic protons and electrons imbedded in a magnetic field. It is the gyration of the charged electrons in the magnetic field that produces the radio radiation that we detect.

Even though we do not know what caused the original explosion, we understand enough about the radiation of these gyrating electrons—synchrotron radiation, as it is called—to estimate how much energy is present in the motions of the electrons and in the magnetic field. For strong radio sources, the amount of energy is astonishingly large, equivalent to the total radiation that millions of solar-type stars would release in their entire lifetimes. The region that emits the radio radiation can also be extremely large. Many radio galaxies have two radio-emitting regions placed symmetrically, far outside the apparent optical limits of the galaxy. Separations of 100 kiloparsecs are not unusual.

From the characteristics of these radio sources, we can estimate how long they are likely to last. One approach is to compare the amount of

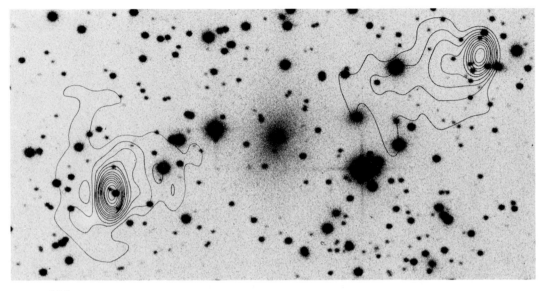

FIGURE 22.5.
The radio source Cygnus A is one of the brightest in the sky, even though the radiation comes from a faint, distant galaxy. In this Hale Observatories photograph, the galaxy is near the center. The superimposed contour lines outline the very extended region from which the radio radiation comes. (Courtesy of A. Moffett. Reprinted from A. Sandage, M. Sandage, and J. Kristian, eds. *Galaxies and the Universe.* Copyright © 1975 by the University of Chicago.)

energy with the rate at which it is being radiated. We cannot be sure that the radiation is always emitted at a constant rate, but such a comparison at least gives us an order of magnitude for the lifetime. It is a few million years. This length of time is also the time that it would have taken a large radio source to expand to its present size if it were moving at, say, a tenth the velocity of light, which is a reasonable estimate for the speed produced by such a violent explosion.

Like so many astronomical times, this is long in human years but quite short compared to the ages of the galaxies. The question then arises, is the formation of a strong radio source an event that every galaxy goes through on rare occasions, or are there a few galaxies that show this peculiarity in a chronic way? With our present information it is impossible to tell. What we can say, however, is that we could account for the number of present-day radio sources if every giant galaxy became a radio source about once every billion years.

The latter hypothesis is an appealing one, because many radio sources are giant elliptical galaxies that show no other peculiarity; they look exactly the same as other galaxies that have no radio emission. This means that the stellar part of the galaxy—most of its mass, that is—is not in any way affected by the tremendous explosion that has taken place at its center. When large amounts of gas and dust are present, however, the

FIGURE 22.6.
The peculiar galaxy M82, photographed through a filter that accentuates Hα, the red Balmer line of hydrogen. M82 is probably an edgewise disc in the center of which an explosion has taken place. (Hale Observatories.)

effects of the explosion are striking. A strong radio galaxy of that type has a form that often defies classification, except by the catch-all word "peculiar."

It is easy enough to understand why stars and gas should behave so differently in the presence of an explosion. The explosion presumably sends out a blast of hot gas. When this encounters other gas, it collides with it violently, so that the arrangement of the gas in the galaxy is completely disrupted. Stars, on the other hand, are so small that they offer very little target to the expanding gas, most of which streams by with no effect on them at all.

The disrupted appearance of some radio galaxies, along with their large internal motions, originally led some astronomers to believe that they were actually pairs of galaxies in collision. More careful analysis has shown this to be a very unlikely explanation, however; the event is almost certainly a central explosion, whatever its basic cause may be.

Although strong radio sources are rare, many galaxies have weak but detectable radio sources at their centers. Among them is the Milky Way, whose center is, in fact, most easily located by means of the small, intense radio source that is located there. (At one time it was thought that considerable quantities of gas were also streaming out of the Galactic center, but it is no longer clear that this interpretation was correct.)

The Milky Way is thus a radio galaxy, although a weak one on an absolute scale, and many other galaxies show small or moderate amounts of radio emission. The strong radio sources are the spectacular ones, but radio emission by galaxies is a common phenomenon.

GALAXIES WITH BRIGHT CENTERS

Another type of peculiar galaxy shows us the disturbance at its center by means of optical emission. These galaxies, called Seyfert galaxies after the astronomer who first studied them, have very strong bright lines in their spectra. This emission is confined to a very small central region, in some a few parsecs or less, but it is strong enough to make the center of the galaxy stand out as an intense bright spot.

Even more peculiar is the width of the emission lines. They seem to indicate rapid motions toward and away from us, as if the gas were expanding out of the nucleus at thousands of kilometers per second. Again, we do not know the reason for this behavior, except that some energetic event is going on at the center of the galaxy. Some Seyferts are radio

FIGURE 22.7.
NGC 4151, a Seyfert galaxy. It looks like a spiral galaxy with an unusually bright center. (Hale Observatories.)

sources, along with their optical peculiarity, but others are not. Some look like ordinary spiral galaxies, except for the intense bright spot at the center, but some Seyfert galaxies are totally peculiar in their form.

The Seyferts are members of a broader class of galaxies that have bright centers. The most striking members of this class are the N galaxies ("N" for nucleus), in which the center is so bright that it completely dominates the appearance of the galaxy.

QUASARS

The most violent and energetic objects of all are the quasars. They are the most luminous objects in the universe. Optically, where we most easily observe them, they put out more light than giant galaxies; on the conventional scale of absolute magnitude, they range up to about −26. In the radio part of the spectrum they are also among the most luminous of sources; even though they are so far away, quasars are among the brightest-looking objects that a radio telescope sees.

In spite of their tremendous brightness, quasars are small objects. Their star-like appearance shows that they are smaller than galaxies, but their real dimensions come as another surprise: fluctuations in brightness show that many quasars are much less than a parsec in diameter. This takes some explanation—how can changes in brightness tell us the size of an object? The answer is based on the time that it takes light to travel. If an object emits a burst of light but its size is one light year across, then the part of the burst that comes from the far side will reach us a whole year later than the part from the near side. This means that a burst of light emitted as a single, simultaneous event would arrive here spread out over a year of time. Any fluctuations that were much faster than a year would then be completely washed out.

In fact, many quasars vary in brightness, and some of them change strikingly in a month or less. This means that their size must be a light month or less, since otherwise the brightness changes would be smeared out by the differences in light-travel time. Since even a light year is equivalent to only a third of a parsec, the source of the quasar's light must be concentrated within a region a fraction of a parsec in size. Yet the quasar is more luminous than a galaxy tens of thousands of parsecs in diameter.

The spectrum of a typical quasar shows a continuum plus a number of bright lines. The lines are similar to those that we see in the spectra of gaseous nebulae, indicating that they are emitted by a gas of low density. They do not vary in brightness, whereas the continuous spectrum does. It thus appears that the bright lines come from a region that is more extended than the tiny center that puts out the continous spectrum. It is hard to estimate the size, since quasars are so far away that the limited

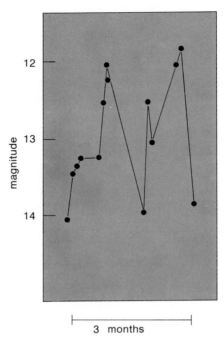

FIGURE 22.8.
Changes in the brightness of the quasar PKS 1510–089
during a 3-month period in 1948. (Adapted from M. and
W. Liller, *Astrophysical J.*, **199** (1975), 134.)

resolving power of our optical telescopes makes even a kiloparsec look like a stellar point.

It is natural to ask what else the spectral lines reveal about the nature of quasars and the physical conditions in the gas from which the lines were emitted. Unfortunately, the answer turns out not to be very revealing. The emission spectrum of a quasar is very much like that of any other region of interstellar gas whose atoms are excited to emit radiation. This was to be expected, however, since the basic physical situation is not very different from that in a bright nebula. In either situation, energy is fed into the gas and heats it up; atoms collide, ionize, and recombine; and the emission lines are what one would expect from a gas where those processes are going on. Unfortunately for our physical study of quasars, the bright-line spectrum depends very little on what the original source of the energy is, and thus the basic mystery of quasars remains.

There is one way, however, in which quasar spectra do give us an opportunity to see something new about the behavior of an excited gas. Many quasars are extremely distant, and at their location in the expanding universe, all objects have a very high velocity away from us. As a result the Doppler shift in their spectra is very large—so large, in fact, that ultraviolet lines are shifted into the observable part of the spectrum. This means that we see in many quasars spectral lines whose rest wavelengths (the wavelengths that they would have in a stationary object) are so short that they would be blocked by the Earth's atmosphere if they were not redshifted. Some of these lines reveal the physical conditions in the gas better than the spectral lines that are normally visible.

THE PROBLEM OF QUASAR REDSHIFTS

We encountered the phenomenon of the redshift when studying galaxies. It is a general property of the universe that distant objects appear to be moving away from us with a velocity that is proportional to their distance. Some faint, distant galaxies have been observed with wavelength shifts that are half or more of the rest wavelength. For many quasars the redshifts are much larger, however, so that the observed wavelengths of the spectral lines are several times their rest wavelengths.

For such large redshifts we must reexamine the Doppler effect. Whereas at small velocities the shift is proportional to velocity, the wavelength shift for higher velocities is not so simple. According to the theory of relativity, the velocity of an object cannot be as great as that of light; but its redshift has no such limit. If velocities of recession are close to the velocity of light, the corresponding redshifts can be very large. Thus the most distant quasars that have been identified are receding from us with a velocity a little more than 90 per cent of the velocity of light, but their redshifts are greater than 3. That is, the shift in wavelength is more than 3 times the rest wavelength, so that, with the addition of the shift, the observed wavelengths of the lines are more than 4 times their rest wavelengths.

It is natural to ask how we can recognize spectral lines at all, when they are shifted into a completely different part of the spectrum. How can we assert that a certain line recorded in the visible spectral range is really an ultraviolet line with a large redshift? The answer is that we would be completely unable to do this for a single line, but groups of lines have recognizable patterns. More specifically, since the redshift within the spectrum of an object is everywhere proportional to wavelength, it turns out that the *ratios* of the wavelengths of lines are not changed by the redshift. Thus if two observed lines have the same wavelength ratio as some well-known pair of lines, then it is a good bet that they are indeed that pair of lines. When several lines are observed, the mutual ratios make the identification quite certain.

Although the redshifts in the spectra of quasars are an undoubted fact, some astronomers have considered their interpretation to be quite controversial. According to the conventional view, which is the one that has been presented here, the redshift that we see is the "cosmological" redshift: that is, in an expanding universe in which more distant objects have larger redshifts, the quasars are extremely distant objects. It follows then, from their apparent magnitudes and their distances, that quasars must have very high luminosities. Yet the speed of their changes in brightness shows that they are very small in size. It is at this point that some astronomers have balked; they have refused to believe that such a stupendous amount of energy could have been released in such a tiny region. Instead, they have looked for other ways of interpreting the redshifts of quasars, in the hope of somehow explaining them as nearby objects of much lower luminosity.

According to one version of the "local hypothesis," the quasars are small objects of an unknown nature that are actually moving with respect to their surroundings at the high velocities that their spectra indicate. Since all the motions are away from us rather than toward us, they must have all been shot out of some nearby center; the most plausible candidate is the center of the Milky Way. In another version of the local hypothesis, the quasars are peculiar objects scattered among the nearby galaxies, and their redshifts arise not from velocity but from some new and unknown principle of physics.

Some astronomers claim to have shown observational evidence of the local nature of quasars, in the form of associations between high-redshift quasars and neighboring normal galaxies of low redshift. In some cases it is asserted that faint bridges can be seen that demonstrate an actual physical connection between a quasar and a galaxy. Another line of argument has cited the frequency with which such associations occur and argued that the observed pairings are far too frequent to be random coincidences.

Most astronomers, however, are quite unimpressed by the arguments in favor of the local hypothesis. The supposed bridges look dubious, and the statistical arguments about what is and is not random are open to dispute. Even more important, the local hypothesis merely replaces an enigma by another one that most astronomers consider even more unpalatable. Although the large energies of quasars are puzzling, redshifts of an unknown cause seem downright incredible.

At the bottom of the controversy is a basic difference in scientific philosophy and taste. When a phenomenon defies immediate explanation, should we merely note it as a puzzle for which we still lack some of the clues, or should we seize upon it as the entering wedge of a scientific revolution? At this point every scientist has to form his own judgment. The conservative majority warns us that a fundamental revision of physical principles is a serious and costly step that should not be undertaken lightly. It should be considered only when there is strong and clear evidence that the fault is in the basic system rather than in the details of a single problem, or else when the introduction of a new concept would explain or pull together a great many problems that had until now lacked coherence or explanation.

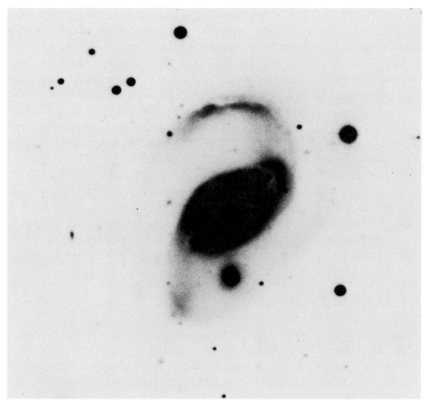

FIGURE 22.9.
The spiral galaxy NGC 4319 has a redshift of 1,800 km/sec; just below it is the peculiar object Markarian 205, whose redshift is 21,000 km/sec. A faint bridge seems to join the two, and some astronomers have used it to argue that they are physically connected and therefore at the same distance. Others argue that the "bridge" is merely a photographic effect caused by the overlapping of the images of the two objects. (Hale Observatories, courtesy of H. C. Arp.)

To this the dissidents would retort, how can you ever discover anything new if you reject radical suggestions just because they are radical? The answer is that it is indeed a matter of personal judgment. Each individual scientist moves in the direction that he thinks is most likely to be productive, and he gives his attention to work that he thinks is well-founded and therefore worthy of his attention.

Thus science, like all other intellectual activities, has aspects in which subjective judgments play an important role. What distinguishes science, however, is that one hopes—and presumes, indeed—that the subjective stage is confined to first steps and tentative choices of direction. Conclusions, by contrast, are subject to a more stringent criterion, according to which verifications should be available in whatever abundance might be desired.

So it is with quasars. They seem to fit best into the fabric of science when interpreted as distant objects of extremely high luminosity. As such, they lie on a continuous sequence that also includes radio galaxies, Seyferts, and N galaxies. All these objects are characterized by a violent release of energy at their centers. Each class of these objects has some properties in common with other classes, and there seem to be no sharp divisions between them at all. An N galaxy, for instance, is like an exaggerated Seyfert galaxy; but it is also like a low-violence quasar—and indeed, many N galaxies would look like quasars if they were too far away for us to see the faint disc part of the galaxy.

As our working hypothesis, then, we take quasars to be the most luminous, distant objects of the universe. The source of their stupendous energies is still a mystery, but so is that of the more prosaic radio galaxies. Just as the last dozen years have brought us this much insight, so we can hope that the next dozen years can clarify the physical nature of quasars and bring them into the comfortable realm of settled science.

THE NUMBER OF QUASARS

Although they are physically the most luminous objects of all, quasars are hard to find. Out of the very large number that are within reach of our telescopes, only a few hundred have been catalogued and studied. The trouble is that optically they look very faint and are superficially indistinguishable from faint stars. Suppose that a photograph taken with a large telescope shows a thousand stars. Most of the dots on the photograph represent ordinary stars, most of them relatively close to us; but sprinkled among them are a few dots that look just like nearby stars but are in reality distant quasars. How are we to distinguish them?

The first part of the answer has already been stated: in fact, very few quasars have yet been identified. Most of the known ones were found first as radio sources. As described earlier, when the radio position became so accurate that it pointed to an object that otherwise looked stellar, that object could then be investigated spectroscopically. If it showed emission lines with a large redshift, it was a quasar.

The great majority of quasars do not put out strong-enough radio emission to be found in this way, however. What we can do instead is to make use of their optical properties. Along with its peculiar emission-line spectrum, a quasar also has a continuous spectrum that is different from those of most stars, and this gives it colors that are also peculiar. Most quasars are rather blue, and they are even more striking in the ultraviolet part of the spectrum—the "near" part of the ultraviolet, that is, which we are able to observe through the Earth's atmosphere. In particular, photographs taken in ultraviolet light can be compared with those taken in other parts of the spectrum; the objects with an ultraviolet excess are good quasar candidates. The final confirmation requires a spectrum, however, and this is the slow stage. Just because a spectroscopic observation spreads the light out instead of concentrating it at a point, the observation of spectra takes a long time. Thus the smallness of our detailed catalog of quasars is a matter of economics rather than principle. A large part of the observing time of the few large telescopes—the only ones that can profitably be used for such faint work—goes into quasar research; but even so, progress is slow.

A few efforts have been made to look for faint quasars in a systematic and complete way, and the results indicate that they are very numerous. Estimates are that the total number that can be detected with our large telescopes may be more than a hundred thousand.

Most of these quasars are very distant, however; in the region closer to us, quasars are quite rare. Neighboring quasars are about a billion parsecs apart. If we compare this with the typical spacing of galaxies, we find that quasars are almost a billion times rarer than giant galaxies. Farther away, however, quasars are quite numerous. In fact, most of the quasars in our catalogs have large redshifts. (Astronomers most commonly cite the redshift as the most convenient indicator of the distance of a quasar.) A large fraction have redshifts around 2—that is, a wavelength shift that is twice the rest wavelength—and only a minority have redshifts of 1 or less.

It is difficult to make an assessment of the true number of quasars at different distances from us, since we must take carefully into account the way in which the likelihood of discovery depends on the distance itself. When such estimates are made, however, we find that quasars really are very much more common—per unit volume of space, that is—in the distant parts of the universe than they are right around us.

What does this mean? Is our place in the universe really a special one, with the quasars all carefully arranged to be very far from us? The answer, of course, is no, and the paradox is resolved easily enough. Light takes time to travel, and the light from distant quasars has been in transit for a considerable fraction of the age of the universe. We see a quasar as it was when its light started on its long journey to us, and thus we observe the distant quasars as they were in the distant past. Apparently, then, the high number of distant quasars tells us that the quasar phenomenon was more common in the past than it is now. The number of quasars with large redshifts is telling us something about the history of the universe.

Although this picture has filled in rapidly, it is still very incomplete. Quasars have been found with redshifts up to about 3.5. Larger redshifts should also be found, but only if they exist in appreciable numbers. There is an indication, however, that the dearth of large redshifts may be real. This would mean that the universe made quasars at a rate that started out small, peaked at a time that corresponds to a redshift between

2 and 2.5, and has tapered off since. At the time of maximum activity, quasars may have been about as common as giant galaxies. Did every giant galaxy once go through a quasar stage? Only future research can tell. Astronomers are confident that it will.

REVIEW QUESTIONS

1. What do we know about radio galaxies, and what do we not know about them?

2. What is a Seyfert galaxy? How does it differ from a normal galaxy? What reason is there for believing that Seyfert galaxies may be related to quasars?

3. How does a quasar look to us? How are quasars formed?

4. What are the reasons for believing that a quasar involves a tremendous release of energy in a quite small volume of space?

5. Outline the arguments in the "redshift controversy," and show how interpretations of facts may be related to philosophical biases.

6. Why is it appropriate to say that most quasars are a relic of the past? Explain how we see into the past.

23 | THE UNIVERSE

Cosmology is the study of the universe, taken as a single entity. We regard galaxies merely as points in the universe; our problem is the behavior of the whole. Even so, the galaxies are the principal actors in the play; it is their positions and motions that we observe, and they contain a large part, if not all, of the mass whose gravitational force governs the motions.

From our position in one of those galaxies, we look out at the universe. Galaxies are everywhere. Unlike the foreground stars, whose number drops sharply outside the disc of the Milky Way, the galaxies go on undiminished in number, as far as we can see. The limit is set not by the universe but by the sensitivity of our telescopes and detectors. Whatever the structure of the universe may be, it is clear that we see only a small part of it. From this limited part, we must try to construct a picture of the whole. To do so, we are forced to assume that remoter regions, which we are unable to observe directly, are the same—at least in their average properties—as the part that we do observe. In other words, our part of the universe is distinguished in no way, except for local details, from any other part. An observer anywhere else would see the same over-all picture.

This assumption, that the universe would look the same from all observing points, is called the *Cosmological Principle.* It is an act of faith, since we can never verify its correctness, but it is a reasonable step nevertheless. If we are to study cosmology at all, we must make some assumption about the parts of the universe that are too distant to observe; and

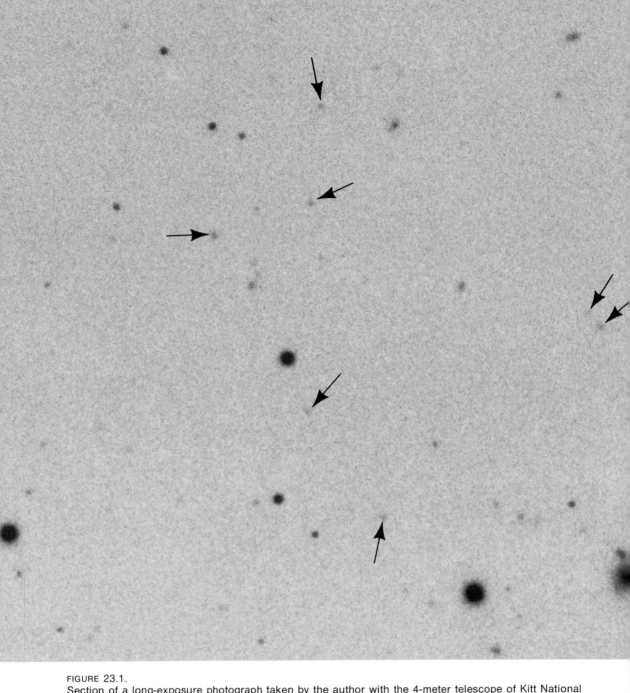

FIGURE 23.1.
Section of a long-exposure photograph taken by the author with the 4-meter telescope of Kitt National Observatory. Most of the faint galaxies, some of which are marked with arrows, are billions of parsecs away. The light that reaches us today left them before the Earth existed.

in the absence of information, the Cosmological Principle chooses the simplest hypothesis.

This largest scale of world-view, in which we confront the universe as an organic whole, is inspiring; but it can also be confusing, perplexing, and paradoxical. Because cosmology, by its very nature, takes us into larger realms of space and time than anything in our previous experience, it introduces ideas and problems that also go beyond our previous experience. The expanding universe, the beginning of time, curved space, a universe that is finite in extent, or, for that matter, one that is infinite: these have a flavor of fantasy rather than science, but they are the stuff that cosmology is made of.

THE EXPANDING UNIVERSE

Even within our own corner of the universe, the first problem of cosmology immediately thrusts itself before our attention. The redshift of distant galaxies is a well-known phenomenon. The more distant a galaxy is from us, the faster is its velocity of recession; and the velocity is in fact proportional to the distance.

At first glance, this picture, in which everything is moving away from us, would seem to violate the Cosmological Principle, by fixing a unique center for the universe. Even worse, it violates the Copernican principle, by placing *us* at the center of everything. Fortunately, the paradox is easily resolved. In this particular kind of motion pattern, where velocity is proportional to distance, *every* observer will see the same picture, in which everything else is moving away from *him*. Consider an observer who lives in a distant galaxy; we see him moving rapidly away from us. But motion is only relative; he sees us moving away from him. Similarly, if we use the relative motions to transform our view of any other galaxy into this other observer's frame of reference, we find that invariably he sees it as moving directly away from him. Figure 23.2 shows in detail how this works for a few galaxies.

Thus the pattern of motion of the galaxies is such that everything is moving away from everything else. We can describe this, in a word, as an *expansion,* because this is just what happens among the parts of a body that is expanding. For a simple example, consider a balloon that has small dots marked all over its surface, and suppose that we slowly blow up the balloon. As the surface of the balloon gets larger, the separation between any two dots increases correspondingly. If we were to follow the motion in detail, we would find that the speed at which two dots are moving away from each other is indeed proportional to the distance between them.

The "expanding universe" is a convenient phrase to describe the pattern of motions of the galaxies, but we must be careful not to read into

460

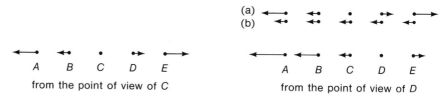

FIGURE 23.2.
Left: an expanding set of points, as seen from point *C*. The arrows indicate the relative velocities. *Right:* The same motions, as seen from point *D*. In the upper pair of rows are shown (a) the motions relative to point *C*, and (b) the corrections needed to refer the motion to point *D* instead.

it meanings that do not belong there at all. In particular, it would be quite misleading to say that "space" is expanding, as if the galaxies were held on some ethereal framework that drags them along as it expands. To repeat, "expansion" is just a word that conveniently describes how this whole collection of objects moves. And it should be clear from this that the expansion of the universe does not imply that objects themselves are getting larger or that there is an expansion within each galaxy; again, it is only the motions of whole galaxies that are being described.

Although we must be careful not to endow the expansion of the universe with false mysteries, the real mystery is intriguing enough. That each galaxy has a motion should not be surprising; after all, Newton's first law says that motion is a perfectly natural thing and that it continues except as forces act to change it. The surprising thing is the pattern of the motions: all the galaxies moving away from each other. How did the universe get that way? The ultimate answer is unknown, and may indeed be unknowable, but we can follow the train of clues a long way, and it takes us into new and strange realms.

THE BEGINNING OF TIME

Since we see the galaxies moving away from each other, in times past they must have been closer together. In the more distant past, moreover, they must have been extremely close together. If, in fact, we assume that each galaxy has always moved with its present velocity, then there must have been a time in the past when all separations were zero, that is, when the galaxies were packed infinitely close together. From our best present-day observations of velocities and distances, we calculate this time to be about 20 billion years in the past.

Since we know only the present-day rate of expansion of the universe, the real age of the expansion is uncertain; it depends on how the expan-

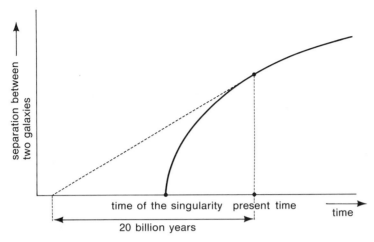

FIGURE 23.3.
The expansion of the universe, shown as a graph. Note how the expansion slows down, and how the present rate of expansion (slope of the curve) determines a characteristic time scale.

sion behaved throughout the past. We take an important step, however, by merely establishing the time-scale of 20 billion years. The actual age may be somewhat greater or less, but at least we know that 20 billion years is a time in which the universe has changed seriously. This means that a million-year interval is only a twinkling in the eye of the universe; events of that duration can have happened again and again in the history of the universe. If, on the other hand, some process would take 10^{12} years to develop, we can forget about it; it is too slow to matter. Thus the time-scale of the Universe puts all other time-scales in perspective.

This 20-billion-year time-scale is, of course, similar to the time-scale that we have already established for the ages of the oldest stars. It is quite significant that we find it again here, because it implies a connection, a suggestion that the beginning of stellar evolution may be an echo of a more basic beginning in the universe itself.

The next step after finding the time-scale is to ask what actually *did* happen: how have the galaxies moved in the past? There is indeed a force acting on them: their gravitational attraction for each other. Its tendency is to slow the expansion down. Each galaxy feels the attraction of all the other galaxies from which it is receding, and this attraction tends to slow its motion down. The over-all effect of this self-attraction of the universe depends on its density and on its rate of expansion; we shall return to this problem later as one of the basic issues of cosmology. What we can say now, at least, is that any change in the motions of the galaxies has been one of slowing down. In the past, galaxies were moving as fast as they do now, or even faster; so the age of the expansion may be rather less than 20 billion years, if it indeed went faster in the past.

If we follow the history of the universe backward, we seem inevitably to reach a point in the past when the universe was infinitely compressed, when its density was everywhere infinite. The material had motion, however; and it has expanded from that time on, condensing into stars and galaxies as it went. It is natural that astrophysicists should refer colloquially to the initial event as the *big bang*, and this term has in fact become its serious name.

Mathematically, the moment of infinite density is called a singularity, in recognition of the impossibility of describing the properties of that moment in ordinary terms. Physically, it is the beginning of time. It is meaningless to ask what happened before that; everything that we now observe in the universe dates from after the singularity. Anything that "happened" before that can have no effect on our observation or our experience; therefore its very existence is meaningless.

The idea of a beginning of time is philosophically difficult, and it can also be intellectually or emotionally upsetting. Many have tried to rationalize or to deny it. Some religious scientists, for instance, have identified the singularity with the Creation. A quite different alternative was suggested two decades ago by a group of British astrophysicists. Their idea was to avoid the singularity by postulating that the universe always remained the same, in its average over-all properties. This was the *steady-state theory*. It did not deny the expansion of the universe; it said instead that the universe has always been expanding and always will. Since an uncompensated expansion would lower the density of the universe, thus violating the steady state, this theory had also to postulate that new matter continually appears at random, everywhere, at a rate just sufficient to keep the density constant. The continuous creation of matter was a new and heretical principle of physics, but it would save us from having to believe in a singular beginning.

The steady-state theory has now been laid to rest, as a result of clear-cut observations of how things have changed systematically with time—the very much larger number of quasars in the distant past, for instance. It serves now as an example of the lengths to which a philosophical system can stretch itself, in the absence of a sufficiently clear factual picture.

Thus we are left, willingly or not, with a universe that began with a singularity. The last lingering doubts about the correctness of this picture were erased, in the mid-1960's, by the spectacular discovery that we can look into the remote depths of the universe and actually see the remnants of the big bang itself. Coming from all around us is faint radio radiation, characteristic of a black body at a temperature of 3 degrees. As strange as it may seem, this pale, cold relic is the radiation emitted by the material of the universe when it was very young and was still compressed into a dense, hot gas.

To understand this, we must remember that radiation takes time to travel, so that in a sense everything that we see is in the past. We see each object as it was when the light left it; and the farther away the object, the longer in the past the light left it. For most purposes this makes no practical difference; only in cosmology does it really matter. Thus we see the Sun as it was eight minutes ago and the Andromeda spiral galaxy as it was two million years ago; but neither object has changed during the light-travel time, except in unimportant details. When the light-travel time is comparable to the age of the universe, however, we see a picture that originated far in the past, and in this younger era the universe looks different from what we see around us today. As we saw in the preceding chapter, the distant quasars show this effect; they are so numerous because we see them as they were at an epoch when the quasar phenomenon happened more frequently than it does today.

Imagine, then, looking even farther into space, and therefore looking even farther back in time. As we go farther back, we see a denser and denser universe; and if we look deep enough into the past, we should eventually see an era when the universe was so dense that it was opaque. This is where the background radiation originates. It marks the epoch when the universe had expanded and thinned just enough to stop being opaque. The radiation that had been emitted at earlier times encountered a denser gas and was reabsorbed; but after this epoch of "decoupling," the radiation found itself passing through a much thinner gas—thinning more all the time, as the universe expanded. Never again could it be absorbed, and it is still traveling. The radiation that long ago originated here is now far, far away; and conversely, what we see here now is radiation that was emitted, back then near the beginning of time, at a point so far away that the radiation is only now reaching us. In the expanding universe, such remote regions are moving away from us very fast, and their radiation thus has a very large redshift, so large that it completely changes the quality of the radiation. Thus the "fireball radiation" looks to us as if it is coming from a cold source. The ultraviolet radiation from the hot gas is redshifted all the way into the microwave radio spectrum, where it matches the radiation of a body whose temperature is only 3 degrees.

The 3-degree background is a pale signature left behind by the big bang, but the big bang has also made a much more important imprint on the present-day makeup of the universe. The chemical composition of most of its material—about three quarters hydrogen and one quarter helium—was very probably determined by the temperatures and densities at an even earlier time. The background radiation decoupled when the universe was about a million years old and the temperature of the gas was a few thousand degrees. Even earlier, the temperature was much higher. When the universe was only a few *seconds* old, the temperature was many billions of degrees. The gas was a mixture of radiation and elementary particles, such as protons and neutrons. At that temperature, collisions caused nuclear reactions that changed neutrons into protons, and vice versa. As the universe expanded and the gas cooled, however, these high-energy reactions were no longer possible; and the number of neutrons, relative to protons, was fixed at a value that was determined by the density and by the temperature, which was then a few billion degrees. Nuclei continued to collide, however, and all the neutrons soon fused with protons to give deuterons, which in turn quickly fused into helium.

At this point nuclear changes ceased. Hydrogen and helium are both very stable except at temperatures much higher than then prevailed, and the temperature continued to drop as the expansion continued. Even the tiny remaining fraction of deuterium was no longer able to fuse, and it has remained as another relic of the first few minutes of history.

There is no way in which the heavy elements could have been made in the big bang, however. We presume that we must look instead to the stars as the furnaces in which the nuclei heavier than helium were forged. This picture has a serious gap, however. All the stars that we know, no matter how old, have some amount of the heavy elements in them. Where did those heavy atoms come from? Are they the ashes of a mysterious "first generation" of stars, born of pure hydrogen and helium, that lived briefly, created the first of the heavy nuclei, and then vanished from the face of the universe? Like cosmology, cosmogony—the origin of things—is full of unanswered questions.

The whole scenario of the early history of the universe is a lively one, and a plausible one too; but it takes a very long extension of present-day knowledge to trace the universe back from its present age of 20 billion years to the first few minutes of its youth. A lot of things could go wrong with our reasoning, in taking such a large step. It may indeed be foolishly naive to view the development of the universe in such a simple way; yet scientists do so, partly in the hope that the answer really will turn out to be simple, and partly out of the pragmatic view that the simple idea is the one to try first. In this case both views appear to be vindicated, since the picture fits together so well. The time-scale agrees with the ages of the stars, the background radio radiation confirms the reality of the big bang, and the helium abundance of the universe is what the same events should have produced. Although a great deal is still unknown, this is one of the cases where the interlocking of our scientific structure suggests that it is soundly built.

THE EXTENT OF THE UNIVERSE

Our optical view of the universe is of course limited by the ability of our telescopes to detect faint light, and we have just seen how the view of radio telescopes encounters in every direction the opaque wall of the dense, distant past. What lies beyond? Does the universe extend for an infinite distance?

A first and a reasonable-sounding answer would be that it *must* be infinite; otherwise there would have to be an edge somewhere, and this would lead to all kinds of absurdities. Reasonable though it may sound, this argument is wrong. The existence of an edge is not the only alternative to infinite extent. An obvious counterexample is the surface of the Earth, or of any other sphere. One can travel forever without encountering an edge, yet the total surface is finite in size.

Clearly, the trick here is that the surface of a sphere is curved, so that a "straight" path curves back around the surface and eventually returns to its origin. But what has this to do with the universe? Can *space* be curved, and close around on itself in the way that the surface of a sphere does? The answer is that space could perfectly well be curved. Whether it really *is*, we can find only by observing it carefully.

Of all the perplexities in the wonderland of cosmology, curved space is perhaps the most curious. It cannot be *visualized*. Since space spans the three dimensions that we can perceive, we could see it as curved only by looking in more dimensions—and it is nonsense even to talk of this. The problem becomes clearer, however, if we study an analogous situation that we *can* visualize.

Imagine a sphere, on whose surface there live creatures who can perceive only the two dimensions that lie along that surface. The surface is their whole world; all motion is along the surface, and light travels only along the surface. Nothing outside or inside of it has any influence on them; therefore the third dimension, perpendicular to the surface, has no reality for them. How would one of these creatures react if we suggested that he lives on a curved surface? He would ask just the same questions that we do: How can my space be curved? What does it curve *in*? How can you say that it curves in another dimension, and then say that that dimension has no reality for me? Perhaps in sympathizing with his perplexity, we can better understand our own situation.

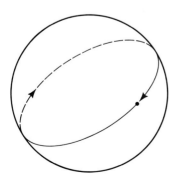

FIGURE 23.4.
The "straightest" path on a sphere is a "great circle"; if extended, it goes completely around and returns to its starting point.

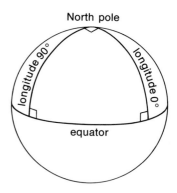

FIGURE 23.5.
This triangle, on the surface of the Earth, has three right
angles, so that the sum of its angles is 270 degrees.

To carry the analogy a step further, how could the surface creature
hope to determine whether his surface is curved or not? (Let us suppose
that his world, like our universe, is too large for him simply to travel
around it.) A simple answer is that the rules of geometry are different on
a curved surface from those of the familiar Euclidean geometry of flat
surfaces. Take, for example, one of the most familiar rules: the sum of
the angles of a triangle is 180 degrees. On the surface of a sphere this is
not so; the angles always add up to more than 180 degrees, depending on
the size of the triangle. Figure 23.5 shows a spherical triangle that has
three 90-degree angles.

Interestingly, there is another kind of surface on which triangles behave the opposite way. On a saddle-shaped surface, the angles of a triangle always add up to less than 180 degrees; again the deviation from
180 degrees increases with the size of the triangle. Mathematicians refer
to the sphere and the saddle as surfaces of positive and negative curvature, respectively.

Looking at these surfaces in three dimensions, we can easily distinguish surfaces with positive, negative, and zero curvature (zero curvature
meaning simply "flat"), just by looking at the directions in which they
curve. The two-dimensional creature, however, cannot do this, because
he is confined to the surface; all that he can do is study the geometry.
This he could do in principle by measuring triangles; if he could study
large enough ones, the sizes of the angles would tell him the curvature of
his surface. If the curvature turned out to be positive, he would probably
conclude, correctly, that his surface was closed and finite in extent.

This is just our situation in investigating the curvature of our own
space: we cannot see the curvature of space directly, because to do so is
beyond our power of perception; but we can determine the curvature
indirectly, by studying the geometry.

It may come as a surprise that geometry is to be considered an observational science, when it has been a staple of our schools for centuries
to develop Euclidean geometry as an exercise in pure thought. Euclid
begins with his axioms and postulates, all of which seem clearly accept-

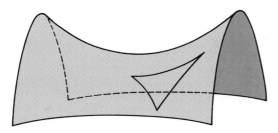

FIGURE 23.6.
On a surface of negative curvature, such as a
saddle or the horn of a trumpet, the angles of a
triangle add up to less than 180 degrees.

able, and then he *proves* that the sum of the angles of a triangle is 180 degrees. But the cosmologists now tell us that it might be otherwise and that we can find the real answer only by looking. What is wrong here? The problem turns out to center on one of Euclid's postulates of plane geometry, which says, "Through a point not on a given line, there is one and only one line that never meets the given line." This "parallel postulate" has always seemed a little bit mysterious and in a class by itself. The other postulates state much more obvious things, like, "Through two points there is one and only one line." The parallel postulate, however, makes a statement that could be verified only by following the lines for an infinite distance in both directions. We cannot do this; so how do we know that the parallel postulate is true?

During the nineteenth century, mathematicians began to ask an interesting question that is related to this one. What would happen if we threw out Euclid's parallel postulate and substituted a different hypothesis instead? Suppose that we assume that all lines intersect, so that there is no such thing as a parallel. What we get then, it turns out, are rules of geometry that apply to a surface of positive curvature, like a sphere. In particular, the angles of a triangle add up to more than 180 degrees, and the excess increases with the size of the triangle.

Alternatively, we could assume that there are many parallels: that through the external point there is a whole range of directions within which no line intersects the given line. If we followed through and derived the theorems of this geometry, it would come out to be that of a surface of negative curvature, where the angles of a triangle add up to less than 180 degrees.

It is easiest to state these properties in terms of two-dimensional or "plane" geometry. They generalize equally well into the "solid geometry" of three dimensions, however. The rules of geometry still depend on whether we adopt the parallel postulate or one of its alternatives.

These alternative geometries are called non-Euclidean geometries, but mathematically they are just as valid as that of Euclid. In each the conclusions follow by logical deduction from the postulates—which is all that

468

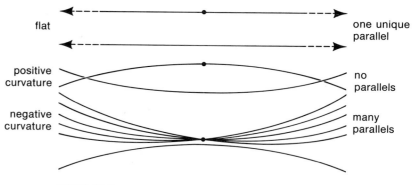

FIGURE 23.7.
Parallelism in geometries of different curvatures.

mathematics sets out to do. The task of physics is different, however; its aim is to find out how the real world *is*. Geometries of positive, zero, and negative curvature are equally valid mathematically; but we must find out which one applies to the real universe, and we can do this only by observation.

This is not an easy task at all. As everyone knows, Euclidean geometry does work very well; it is a very accurate description of how the world actually is. Why, then, do we not accept that the universe obeys Euclidean geometry and thus has zero curvature? The answer is that *all* geometries behave like Euclidean geometry within small distances; it is only for very large distances that the effects of curvature will show up. We see a similar effect on the surface of the Earth; although it is a curved surface, plane geometry describes it quite accurately, as long as we confine ourselves to a small area; for larger areas, however, we have to take the curvature into account. So it is in studying the geometry of the universe: small regions will follow Euclidean geometry very closely, and we can hope to find curvature effects only by studying as large a region as possible.

Because we need to cover large distances, we cannot use the particular geometric rule we have already considered; we cannot travel around sufficiently large triangles and measure their angles. Instead we must resort to some type of measurement that can be made completely from the single point from which we observe. One such measurement derives from the rule for the volume of a sphere. In a flat space, the volume is proportional to the cube of the radius, no matter how large the sphere is. In a space of positive curvature, however, the volumes of large spheres go up less rapidly than the cube of the radius. (Similarly, on the curved surface of the Earth the area included in a circle increases less rapidly than the square of its radius.) Correspondingly, in a space of negative

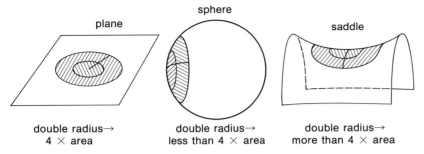

FIGURE 23.8.
The rule for the area contained within a circle depends on the curvature of the surface on which the circle lies.

curvature the volume of a sphere goes up more rapidly than the cube of the radius.

Fortunately, we can in principle measure volumes by direct observation, without traveling around. If galaxies are indeed distributed uniformly in space, then the number of galaxies out to a certain distance from us should be a measure of the volume contained within a sphere of that size. If we then count the galaxies out to twice that distance, we can ask whether there are just eight times as many (since $8 = 2 \times 2 \times 2$), or whether there are more or fewer. Given large enough volumes, this ought to provide us with a test of the curvature of space.

In practice, however, it has not yet been possible to apply this test successfully. There are three difficulties. First is the difficulty of the observations themselves. To find curvature effects, we have to look very far into the distance, and such distant galaxies are extremely faint. The second problem is the irregularity of the distribution of galaxies. Our premise is that we can measure the volume of a large region by counting the number of galaxies in it; what happens, however, if different regions have different numbers of galaxies? On a small scale, we know that galaxies cluster; on a larger scale, can we trust them to be uniform? Suppose, for instance, that there are nonuniformities as large as the whole region in which we attempt to determine the number of galaxies. Our counts will then refer not to the curvature of the universe, but rather to the nonuniformity of the region that we have studied.

The third difficulty is evolution. We see distant galaxies as they were a long time in the past; yet all our methods of measuring distance assume that we know the luminosities of the galaxies. This means that we must know how the brightness of a galaxy has changed with time, and we must allow for this change in measuring the large distances that look far back in time. Evolutionary corrections can be calculated in principle, by knowing what kind of stars the galaxy is made of and calculating how they change with time. Such calculations are uncertain, however, and that uncertainty is a major obstacle in applying this and other cosmological tests.

It therefore seems too difficult, for now, to find out the curvature of the universe by direct observation. What we shall do instead is to look at other properties of the universe, which can eventually be related to its curvature, and therefore to the question of whether it is finite or infinite in extent.

THE DYNAMICS OF THE UNIVERSE

The universe began with a "bang," and it is still expanding today, but what will be its future? Will it expand forever, until the galaxies are almost infinitely far apart? Or will the expansion slow and stop, and will the galaxies then fall crashing back upon each other? We can attempt to answer these questions by studying the dynamics of the expansion.

As we have seen, the galaxies of the universe attract each other, and this attraction slows down the expansion. The crucial question is, how strong is this deceleration; is it enough eventually to stop the expansion? The self-attraction of the universe depends on its density; the more mass per unit volume, the stronger the self-attraction. On the one hand, therefore, we need to know the density of the universe. The other vital quantity to know is its current rate of expansion. Clearly, the faster it is going, the stronger the force that would be needed to stop it. The problem is very similar to that of a projectile shot upward against the force of gravity. If it has a low velocity, it will reach a maximum height, stop, and come down again; but if its speed is high enough, it will escape from the Earth completely and never come back.

So it is with the universe. If its velocity is too high for its gravitation, it will expand forever; but if gravitation outweighs expansion, the universe will turn around and fall together again. Cosmologists state this question in terms of the *critical density,* which is the density of matter whose gravitation would be just sufficient to put an eventual stop to the expansion. Is the actual density of the universe greater or less than the critical density?

We can attempt to answer this question by adding up the density of all the known material of the universe, which of course means the galaxies. When we do this, the total adds up to only a few per cent of the critical density. If the only mass is that of the galaxies, the universe will expand forever.

This may not be the final answer at all, however. How do we know that this summation of galaxy masses gives us a correct total for the universe? In the clusters of galaxies, where there is an opportunity to mea-

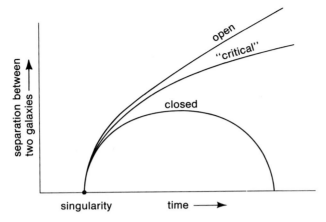

23.9.
The expansion of "model universes," shown graphically. All models decelerate, but only the closed models fall back. In the closed models, time has an end as well as a beginning.

sure the total gravitating mass, we saw that the total was far more than we could account for in the galaxies. Might there not be a similar factor, or even a greater one, for the universe at large? The answer is that we simply do not know. The universe has at least as much mass as we see in the galaxies, but it could have considerably more.

How then can we hope to settle this crucial question of whether the universe will go on expanding forever? Fortunately, there is a direct observational approach. Just as the universe will expand less rapidly in the future, so it must have expanded more rapidly in the past. But the past history of the universe is something that we can see directly, by looking into the distance. Because of the time that light takes to travel, we see the distant parts of the universe as they were in the past, when they emitted the light that is reaching us today. Thus, by looking at the redshifts of distant galaxies, we can attempt to see how fast the universe was expanding in the past. From this, we can hope to measure the deceleration and predict what the eventual fate of the expansion will be.

The crucial cosmological test therefore becomes the relation between the redshifts of galaxies and their magnitudes, since the apparent magnitude of a standard galaxy is our best indicator of distance. Astronomers are working strenuously on this problem, but the results are still far from giving the answer. Observationally the difficulty is finding distant-enough clusters of galaxies—only for clusters can we measure distances reliably enough—and measuring their redshifts.

Here again there is a serious difficulty in interpreting the redshift-magnitude relation, one caused by the evolution of the galaxies themselves. All our calculations of distance depend on assuming a standard

472

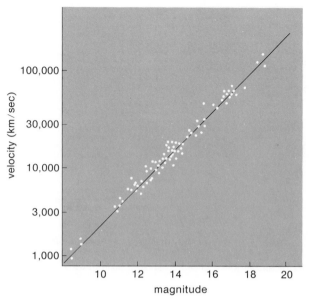

FIGURE 23.10.
The redshift-magnitude relation, as measured from the
brightest galaxy in each of 84 clusters. The data are fitted
here to a straight line. In principle, the nature of our
universe can be deduced when more extensive
observations have been made, from the curvature of the
upper end of the line. (Adapted from A. Sandage,
Astrophysical J., **178** (1972), 12.)

luminosity for a galaxy—but what if this has changed with time? It almost
certainly has; and the size of the evolutionary correction is distressingly
similar to that of the deceleration effects that we are studying.

Thus the answer to this basic cosmological question is still beyond
our reach. In principle we can determine the past history of the expansion—and therefore its future, too—by measuring the redshifts of faint,
distant galaxies. In practice, however, the measurements are difficult,
and the existing results do not reach far enough. And even with adequate
observations, we will have to know more about the history of galaxies if
their redshifts are to tell us the true nature of the universe.

UNIFIED COSMOLOGY: GRAVITATION AND CURVATURE

We have seen the two great problems of cosmology: the curvature of
space and the future of the expansion. Each can be solved only by observing the faint, distant reaches of the universe, and for each the
present-day observations are still inadequate. Very fortunately, a deeper

study of physics shows that the two problems are directly connected, so that a solution to one of them will also answer the other. The link between them is gravitation. In modern physics, gravitation is described by Einstein's theory of general relativity, which connects the phenomenon of gravitation with the properties of space. General relativity predicts that the curvature of space should be related to the density and motion of the matter in it, so that gravitation becomes a natural property of space. Thus the curvature and the decleration are each determined by the same quantities.

General relativity is an extreme example of the paradox of science: we create a simple picture by introducing concepts that may themselves be very difficult to grasp. In its logic, general relativity displays a beautiful simplicity, explaining the whole phenomenon of gravitation as a consequence of something more fundamental. This more basic starting point, however, is full of ideas that are so far from everyday thought and experience that they strain the flexibility of our mental processes. Viewing gravitation through the eye of general relativity makes gravitation a simple thing—but to do so we must think in terms of a curved four-dimensional continuum that joins space and time into a single manifold.

How can one claim that a picture is simpler, when it is so much harder to understand? The answer is in a premise that underlies all of science. We call an explanation "better" when it starts with fewer hypotheses. The aim of science is to explain the behavior of the universe as a logical consequence of as brief a set of rules as possible. The rules themselves we cannot fathom; all we can say is, "That's the way it is." Thus it is no simplification at all to say that a body falls because it seeks its natural abode at the center of the Earth. One statement has merely been replaced by another. The law of gravitation is a great step forward, however. By a single hypothesis it explains the weight of an elephant, the fall of an apple, and the shape of the orbit of Mars. This is a great logical economy, but already the cost is evident. Any high-school student can understand the fall of the apple, but one must be an expert in integral calculus to calculate the orbit of Mars. Even so, Newtonian mechanics can regard gravitation itself only as an inscrutable property of nature. This is where general relativity makes a large step forward. By introducing a deeper view of space and time, it is able to consider gravitation as just another form of acceleration, one that is inherent in the way that matter interacts with the space around it. The problem, however, is that the logical simplicity of general relativity becomes evident only in the context of the tensor differential geometry of four-dimensional spacetime.

When applied to the cosmological problems of expansion and curvature, general relativity gives an amazingly simple answer. If the density of the universe is so great that its self-attraction will win out over the expansion, and cause the galaxies eventually to fall back together again, then the curvature of space is positive and the universe is closed in extent. If, on the other hand, the expansion will win out over gravitational attraction and run unbridled forever, then space is negatively curved and the universe is open and infinite. Just on the dividing line, at the critical density that forever slows the expansion but would take an infinite time to stop it, space would be flat.

It is interesting that this connection was used to predict the expansion of the universe long before it was actually observed. One of Einstein's applications of general relativity was to the structure of the universe. Not being aware of the expansion, and looking only at the mean density of matter in the universe, he necessarily concluded that space must be positively curved and finite in extent. Furthermore, the time for the universe to fall in on itself was distressingly short. Einstein therefore added to his equations a "cosmological term," which counteracted the self-gravitation of the universe and allowed it to remain static without collapsing. But in doing so he spoiled the beautiful simplicity of his theory, by introducing an assumption that was *ad hoc*: its purpose was simply to avoid the prediction that a static, self-gravitating universe would collapse.

Fortunately, other workers in general relativity soon showed that there was another way out, without introducing the "cosmological term." They found that there were solutions of Einstein's equations that corresponded to a universe whose framework continually expands. To some contemporaries this alternative seemed equally fantastic—but a few years later the redshift was discovered observationally, and modern cosmology was born.

Given the difficulty of all the direct cosmological tests, which require observations of very faint galaxies at very great distances, it may well be that measurements of the mean density of the universe offer the best chance to distinguish its real nature. Is the mean density of the universe greater than the critical density, or is it less? According to relativistic cosmology, the answer to this question determines not only whether the expansion will stop and fall back, but also whether the universe is closed or open. As we have seen, the observational answer to this question is still quite uncertain. The masses of galaxies, with a reasonable allowance for "missing mass," add up to about a third of the critical density; and this result is so uncertain that the correct answer could go either way. Once again the cosmological answer eludes our grasp.

Cosmology, by its very nature, introduces us to realms, problems, and concepts that range beyond our previous experience. The answers to many of the most important questions are still obscure, in spite of seeming tantalizingly close. Indeed, many of the answers have appeared to be so close for so many years that they lead one to doubt Einstein's kindly precept: that nature is only sophisticated, but not really malicious. In the flow of intellectual history, however, it is a great step for us even to have asked these questions.

REVIEW QUESTIONS

1. Explain why the motion of all distant galaxies away from us does not mean that we are at the center of the universe.

2. What is meant by the "singularity?" What does it mean to say that it was the beginning of time? What happened before that?

3. Describe and explain the 3-degree background radiation.

4. By what steps do we believe the big bang to have produced hydrogen and helium?

5. What is meant by curved space? Distinguish between sense and nonsense in describing its properties and the ways in which we might observe them. In what sense is Euclidean geometry still true?

6. How is the future of the universe connected with its present density and rate of expansion?

7. "When we study the redshifts of distant galaxies, we look into the past in order to predict the future." Explain.

8. Why is the evolution of galaxies such a serious problem for observational cosmology?

In the beginning was the singularity. From it the arrow of time took flight, and the history of the universe began to write itself. At first, change was very rapid. The clashing and merging of nuclear particles was over in a few minutes, leaving the hydrogen and helium that have made up the cosmos ever since. Then, as the fiery hot gas expanded, minutes became years, and then thousands of years. But the gas cooled as it went, and after the first million years it was cool enough and thin enough to be transparent. The first act of history was over, leaving behind the afterglow that we now see as the 3-degree background radiation.

The density of the infant universe was not everywhere the same, however; and in the denser regions gravity soon began to assert itelf, slowing down the expansion inside each of these regions. This scene was played more slowly. Perhaps a billion years passed, as the material of each condensation slowed, stopped, and began to fall back on its center. This was the birth of galaxies.

Then in smaller regions, but more rapidly, the material within each galaxy began to condense even more, and stars were born. The first burst of star formation produced many stars that were large and bright and fast-burning. These lived and died, and then enriched the surrounding gas with the products of their nuclear burning.

The material of each galaxy continued to fall together, until the motions were fast enough to prevent further contraction. The stable, settled era of cosmic history had begun. Each galaxy has gone on since; stars are born and die, and occasional explosions ruffle the interstellar calm, but now the long past flows into a future that is figuratively and perhaps literally infinite.

Within this universe our own history has unfolded. Some 4½ billion years ago, when the universe was already more than half its present age, the continuing processes of star formation produced a certain middle-sized star in an outer spiral arm of an intermediate-type galaxy within a small group of galaxies. This was our Sun. Not all the material went into the Sun, however. The newly born star was unable to contain all the angular momentum of its protostellar cloud, and the remainder of the cloud flattened down into a spinning disc of dust and gas. Condensations formed, merged, and grew, until most of the material of the disc was collected in a few solid bodies. These were the planets.

On one planet, the Earth, the temperature was particularly favorable for the growth of complicated molecules. Some of these joined into chains that had the marvelous ability to duplicate themselves; some three billion years ago, life was born. For most of its history the evolution of living forms was slow. It was only half a billion years ago that the oceans began to fill with creatures that resemble the fishes of today; then hundreds of millions of additional years passed before plants and animals emerged from the ocean to cover the land. Only 60 million years ago, the giant reptiles disappeared and left the Earth to the mammals. A few million years ago, one line of tree-dwellers came down and began using rocks as tools. Late within the last million years, their brains enlarged rapidly, and communities developed new skills in finding food. Ten thousand years ago, a few tribes began replanting the seeds of some newly evolved, richer grasses, and settled civilization began. For all but the last 200 years of this time, power came from the muscles of men and animals, and development was slow. Then the first industrial revolution brought large-scale manufacturing and easy travel. Now, in the most recent two decades, has come the second industrial revolution, of electronics and nuclear energy, and another clan of scientists is moving toward a future third revolution, which will be based on an understanding of biological processes. Unlike the universe, with its frenetic beginning and subsequent slowing down to a majestic calm, our own history began slowly but has evolved with a dizzying acceleration. Where it leads, we can only wonder.

Indeed, it is this tendency to wonder that distinguishes Man from other organisms that are equally complex and function in similar ways. What makes us really different is not just the capacity to think, but rather the involuted way in which we use that capacity. The cat in the jungle asks, "How can I get what I want?" but Man also asks, "Who am I?" Science does its utmost to answer one aspect of this question, but the question goes on reverberating in other keys. Physical science can only explain what the scene is and how we have come into it. The part that we play in that scene is in our own hands. Ten thousand years after our ancestors began to build towns, we learned the source of the Sun's energy; and a mere decade later we were capable of using it to destroy whole cities. The end of our own world may in fact be near; but even so, we should not confuse our small world with the universe, in which our fiery apocalypse would be less than the daily pulsation of an RR Lyrae star. The difference is only a shift of perspective, and the most important human role for astronomy is to develop that perspective.

The perspective is a modern one, however; it has looked very different in the past. The Bible begins, "In the beginning God created the heavens and the Earth," and a few verses later comes the divine creation of Man. Today no serious thinker would take this as anything but a highly figurative account, but past centuries held the word of the Bible to be literally true. The Copernican revolution led to the "war between science and religion," and three centuries later the theory of evolution brought that war to a climax. In the century that has passed since then, science has shown us much about the world into which we came, and it has also bred a technology that offers us a large control of our surroundings. Willing or unwilling, we are faced with the question of how we will use that control. Although few scientists would seek the answer to such questions in revealed religion, even fewer would deny that human existence poses a dimension of human questions whose thrust is outside the directions in which science looks. "Who am I?" asks Man, and it would be foolish to answer the question as a scientific exercise. Yet it would be equally foolish to ignore the answers that science does give. We are part of a physical universe, and our existence is a result of its physical workings. It is only in this setting that we can properly approach the questions of consciousness, feeling, and choice that make up so much of the human outlook.

Another account in the Bible opens, "In the beginning was the Word." This phrase can apply equally to science. As the universe unfolds, in space and time, its actions express the set of underlying laws that make it what it is. The task of science is to find those laws and to probe their workings. In the universe Man is small, but in this task he is great. Here, on a small planet circling an undistinguished star, far out in one galaxy among millions, we unravel the secrets of nature at a rate whose progress is bewildering. This, indeed, is the twofold wonder of astronomy: the wondrous structure of the universe, and the wondrous ways in which we come to understand it. This is the place of Man in the universe.

STAR MAPS

On the following four pages are star maps that show the appearance of the sky at four different sidereal times, as seen from latitude 35° north. (This is the latitude of North Carolina, Oklahoma, and central California. Observers farther south will be able to see some southern stars that are not shown here; those farther north will miss the southern extreme of the maps.)

Each map shows the entire sky, with the zenith near the center and the horizon all around the edge. In the projection that is used here, constellations near the horizon are exaggerated in size; but that is how the psychology of our perception actually makes them look. All stars are shown that are easily visible on a dark night. The constellations are connected by lines in a simple way, and their names are given in capital letters. First-magnitude stars are labeled with small letters, along with Polaris, the eclipsing star Algol, and the long-period variable Mira (which is visible only at times). Also labeled are the Pleiades and Hyades clusters, the galaxy M31, and the globular cluster Omega Centauri.

To use the maps, first decide which one you need, according to the date and the time of night. It will help very much beforehand if you know which way is north. Whatever the direction in which you are facing, turn the map so that the part of the horizon at which you are looking is at the bottom. Remember then that stars in the middle of the map are above you, and stars near the top of the map are behind you.

The crucial step is to make the first identification. For this, choose a distinctive configuration that is easy to recognize. Toward the north, the "Big Dipper" in Ursa Major and the "W" (or "M") of Cassiopeia are good starting points. Use your map to find which way they are turned, and begin with whichever one is higher; then identify the other one too, unless it is too close to the horizon. Note Polaris, about halfway between Ursa Major and Cassiopeia; the last two stars in the Big Dipper point at it. If possible, move from these constellations directly to others. For instance, if the Big Dipper is high, follow the curve of its handle to Arcturus. If Cassiopeia is high, move from it to Perseus and then down Andromeda to the "Great Square" of Pegasus. You can also look for distinctive groupings in the southern half of the sky, or overhead. Easy ones to recognize are Orion (note the three stars of his belt), the Great Square of Pegasus, the "sickle" of Leo, and the "Northern Cross" of Cygnus, with Vega nearby.

In star-finding, planets can be confusing, since their motion makes it impossible to include them on printed star maps. Fortunately the planets stay close to the ecliptic (which is marked on the maps), and some of them are easy to recognize in special ways. Mercury is no problem, since it can barely be glimpsed in the sunset or sunrise. Nor should Venus cause much difficulty; when it is visible, it is incomparably brighter than any star. Jupiter is similar, though not so dazzling as Venus. Mars and Saturn are easily mistaken for first-magnitude stars, however (except for a few months around Mars' opposition, when it is very bright).

One way to recognize planets is by their lack of twinkling. The other is to know beforehand where they are—although knowing this is not easy for Venus and Mars, which move rapidly among the stars. Jupiter and Saturn move slowly, however. Jupiter moves by about one zodiac constellation per year: Aries in 1976, Taurus in 1977, etc. Saturn spends about $2\frac{1}{2}$ years in each zodiac constellation; in 1976 it is in Cancer.

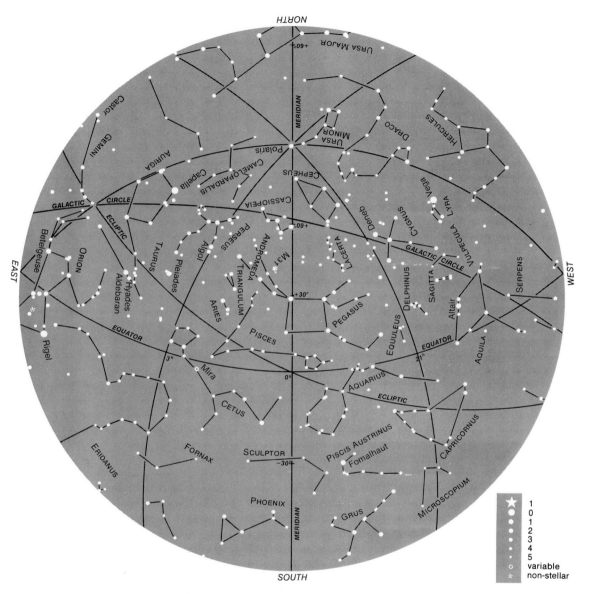

STAR MAP 1.

The autumn stars. This map shows the appearance of the sky at each of the following times:

early September	1 A.M. (2 A.M. Daylight Time)
late September	midnight (1 A.M. Daylight Time)
early October	11 P.M. (midnight Daylight Time)
late October	10 P.M. (11 P.M. Daylight Time)
early November	9 P.M.
late November	8 P.M.
early December	7 P.M.
late December	6 P.M.

For use of the map, see page 479.

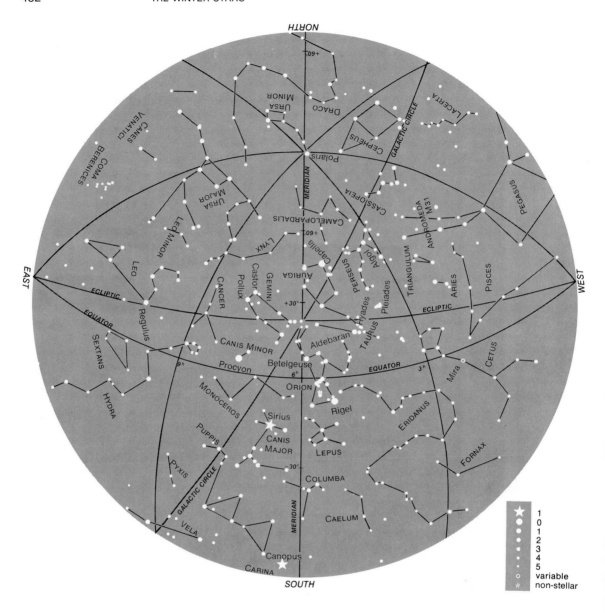

STAR MAP 2.
The winter stars. This map shows the appearance of the sky
at each of the following times:

early January	11 P.M.
late January	10 P.M.
early February	9 P.M.
late February	8 P.M.
early March	7 P.M. (8 P.M. Daylight time)

For use of the map, see page 479.

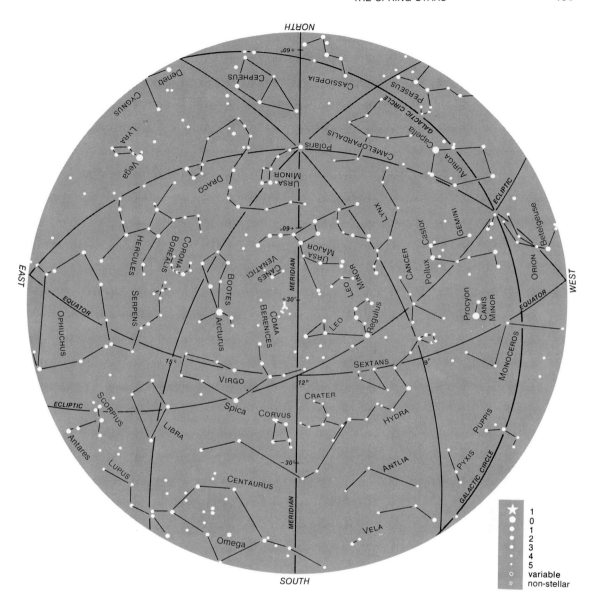

STAR MAP 3.
The spring stars. This map shows the appearance of the sky
at each of the following times:

late March	midnight (1 A.M. Daylight Time)
early April	11 P.M. (midnight Daylight Time)
late April	10 P.M. (11 P.M. Daylight Time)
early May	9 P.M. (10 P.M. Daylight Time)
late May	8 P.M. (9 P.M. Daylight Time)

For use of the map, see page 479.

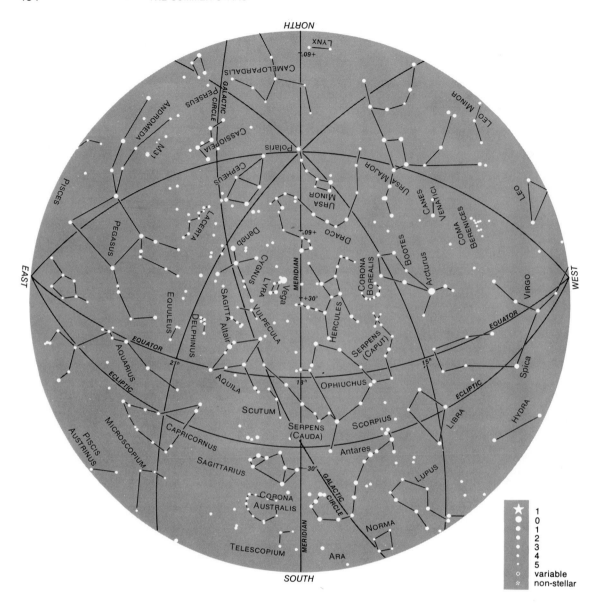

STAR MAP 4.

The summer stars. This map shows the appearance of the sky at each of the following times:

early June	1 A.M. (2 A.M. Daylight Time)
late June	midnight (1 A.M. Daylight Time)
early July	11 P.M. (midnight Daylight Time)
late July	10 P.M. (11 P.M. Daylight Time)
early August	9 P.M. (10 P.M. Daylight Time)
late August	8 P.M. (9 P.M. Daylight Time)

For use of the map, see page 479.

GLOSSARY

absolute magnitude The apparent magnitude that a star would have if it were 10 parsecs away from us. A measure of the luminosity of the star.

absolute temperature Temperature, on a scale of centigrade degrees, measured from the "absolute zero" that is the lowest temperature possible. Symbol is K. Water freezes at 273°K; a very hot day is 310°K.

absorption line A dark line on a continuous spectrum. A narrow range of wavelengths within which the brightness of the spectrum is reduced. Corresponds to a specific atomic transition.

acceleration The rate at which velocity changes. A vector quantity, including both change of speed and change of direction.

airglow A faint permanent aurora that contributes about half the brightness of the darkest night sky. Limits astronomical observation of faint objects.

alpha particle A helium-4 nucleus, containing 2 protons and 2 neutrons. Participant in, or product of, many nuclear reactions.

altitude On the surface of the Earth, height above sea level. On the celestial sphere, angular height above the horizon; called "elevation" in engineering.

amplitude The amount by which a quantity varies. In most cases, amplitude is measured in either direction from the mean value, but in variable-star astronomy it refers to the total range.

Ångstrom The unit in which astronomers most commonly express wavelengths of light. (Its length is 10^{-8} centimeter.) The middle of the visible spectrum is around 5,000 Ångstroms.

angular momentum A quantitative measure of the amount of turning motion in a rotation or a revolution. It is the mass times the transverse velocity times the distance from the center. In any interaction, the total angular momentum of all the bodies remains constant.

annular eclipse An eclipse in which an unobscured ring remains visible around an eclipsed central disc.

apastron The point in a binary-star orbit where the two stars are farthest apart.

aperture The diameter of a telescope: for a reflector, the diameter of the large mirror; for a refractor, the diameter of the lens.

aphelion In an orbit around the Sun, the point at which the object is farthest from the Sun.

apogee In an orbit around the Earth, the point at which the object is farthest from the Earth.

apparent magnitude The magnitude of an object as it looks to us, without any adjustment for its distance. A measure of brightness.

apse (plural **apsides**) An end of the major axis of an ellipse.

apsidal motion Rotation of the direction of the major axis of an orbit, because of perturbations.

ascending node   The point at which an orbit crosses its reference plane in the upward direction. For a body in the Solar System, its northward crossing of the ecliptic.

aspect   The apparent position of a body (as seen from the Earth) with respect to the Sun. Determines its phase.

association (stellar)   A group of stars that were born together and are still visibly associated with each other. Differs from a star cluster, in that an association is in process of dissipating.

asteroid   A small body in orbit around the Sun, usually between the orbits of Mars and Jupiter. Sometimes called a minor planet.

astrology   The superstition of fortune-telling by means of the stars and planets.

astronomical unit   The semi-major axis of the Earth's orbit (average distance from the Earth to the Sun), often used as a unit of distance in the Solar System.

astrophysics   The part of astronomy that deals with the local physical properties of the material of the universe. Often used loosely, in an aggrandizing sense, for all of astronomy, especially for astronomical theory.

atom   A tiny body consisting of a nucleus surrounded by electrons. Each chemical element consists of a specific kind of atom, distinguished by the electric charge of its nucleus.

atomic energy   Energy produced by nuclear reactions.

aurora   A glow of the upper atmosphere, caused by emission by its atoms when high-energy particles strike them.

azimuth   Compass direction in degrees, measured around the horizon from 0° at the north, through 90° at the east, 180° at the south, and so on.

Balmer series   The transitions of the hydrogen atom upward from, or downward to, the second energy level. These spectral lines are in the visible and the near ultraviolet.

band   A series of lines in the spectrum of a molecule. In most bands the lines converge to a limit called the head.

barred spiral   A spiral galaxy that has a straight, bar-like structure across its center.

big bang   The brief period of very high density and temperature at the beginning of the expansion of the universe.

binary   A double or multiple star.

black body   An idealized absorber and emitter of radiation. It absorbs all radiation that strikes it, and emits exactly according to Planck's law.

black hole   A mass so condensed that neither matter nor light can escape from its gravitational field.

blink comparator   A device that allows two photographs to be viewed in rapid alternation, to allow one to look for small changes in position or brightness of stars.

Cassegrain focus   An alternative focus of a reflecting telescope. A small secondary mirror at the top refocuses the light through a small hole in the center of the primary mirror.

celestial equator   The circle that goes around the celestial sphere halfway between the two poles, exactly above the Earth's equator. The zero-point of declination.

celestial mechanics   The theory of orbits in a gravitational field, especially in the situation where several bodies perturb each other's motions.

celestial poles   The two points on the celestial sphere that lie exactly above and below the north and south poles of the Earth.

celestial sphere   An imaginary sphere surrounding the observer and the Earth, large enough that directions seen from the observer are the same as they would look from the center of the Earth. Points on the celestial sphere correspond to directions from the observer.

center of mass   (also called **center of gravity**) Mean position of a set of bodies, considered in proportion to their masses. For two bodies, it is the point at which they would balance, if connected by a rigid rod and suspended in a gravitational field.

cepheid   A type of pulsating variable star. Classical cepheids have evolved away from the upper main sequence and have a period-luminosity relation that allows their distances to be easily measured.

chain reaction   A process that sustains itself when each event sets off others.

chromosphere   The thin, hot transition layer between the Sun's surface and the even-hotter surrounding region of the corona.

circular velocity   The speed that a body needs in order to follow a circular orbit, at a given distance from a center of attraction.

circumpolar regions   The region around the observer's north celestial pole in which stars are always above the horizon, and the region around his south celestial pole in which they are always below the horizon. (For a southern-hemisphere observer, interchange the words "north" and "south.")

cluster of galaxies   A group of many galaxies that are bound together by their mutual gravitational attraction. The word "cluster" is usually reserved for groups that have more than a dozen or so obvious members.

CNO cycle   A sequence of nuclear reactions by which carbon, nitrogen, and oxygen act as intermediaries in fusing hydrogen nuclei to form helium.

color excess   Reddening of an object by interstellar dust, expressed as the difference between its observed color index and the color index that it would have had in the absence of reddening.

color index   The color of an object expressed as a difference of magnitudes measured through different color filters. If otherwise unspecified, color index refers to blue magnitude minus yellow magnitude.

color-magnitude array   A graph in which each star is represented by a point that corresponds to its apparent magnitude and its color. Corresponds to an HR diagram except for shifts of zero point; magnitude differs by the distance modulus, and color differs by the interstellar reddening.

coma   (1) A defect in optical images, which spreads them into cometlike flares. Limits the field of sharp imaging of most reflecting telescopes. (2) The diffuse part of the head of a comet.

comet   A small Solar System body whose ices vaporize when it comes close to the Sun, brightening the body greatly and often forming a tail.

comparison spectrum   A set of emission lines put at the edge of a spectrum in order to serve as a standard of wavelength.

compound   A substance, such as water or salt, in which atoms of different elements are joined together in a regular way. The regularity and the fixed proportions of the different atoms distinguish a compound from a mixture.

conjunction   The situation in which two bodies are as close together in the sky as the inclination of their planes of motion will allow them to be. If no second body is mentioned, then conjunction with the Sun is implied.

conservation law   A rule according to which the total amount of some quantity, such as energy or momentum, remains fixed, even though it may shift from one place or form to another.

constellation   A grouping of stars into a pattern. Also, a fixed section of the celestial sphere, containing such a grouping.

continuous spectrum   A distribution of emitted radiation over a continuous range of wavelength. More specifically, the shape of this distribution.

convection   The transport of energy by rising and falling currents of hotter and cooler material.

Copernicus, Nicholas   Polish astronomer (1473–1543), whose book *De Revolutionibus* marks the beginning of modern astronomy.

Coriolis force   An apparent acceleration of a body that moves across the Earth's surface, such that its direction of motion seems to be deviated. Caused by the Earth's rotation.

corona   The hot, extensive outermost part of the Sun's atmosphere.

coronagraph   A telescope in which the Sun's disc is blotted out so that the chromosphere and corona can be studied.

cosmic rays   High-speed atoms, trapped in the Galactic magnetic field, that strike the Earth from all directions.

cosmogony   The study of the origin of the Solar System or the universe.

cosmological constant   A quantity sometimes included in the equations of general relativity, but usually omitted. Its original purpose was to counteract the self-attraction of the universe.

cosmological principle   The assumption that the large-scale properties of the universe would look the same from any point of observation.

cosmology   The study of the large-scale structure and development of the whole universe.

coudé focus   A fixed point to which the light of a reflecting telescope can be brought by a series of mirrors.

dark nebula   A dusty region that hides the light of stars behind it.

declination   A latitude-like coordinate on the celestial sphere. Measured from the celestial equator toward the poles.

degenerate material   Material that is so compressed that all, or nearly all, its allowed energy states are filled. It strongly resists further compression.

density   Mass per unit volume. Characteristic of the kind of material rather than of the quantity.

descending node   The point at which an orbit crosses its reference plane in the downward direction. For a body in the Solar System, its southward crossing of the ecliptic.

deuterium   Hydrogen whose nucleus is H^2 (a deuteron) rather than H^1 (a proton). The heavier stable isotope of hydrogen.

deuteron   A hydrogen nucleus that contains one proton and one neutron.

differential rotation   A rotation of a nonrigid body in which different parts go around at different rates.

diffraction   The bending of waves at the edge of an object or a sequence of objects.

diffraction grating   An array of regular, finely spaced lines that reflects or transmits light of different wavelengths in different directions.

diffuse nebula   A bright nebula, often irregular in shape. May be either an HII region or a reflection nebula.

disc   (1) The round apparent surface of a body such as the Sun or a planet. (2) The flattened circular region of the Milky Way in which the stars of Population I are concentrated.

dispersion   Spreading of light according to wavelength.

distance modulus   A measure of distance, expressed in magnitudes. It is equal to zero at a distance of 10 parsecs and increases by 5 magnitudes for each factor of 10 in distance.

Doppler effect   The shift in wavelength (and frequency) of light, caused by movement of its source toward or away from the observer.

dwarf   A star of low luminosity. Often used as a relative term, to distinguish the main sequence from stars of higher luminosity.

eccentricity   A measure of the shape of an ellipse: the center-to-focus distance divided by the semi-major axis. If e is the eccentricity of an ellipse, the ratio of minor axis to major axis is $\sqrt{1 - e^2}$.

eclipse   The hiding of one celestial body by another.

eclipsing binary   A double star in which one star periodically hides all or part of the light of the other.

ecliptic   The circle that the plane of the Earth's orbit makes on the celestial sphere; it is the apparent path of the Sun during the year.

ecliptic plane   The plane of the Earth's orbit.

electromagnetic radiation   A broad class of radiation, of which visible light is an example. Varieties, distinguished by wavelength, include radio, infrared, visible, ultraviolet, X-rays, and gamma rays. All travel at the velocity of light.

electron   One of the fundamental particles of physics, with negative electric charge and very small mass. The outer part of an atom is a cloud of electrons, equal in number to the positive charge on the nucleus of the atom.

element   Atomic species. All atoms of a given element have the same nuclear charge and the same chemical behavior.

elements of an orbit   The six numbers that specify an orbit: semi-major axis, eccentricity, inclination, angle of the node, argument of perihelion, and time of perihelion.

ellipse   A mathematical curve that is equivalent to a uniformly stretched or compressed circle.

elliptical galaxy   A round galaxy, usually flattened, that consists of old stars and has very little interstellar material.

elongation   The angular distance of an object from the Sun.

emission line   A bright line in a spectrum, caused by atoms making a specific transition.

energy   One of the fundamental quantities of physics, often converted from one form to another but never created or destroyed. May take the form of motion, heat, or electromagnetic fields or radiation, or may be stored in forces under tension, including the forces in atoms and their nuclei.

energy level   One of the specific total amounts of energy that a specific kind of atom is capable of containing.

ephemeris   A listing of calculated positions of an object for different times.

epicycle   A circle whose center moves around on another circle.

equation of time   The difference, on any day of the year, between apparent solar time, as the Sun is actually observed, and mean solar time, which runs at a uniform rate.

equatorial system   A system of coordinates on the celestial sphere, based on the Earth's poles and equator and the vernal equinox.

equinox   (1) Either of the two times of the year when the Sun appears to cross the Earth's equator. (2) The direction of the Sun in the sky at one of those two times.

escape velocity   The minimum velocity that a body needs, at a given point in a gravitational field, to escape completely from that field.

excitation   The raising of the internal energy level of an atom, by putting energy into the atom.

extinction   The dimming of light by passage through some material that absorbs or scatters it.

extragalactic   Outside our Galaxy.

fission   The breaking of an atomic nucleus into two nearly equal parts. A practical source of nuclear energy, by means of the fission of uranium or plutonium.

flare   A sudden brightening of a small part of the Sun's surface, which becomes intensely hot and emits X-rays, ultraviolet, and radio waves.

flare star   A red-dwarf star that brightens suddenly for a short time, at irregular intervals.

fluorescence   Emission, by an atom, of one or more photons after it has been excited by absorbing a shorter-wavelength, higher-energy photon.

focal length   The distance from the lens of a refracting telescope to the focus, or from the main mirror of a reflector to the prime focus. Determines the scale of the image that is formed at the focus.

focus   (1) A point at which an optical instrument forms an image. (2) Either of two points on the major axis of an ellipse whose positions can be related, by various mathematical definitions, with the shape of the ellipse.

forbidden line   A spectral line corresponding to a transition that an atom is very unlikely to make. Forbidden lines can be strong in the emission spectrum of low-density gas.

free-free transition   An interaction between an atom and a passing electron in which the electron increases or decreases its energy of motion and radiation is absorbed or emitted.

frequency   Number of oscillations per second. For waves, frequency is inversely proportional to wavelength.

fusion   Combination of atomic particles to build larger nuclei. Most commonly, fusion of hydrogen nuclei to form helium. A powerful source of nuclear energy.

Galactic circle   The Galactic equator.

galactic cluster   An "open" cluster, made up of Population I stars, in the Galactic disc.

Galactic equator   The circle that is traced on the celestial sphere by the central plane of the Milky Way.

Galactic latitude   Angular distance of an object from the Galactic equator.

Galactic longitude   An angle measured around the Galactic equator in just the same way that longitude is measured on the Earth. Galactic longitude is zero in the direction of the Galactic center.

Galactic rotation   The revolution, around the Galactic center, of objects in the Milky Way.

galaxy   A large, independent stellar system. When capitalized, the term means specifically our Galaxy, the Milky Way.

Galileo Galilei   Italian astronomer (1564–1642), who made the first important telescopic observations, developed experimental dynamics, and was silenced by the Inquisition.

gamma rays   Electromagnetic radiation of the shortest wavelength, highest frequency, and largest energy per photon. This term refers to any radiation beyond the X-ray part of the spectrum.

geocentric   Earth-centered.

giant branch   A sequence in the HR diagram of a stellar group, occupied for a time by stars that have ended the main-sequence phase of their evolution.

gibbous   Somewhat less than fully illuminated. Used of the Moon or a planet between quarter (that is, half-illuminated) and full phase.

globular cluster   A star cluster of the halo population. Most are quite round and rich.

globule   A small, compact dark cloud of interstellar dust.

granulation   The fine, rapidly changing pattern of cells on the surface of the Sun.

gravitation   The tendency of all bodies to attract each other, according to their masses and separation. One of the basic properties of mass.

gravity   The gravitational field at the surface of a body such as the Earth.

greenhouse effect   An excess heating by an atmosphere that allows warming sunlight to enter but impedes the escape of cooling infrared radiation.

Greenwich time   Standard time at the meridian of longitude 0°, which passes through Greenwich Observatory. Used when a worldwide standard of time is needed.

Gregorian calendar   The calendar that is in almost universal use today, using the leap-year rules introduced by Pope Gregory XIII in 1582.

H I region   A region in which the interstellar hydrogen gas is cool and not ionized.

H II region   A region, surrounding one or more hot stars, in which the interstellar hydrogen gas is hot and ionized.

half-life   The length of time in which half the atoms in a given radioactive sample can be expected to disintegrate.

halo (Galactic)   The little-flattened, slowly rotating component of the Milky Way, consisting of stars of Population II.

heavy elements   A term that is often used loosely in astronomy to refer to all elements except hydrogen and helium.

heliocentric   Sun-centered.

helium flash   The explosive start of helium burning in degenerate material.

Hertzsprung-Russell diagram   The full name of the HR diagram.

high-velocity star   A star whose velocity relative to the Sun is very large. Most high-velocity stars actually go around the Galactic center more slowly than the Sun, and most are members of the halo population.

horizon system   A system of coordinates on the celestial sphere, based on the zenith, the horizon, and the celestial meridian.

horizontal branch   A region of the HR diagram of Population II, inhabited by stars that have already gone through the red-giant phase of evolution.

hour angle   A coordinate on the celestial sphere, in the equatorial system. It measures the distance of an object from the meridian.

HR diagram   The fundamental diagram of stellar astronomy, in which absolute magnitudes or luminosities of stars are plotted against spectra, colors, or temperatures.

hyperbola   A geometric curve whose two arms extend infinitely. The hyperbola, along with the ellipse, is a member of the class of curves called conic sections. Unbound orbits have the shapes of hyperbolas.

image tube   A photoelectric tube in which electrons emitted by a light-sensitive surface produce a brighter image than the original one.

inclination   The angle between a given plane and a reference plane.

inertia   The resistance of a body to changes in its motion. One of the basic properties of mass.

inferior conjunction   The conjunction of an inferior planet at which it is between the Earth and the Sun.

inferior planet   A planet that is closer to the Sun than the Earth is; specifically, Mercury or Venus.

infrared   The part of the electromagnetic spectrum that is intermediate in wavelength between visible light and radio.

interferometer   An arrangement of two or more receivers of radiation, in which high resolving power is achieved by comparing the radiation received simultaneously at different points.

ion   An atom that has fewer than its normal number of electrons (positive ion), or else more than its normal number (negative ion). In astrophysics, "ion" usually refers only to an atom from which one or more electrons have been removed.

ionization   The removal of one or more electrons from an atom, making it into a positive ion. (The term is normally used in astrophysics to refer only to the removal of electrons, not to their addition.)

ionosphere   The layers, in the Earth's upper atmosphere, in which ions and electrons absorb or reflect some kinds of radio waves.

irregular galaxy   A galaxy containing a mixture of old stars, many young stars, and a large amount of interstellar material in a chaotic arrangement.

isotope   A variety of a chemical element. The nuclei of different isotopes have different atomic weights. Most elements include nuclei of several different isotopes, each having the same number of protons (hence the same chemical identity) but a different number of neutrons.

Jovian planet   One of the large, Jupiter-like planets of the Solar System; specifically, Jupiter, Saturn, Uranus, or Neptune.

Julian calendar   The calendar established by Julius Caesar in 46 B.C., replaced in modern times by the Gregorian calendar.

Kepler, Johann   German astronomer (1571–1630), who discovered the laws of planetary motion.

kiloparsec   1,000 parsecs.

libration   Small periodic changes in the orientation of the Moon's face toward the Earth.

light curve   A graph of the magnitude of a variable star, plotted against time.

light year   The distance that light travels in a year. About a third of a parsec.

limb   The edge of the apparent disc of a body such as the Sun.

limb-darkening   Nonuniform brightness of the apparent disc of a body, with the edge not as bright as the center.

line of apsides   The direction of the major axis of an orbit.

line of nodes   The line along which the plane of an orbit intersects a reference plane.

line profile   The detailed distribution of brightness within a spectral line.

local group   The small group of about two dozen galaxies of which the Milky Way is a member.

logarithmic scale   A scale in which equal intervals represent equal factors. The magnitude scale is a logarithmic scale.

long-period variable   A cool pulsating star whose brightness varies in a nearly regular way, with a period of about a year.

luminosity   Intrinsic brightness, on an absolute scale.

luminosity function   The number of stars of each absolute magnitude, per unit volume of space.

Lyman series   The spectral lines of hydrogen that arise from transitions between the lowest energy level and higher levels. Falls in the ultraviolet part of the spectrum.

Magellanic Clouds   A pair of irregular galaxies, about 60 kiloparsecs away, that are companions of the Milky Way.

magnetosphere   A region around a planet in which its magnetic field strongly affects the motions of charged particles and holds off the solar wind.

magnitude   In stellar astronomy, a logarithmic measure of the brightness of an object. Each magnitude corresponds to about 2.5 times fainter; larger numbers refer to fainter objects.

main sequence   The diagonal line through the HR diagram on which most stars lie; the principal phase of stellar evolution.

major axis   The longest diameter of an ellipse.

mantle   The outer part of a planet's interior, beneath the crust and outside the core.

mare (plural **maria**)   A flat, dark plain on the surface of the Moon, usually round in outline.

mass   One of the fundamental quantities of physics, which manifests itself by the two properties of inertia and gravitation. A measure of the quantity of material; unlike weight, it is independent of the force of gravity at the place where it is measured.

mass-luminosity relation   For main-sequence stars, a close correspondence between the mass of a star and its luminosity.

mean solar day   The length of the uniform "civil" day to which our clocks are set. "Apparent time," as measured by a sundial, runs alternately fast and slow during the year.

megaparsec   A million parsecs.

meridian   (1) On the celestial sphere, the north-south line through the zenith. (2) On the Earth's surface, a north-south line of longitude.

Messier   An eighteenth-century astronomer, who assigned numbers to about 100 galaxies, nebulae, and star clusters. The numbers, preceded by an M, are still commonly used.

metal abundance   The proportion of heavy elements in the chemical composition of a star.

meteor   A tiny body that flashes across the sky, burning up by friction as it strikes the Earth's upper atmosphere. A "shooting star."

meteor shower   A group of meteors, all belonging to the same swarm; they appear at a particular time of year and seem to radiate from a particular part of the sky.

meteorite   A body from outside the Earth that has struck the Earth's surface.

microwave   A term that distinguishes the shortest-wavelength, highest-frequency part of the radio spectrum, up to the boundary between radio and infrared.

Milky Way   The galaxy in which we live. On the celestial sphere, the bright band around the sky that we see because we are looking nearly edgewise through the disc of the Milky Way.

Mira star   A long-period variable; that is, a slowly pulsating giant M star.

molecule   A structure formed by joining specific atoms in a specific way. The fundamental concern of chemistry.

momentum   A quantity that measures amount of motion: the mass of a body times its velocity.

N galaxy   A galaxy with an extremely bright center, probably intermediate in its properties between a Seyfert galaxy and a quasar.

nadir   The point on the celestial sphere that is directly below us, opposite to the zenith.

nebula   Usually, a bright cloud of interstellar gas. The term "dark nebula" is still occasionally used for dust clouds, however. The word "nebula" is no longer used to refer to galaxies.

neutrino   One of the fundamental particles of physics. Neutrinos are produced by many nuclear reactions, but otherwise they interact very little with other particles.

neutron   One of the fundamental particles of physics. It has no charge, but has almost the same mass as a proton. Left to itself, a neutron will decay into a proton and an electron (plus a neutrino).

neutron star   The most compressed stellar configuration, whose interior consists not of atoms but of neutrons.

Newton, Isaac (1642–1727)   English physicist and mathematician, the founder of modern mechanics. His laws of motion and law of gravitation first explained why the planets move as they do.

Newtonian focus   An alternative focus at the top end of a reflecting telescope. A small flat mirror diverts the converging beam of light outside the telescope tube.

NGC   The New General Catalogue, a nineteenth-century listing of nonstellar objects. NGC designations are still commonly used.

node   One of the two points at which an orbit crosses a reference plane.

non-Euclidean geometry   A set of geometrical rules that differ from those of flat surfaces.

nova   A star that suddenly brightens by a very large amount when an explosion occurs on its surface.

nucleosynthesis   The building up of atomic nuclei by nuclear reactions; the processes by which the various kinds of nuclei were formed.

nucleus   The central body of an atom, around which the electrons move.

objective   The large light-collecting element of a telescope: the lens of a refractor, or the main mirror of a reflector.

opacity   Resistance to the passage of radiation. Inside a star, opacity controls how rapidly the radiation flows outward.

open cluster   A star cluster of Population I, usually in the disc of the Milky Way. Also called a "galactic cluster."

opposition   The aspect, or time, at which an object is opposite the Sun in the sky.

orbit   The regular path of an astronomical body.

parabola   An open curve whose two arms extend infinitely, becoming closer and closer to parallel. The limiting case between a very long ellipse and a very narrow hyperbola.

parallax   An apparent displacement of an object because of the motion of the observer. The Earth's orbital motion produces annual heliocentric parallax of nearby stars. In stellar astronomy, the term "parallax" is often used in a more general way, to mean the reciprocal of distance.

parsec   The basic unit of distance, at which a star has a parallax of one second of arc. Equal to 3.08×10^{18} centimeters, or 3.26 light years.

peculiar motion   The motion of an object relative to the average motion of its system or its surroundings.

penumbra   The part of a shadow in which not all the light of the source is blocked.

periastron   The point of closest approach, in a double-star orbit.

perigee   The closest point to the Earth, in the orbit of an Earth satellite.

perihelion   The closest point to the Sun, in an orbit around the Sun.

period-luminosity relation   For cepheid variables, the relation between the period of brightness-variation of the star and its luminosity or its absolute magnitude.

perturbation   A force that disturbs an orbit; also, the effect of that force on the orbit.

photoionization   The removal of an electron from an atom by absorption of a photon that has more than enough energy to ionize the atom.

photometry   The accurate measurement of brightness.

photon   A tiny packet of radiation, whose energy is proportional to the frequency of the radiation, or inversely proportional to the wavelength. Each photon is absorbed or emitted by a single atom.

Planck's law   The formula for (or graph of) the amount of radiation at each wavelength, from a body of a given temperature.

planet   One of the nine large bodies that go around the Sun. (The distinction between planets and asteroids is an arbitrary one, based on size.) Outside the Solar System, a nonluminous body going around a star.

planetarium   A domed theater in which images of the stars can be projected onto the ceiling.

planetary nebula   A bright nebula consisting of a shell of gas thrown out by a star that is between the red-giant and white-dwarf stages of its evolution.

plate tectonics   The process by which the surface of the Earth reforms itself by means of the lateral motion of the "plates" that carry the continents.

polarization   Selection of electromagnetic waves whose vibrations are in one plane rather than another.

population   A general family-type of stars, according to chemical composition and age.

Population I   The stellar population whose chemical composition is like that of the Sun. May contain stars of all ages.

Population II   A stellar population with lower metal abundance than that of Population I. Contains old stars only.

positron   A particle similar to an electron, but with a positive charge. A positron usually exists only a short time before encountering an electron, whereupon the two annihilate each other.

precession   A slow wobbling of the axis of a rotating body. The Earth's axis precesses because of the attractions of the Moon and the Sun, causing the apparent positions of the stars to change.

prime focus   The direct focus of a reflecting telescope, in the middle of the incoming beam of light.

prism   A piece of glass of triangular cross-section, which disperses light by bending different wavelengths by different amounts.

prominence   A region of bright gas protruding beyond the limb of the Sun.

proper motion   The small change in a star's position that is due to its motion across the line of sight.

proton   One of the fundamental particles of physics, with positive charge and much greater mass than the electron. Atomic nuclei are made of protons and neutrons.

proton-proton chain   A series of nuclear reactions that starts with the fusion of two protons, and finally results in the fusion of four hydrogen nuclei to form a nucleus of helium.

protostar   A configuration of material that is on its way to becoming a star.

Ptolemy, Claudius   Ancient astronomer (about A.D. 150), whose *Almagest* set forth the Earth-centered picture of planetary motions that was accepted until the Copernican revolution.

pulsar   A neutron star, whose strong magnetic field and rapid rotation make its radio radiation look to us like a sequence of brief pulses.

quadrature   The aspects at which a body is 90 degrees on one side or the other of the Sun.

quasar   An object of the most energetic and luminous type in the universe, more luminous than a galaxy but smaller than the center of a galaxy.

Quasars are explosive in nature, but the mechanism is not yet understood.

radar astronomy The observation of objects by aiming radio waves at them and then measuring and timing the reflected waves.

radial velocity The part of a body's velocity that is along the line of sight; measured by means of the Doppler effect.

radio waves The longest waves of the electromagnetic spectrum, with the lowest frequencies. Includes all wavelengths longer than the infrared.

radioactivity The spontaneous breakdown of atomic nuclei that are only slightly unstable.

red giant A star of spectral class K or M, much brighter than main-sequence stars of those classes, but not bright enough to be called a supergiant.

redshift The Doppler shift in the spectra of distant galaxies, caused by the expansion of the universe.

reflection nebula Interstellar dust illuminated by stars in or near it.

reflector A telescope whose main light-gathering element is a mirror.

refractor A telescope whose main light-gathering element is a lens.

relativity The modern theory of mechanics that deals with the behavior of objects at very high velocities or in very strong gravitational fields. "Special relativity" modifies Newton's laws of motion, and "general relativity" states a new theory of gravitation.

resolution The clear separation of objects that are close together in direction.

resonance A simple ratio of periods of two phenomena, so that they occur again and again in the same relationship to each other.

revolution The motion of a body around a distant point.

right ascension The longitude-like coordinate in the equatorial system. Measured around the celestial equator, starting at the vernal equinox.

rotation Turning of a body around an axis within itself.

S0 galaxy A galaxy that consists mainly of a disc of old stars, with no young stars present.

satellite A body that goes around a planet. Sometimes called a moon.

seeing Blurring of astronomical images due to unsteadiness in the Earth's atmosphere.

seismic Related to earthquake shocks.

semi-major axis Half the length of the major axis (usually of an ellipse).

Seyfert galaxy A galaxy with a very bright center that indicates energetic events going on there.

sidereal period The period of a rotation or a revolution, measured with respect to a framework fixed in the stars.

sidereal time Time as measured by the rotation of the Earth with respect to the stars. Equal to the right ascension of objects that are on the meridian.

solar motion The velocity of the Sun, with respect to the average velocity of nearby stars.

solar wind A high-speed flow of gas, outward in all directions from the Sun.

solstice One of the times when the Sun is farthest north of the Earth's equator (summer solstice, about June 21) or farthest south of the equator (winter solstice, about December 21).

spectral class One of a set of broad types recognized among stellar spectra.

spectral type The spectral class and subclass of a star.

spectrogram A photograph of a spectrum.

spectrograph An instrument that produces a spectrum of the light that is put into it.

spectroscopic Related to the use or study of spectra.

spectroscopic binary A double star whose orbital motion can be seen by means of radial-velocity shifts in its spectrum.

spectrum The array of the radiation of an object spread out according to wavelength.

spiral arms Spirally wound regions, in some galaxies, which look bright because of luminous young objects in them.

spiral galaxy A galaxy that has spiral arms in its disc.

star cluster A group of stars that are held together by their mutual gravitational attraction.

statistical parallax A parallax, or distance measure, that is derived from the mean motion of a set of stars across the line of sight.

Stefan's law The rule that the total rate of radiation from a black body is proportional to the area of its surface and to the fourth power of the temperature.

stratosphere A layer of the Earth's atmosphere, beginning about 10 kilometers (6 miles) above the surface, in which motions are horizontal, without vertical mixing.

subdwarf A star that lies slightly below the usual main sequence in the HR diagram. Most subdwarfs are main-sequence stars of Population II.

subgiant A star that lies between the main sequence and the red giants in the HR diagram. Subgiants lie on the post-main-sequence evolutionary track of stars like the Sun.

sunspot A region on the Sun's surface that looks dark, because it is not as hot as its surroundings.

sunspot cycle   The 11-year cycle in which solar activity grows and diminishes, as evidenced most readily by the number of sunspots.

supergiant   A star of very high luminosity. The term is usually reserved for stars that are far off the main sequence, but it is occasionally used to refer to stars at the top of the main sequence itself.

superior conjunction   For Mercury or Venus, the conjunction at which the planet is on the far side of the Sun.

supernova   A star that blows up completely, leaving only a small remnant.

synchrotron radiation   Radiation emitted by electrons of very high energy as they gyrate in a magnetic field.

synodic period   The length of time between the returns of a body to the same apparent position with respect to the Sun.

T Tauri stars   Irregular variable stars that are in a pre-main-sequence phase of their evolution.

tectonic   Mountain-building.

terrestrial planet   A planet that is somewhat like the Earth in its properties. Refers to Mercury, Venus, Earth, and Mars; and Pluto is sometimes included.

tidal force   A stretching force that results from a gravitational attraction that is stronger on the near side of a body than on the far side.

triple-alpha process   The sequence of nuclear reactions in which three helium nuclei fuse to form a nucleus of carbon.

troposphere   The lowest layer of the Earth's atmosphere, about 10 kilometers (6 miles) deep. Stirred by rising and falling currents of air, which are responsible for a large part of our weather.

Tycho Brahe   Danish astronomer (1546-1601), on whose observations Kepler based his laws of planetary motion.

ultraviolet   The part of the electromagnetic spectrum that is at the short-wavelength (high-frequency) side of the visible region. Bounded at even shorter wavelengths by the X-ray region.

umbra   The part of a shadow in which light is cut off completely.

uncertainty principle   The underlying principle of quantum physics, which says that we cannot measure both the position of a particle and its velocity with infinite precision.

Universal Time   A world standard time that is almost the same as Greenwich time, from which it differs only in small scientific details.

universe   The entire cosmos, considered as a unit.

variable star   A star whose brightness changes with time.

vector   A quantity that has both magnitude and direction.

velocity   The rate at which an object moves, including the direction of motion. A vector quantity.

vernal equinox   (1) The time of year when the Sun appears to cross from the south side of the celestial equator to the north side. (2) The apparent position of the Sun among the stars at that time, used as a reference point for right ascension.

visual binary   A double or multiple star whose components can be seen separately.

W Ursae Majoris star   An eclipsing binary that consists of two middle-main-sequence stars going around almost in contact with each other.

wavelength   The quantity that is most often used to distinguish different parts of the electromagnetic spectrum. Physically, it is the distance between successive oscillations of the electromagnetic fields that make up the radiation.

weight   The force with which gravity pulls a body downward.

white dwarf   A star, far below the main sequence in the HR diagram, whose dense interior is slowly cooling off.

Wien's law   Rule for the wavelength at which a black body shines most brightly, which is inversely proportional to its temperature.

X-rays   A short-wavelength (high-frequency) part of the electromagnetic spectrum, between ultraviolet and gamma rays.

zenith   The point on the celestial sphere that is straight overhead.

zodiac   The band of constellations around the ecliptic, which are traditionally 12 in number.

zodiacal light   Light scattered by dust near the plane of the ecliptic, often visible after sundown or before dawn.

INDEX

tion